STUDY GUIDE *for* MEMMLER'S The Human Body in Health and Disease

11TH EDITION

Barbara Janson Cohen

Kerry L. Hull

Professor
Department of Biology
Bishop's University
Lennoxville, Quebec
Canada

Wolters Kluwer Health | Lippincott Williams & Wilkins

Senior Acquisitions Editor: David Troy
Developmental Editor: Dana Knighten
Managing Editor: Renee Thomas
Marketing Manager: Allison Noplock
Managing Editor, Production: Eve Malakoff-Klein
Designer: Doug Smock
Artist: Dragonfly Media Group
Compositor: Maryland Composition, Inc.
Printer: Data Reproductions Corporation

11th Edition

351 West Camden Street
Baltimore, MD 21201

530 Walnut Street
Philadelphia, PA 19106

Printed in The United States

9 8 7 6 5

Library of Congress Cataloging-in-Publication Data

Cohen, Barbara J.
Study guide for Memmler's the human body in health and disease / Barbara Janson Cohen, Kerry L. Hull. — 11th ed.
p. cm.
To be used with the textbook: Memmler's the human body in health and disease, 11th ed.
Includes bibliographical references and index.
ISBN 978-0-7817-6581-7
1. Human physiology—Problems, exercises, etc. 2. Physiology, Pathological—Problems, exercises, etc. 3. Human anatomy—Problems, exercises, etc. I. Hull, Kerry L. II. Cohen, Barbara J. Memmler's the human body in health and disease. III. Title.
QP34.5.M48 2009 Suppl.
612.0076--dc22

2008029604

To purchase additional copies of this book, call our customer service department at (800) 638-3030 or fax orders to (301) 223-2320. International customers should call (301) 223-2300.

Visit Lippincott Williams & Wilkins on the Internet: http://www.lww.com. Lippincott Williams & Wilkins customer service representatives are available from 8:30 am to 6:00 pm, EST.

Preface

The *Study Guide for Memmler's The Human Body in Health and Disease*, 11th edition, assists the beginning student to learn basic information required in the health occupations. Although it will be more effective when used in conjunction with the 11th edition of *The Human Body in Health and Disease*, the *Study Guide* may also be used to supplement other textbooks on basic anatomy and physiology.

The questions in this edition reflect revisions and updating of the text. The labeling and coloring exercises are taken from the illustrations designed for the book. The "Practical Applications" section of each chapter uses clinical situations to test understanding of a subject. Comparing the normal with the abnormal helps a student to gain some understanding of disease prevention and health maintenance.

The exercises are planned to help in student learning, not merely to test knowledge. A certain amount of repetition has been purposely incorporated as a means of reinforcement. Matching questions require the student to write out complete answers, giving practice in spelling as well as recognition of terms. Other question formats include multiple choice, completion, true–false, and short essays. The true–false questions must be corrected if they are false. The essay answers provided are examples of suitable responses, but other presentations of the material are acceptable.

All answers to the *Study Guide* questions are in the *Instructor's Manual* that accompanies the text.

Introduction

Learning about the Human Body

You already have some ideas about the human body that will influence how you learn the information in this textbook. Many of your theories are correct, and this Study Guide, created to accompany the 11th edition of *Memmler's The Human Body in Health and Disease*, will simply add detail and complexity to these ideas. Other theories, however, may be too simplistic. It can be difficult to replace these ingrained beliefs with more accurate information. For instance, many students think that the lungs actively inflate and deflate as we breathe, but it is the diaphragm and the rib-cage muscles that accomplish all of the work. Learning physiology or any other subject therefore involves:

1. **Construction**: Adding to and enhancing your previous store of ideas.
2. **Reconstruction**: Replacing misconceptions (prior views and ideas) with scientifically sound principles.
3. **Self-monitoring**. Construction and reconstruction require that you also monitor your personal understanding of a particular topic and consciously formulate links between what you are learning and what you have previously learned. Rote learning is not an effective way to learn anatomy and physiology (or almost anything else, for that matter). **Metacognition** is monitoring your own understanding. Metacognition is very effective if it takes the form of self-questioning during the lectures. Try to ask yourself questions during lectures, such as "What is the prof trying to show here?" "What do these numbers really mean?" or "How does this stuff relate to the stuff we covered yesterday?" Self-questioning will help you create links between concepts. In other words, try to be an active learner during the lectures. Familiarity with the material is not enough. You w have to internalize it and apply it to succeed. You can greatly enhance your ability to be an active learner by reading the appropriate sections of the textbook before the lecture.

Each field in biology has its own language. This language is not designed to make your life difficult; the terms often represent complex concepts. Rote memorization of definitions will not help you learn. Indeed, because biological terms often have different meanings in everyday conversation, you probably hold some definitions that are misleading and must be revised. For example, you may say that someone has a "good metabolism" if they can eat enormous meals and stay slender. However, the term "metabolism" actually refers to all of the chemical reactions that occur in the body, including those that build muscle and fat. We learn a new language not by reading about it but by using it. The *Study Guide* you hold in your hands employs a number of learning techniques in every chapter to help you become comfortable with the language of anatomy, physiology, and disease.

Addressing the Learning Outcomes

This section will help you master the material both verbally and visually. The first activity, writing out answers to the learning outcomes, will provide you with a good overview of the important concepts in the chapter. The labeling and coloring atlas will be especially useful for mastering anatomy. You can use the atlas in two ways. First, follow the instructions to label and color (when appropriate) each exercise, using your textbook if necessary. You will use the same color to write the name of the structure and to color it. Colored pencils work best, but you may have to outline in black names written in light colors. Coloring exercises are fun, and have been shown to enhance learning.

"Making the Connections": Learning through Concept Maps

This learning activity uses concept mapping to master definitions and concepts. You can think of concept mapping as creating a web of information. Individual terms have a tendency to get lost, but a web of terms is more easily maintained in memory. You can make a concept map by following these steps:

1. Select the concepts to map (6–10 is a good number). Try to use a mixture of nouns, verbs, and processes.
2. If one exists, place the most general, important, or over-riding concept at the top or in the center and arrange the other terms around it. Organize the terms so that closely related terms are close together.
3. Draw arrows between concepts that are related. Write a sentence to connect the two concepts that begins with the term at the beginning of the arrow and ends with the term at the end of the arrow.

For instance, consider a simple concept map composed of three terms: student learning, professors, and textbooks. Write the three terms at the three corners of a triangle, separated from each other by 3–4 inches. Next is the difficult part: devising connecting phrases that explain the relationship between any two terms. What is the essence of the relationship between student learning and professors? An arrow could be drawn from professors to student learning, with the connecting phrase "***can explain difficult concepts to facilitate***." The relationship would be "*Professors* ***can explain difficult concepts to facilitate*** *student learning*." Draw arrows between all other term pairs (*student learning* and *textbooks*, *textbooks* and *professors*) and try to come up with connecting phrases. Make sure that the phrase is read in the direction of the arrow.

There are two concept mapping exercises for most chapters. The first exercise consists of filling in boxes and, in the later maps, connecting phrases. The guided concept maps for Chapters 1 through 7 ask you to think of the appropriate term for each box. The guided concept maps for Chapters 8 through 25 are more traditional concept maps. Pairs of terms are linked together by a connecting phrase. The phrase is read in the direction of the arrow. For instance, an arrow leading from "*genes*" to "*chromosomes*" could result in the phrase "*Genes* ***are found on pieces of DNA called*** *chromosomes*." The second optional exercise provides a suggested list of terms to use to construct your own map. This second exercise is a powerful learning tool, because you will identify your own links between concepts. The act of creating a concept map is an effective way to understand terms and concepts.

Testing Your Knowledge

These questions should be completed after you have read the textbook and completed the other learning activities in the study guide. Try to answer as many questions as possible without referring to your notes or the text. As in the end-of-chapter questions, there are three different levels of questions. Type I questions (Building Understanding) test simple recall: how well have you learned the material? Type II questions (Understanding Concepts) examine your ability to integrate and apply the information in simple practical situations. Type III questions (Conceptual Thinking) are the most challenging. They ask you to apply your knowledge to new situations and concepts. There is often more than one right answer to Conceptual Thinking questions. The answers to all questions are available from your instructor.

Learning from the World Around You

The best way to learn anatomy and physiology is to immerse yourself in the subject. Tell your friends and family what you are learning. Discover more about recent health advances from television, newspapers, magazines, and the Internet. Our knowledge about the human body is constantly changing. The work you will do using the Study Guide can serve as a basis for lifelong learning about the human body in health and disease.

Contents

UNIT I

The Body as a Whole

CHAPTER 1

Organization of the Human Body

Overview

Anatomy is the study of body structure, whereas **physiology** is the study of how the body functions. Anything that disrupts normal body structure or function is a disease, and the study of disease is called **pathology**.

Living things are organized from simple to complex levels. The simplest living form is the **cell**, the basic unit of life. Specialized cells are grouped into **tissues**, which in turn are combined to form **organs**; these organs form systems, which work together to maintain the body.

The systems are the:

- skin, the body's covering, which includes the skin and its associated structures;
- skeletal system, the framework of the body;
- muscular system, which moves the bones;
- nervous system, the central control system that includes the organs of special sense;
- endocrine system, which produces the regulatory hormones;
- circulatory system, consisting of the heart, blood vessels, and lymphatic vessels, which transport vital substances;
- respiratory system, which adds oxygen to the blood and removes carbon dioxide;
- digestive system, which converts raw food materials into products usable by cells;
- urinary system, which removes wastes and excess water; and the
- reproductive system, by which new individuals of the species are produced.

All the cellular reactions that sustain life together make up **metabolism**, which can be divided into **catabolism** and **anabolism**. In catabolism, complex substances are broken down into simpler molecules. When the nutrients from food are broken down by catabolism, energy is released. This energy is stored in the compound **ATP** (adenosine triphosphate) for use by the cells. In anabolism, simple compounds are built into substances needed for cell activities.

All the systems work together to maintain a state of balance, or **homeostasis**. The main mechanism for maintaining homeostasis is **negative feedback**, by which the state of the body is the signal to keep conditions within set limits.

The human body is composed of large amounts of fluid, the amount and composition of which must be constantly regulated. The **extracellular fluid** consists of the fluid that surrounds the cells as well as the fluid circulated in blood and lymph. The fluid within cells is the **intracellular fluid**.

Study of the body requires knowledge of directional terms to locate parts and relate various parts to each other. Several planes of division represent different directions in which cuts can be made through the body. Separation of the body into areas and regions, together with the use of the special terminology for directions and locations, makes it possible to describe an area within the human body with great accuracy.

The large internal spaces of the body are cavities in which various organs are located. The **dorsal cavity** is subdivided into the **cranial cavity** and the **spinal cavity (canal)**. The **ventral cavity** is subdivided into the **thoracic** and **abdominopelvic cavities**. Imaginary lines are used to divide the abdomen into regions for study and diagnosis.

The metric system is used for all scientific measurements. This system is easy to use because it is based on multiples of 10.

Addressing the Learning Outcomes

1. Define the terms *anatomy*, *physiology*, and *pathology*.

EXERCISE 1-1.

INSTRUCTIONS

Write a definition of each term in the spaces below.

1. Anatomy ______________________________
2. Physiology ______________________________
3. Pathology ______________________________

2. Describe the organization of the body from chemicals to the whole organism.

EXERCISE 1-2: Levels of Organization (Text Fig. 1-1)

INSTRUCTIONS

1. Write the name or names of each labeled part on the numbered lines in different colors.
2. Color the different structures on the diagram with the corresponding color. For instance, if you wrote "cell" in blue, color the cell blue.

1. ____________________
2. ____________________
3. ____________________
4. ____________________
5. ____________________
6. ____________________

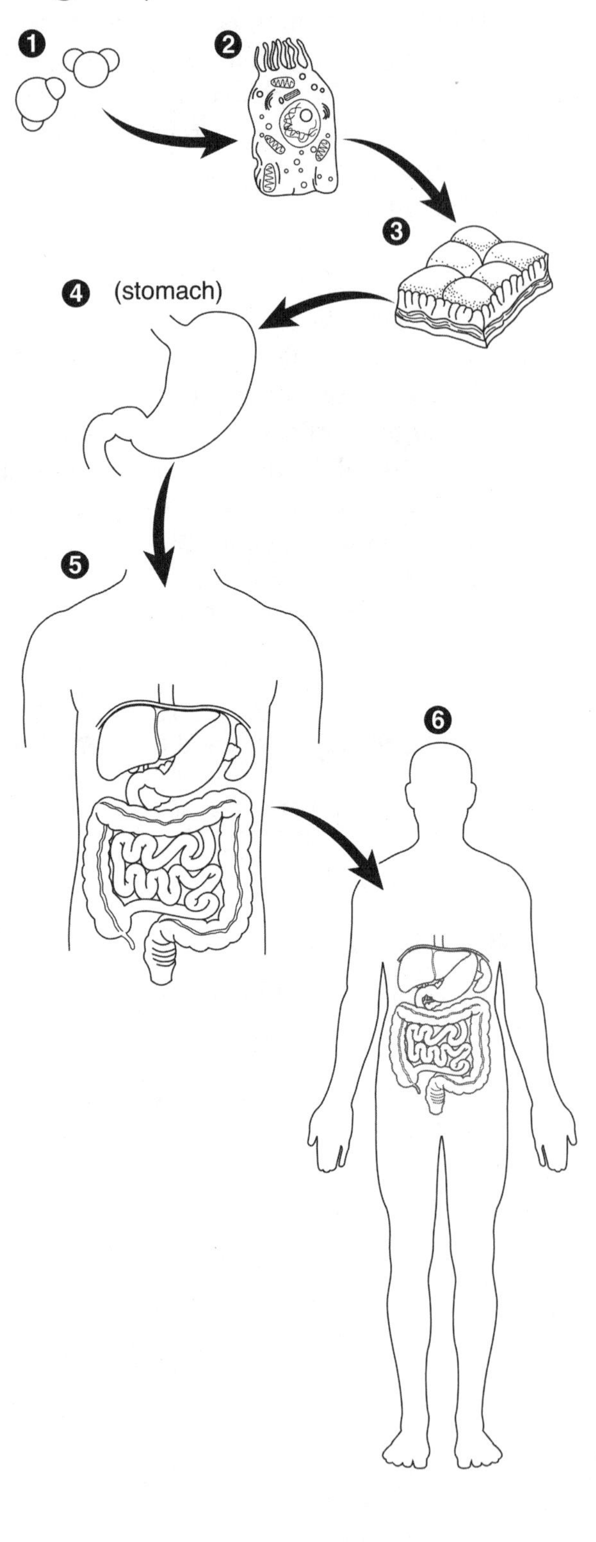

3. List 11 body systems and give the general function of each.

EXERCISE 1-3.

INSTRUCTIONS

Write the appropriate term in each blank.

nervous sytem	integumentary system	cardiovascular system
respiratory system	skeletal system	urinary system
endocrine system	lymphatic system	digestive system

1. The system that processes sensory information ____________
2. The system that delivers nutrients to body tissues ____________
3. The system that breaks down and absorbs food ____________
4. The system that includes the fingernails ____________
5. The system that includes the bladder ____________
6. The system that includes the joints ____________
7. The system that delivers oxygen to the blood ____________
8. The system that includes the tonsils ____________

4. Define *metabolism* and name the two phases of metabolism.

5. Briefly explain the role of ATP in the body.

EXERCISE 1-4.

INSTRUCTIONS

Fill in the blanks in the paragraph below using the following terms: ATP, metabolism, catabolism, and anabolism.

The term ____________ (1) refers to all life-sustaining reactions that occur within the body. The reactions involved in ________ (2) assemble simple components into more complex ones. The reactions of ____________ (3) break down substances into simpler components, generating energy in the form of (4) ______. This energy can be used to fuel cell activities.

6. Differentiate between extracellular and intracellular fluids.

EXERCISE 1-5.

INSTRUCTIONS

Fill in the blank after each statement. Does it apply to extracellular fluid (EC) or intracellular fluid (IC)?

1. includes lymph and blood ____
2. refers to fluid inside cells ____
3. includes fluid between cells ____

7. Define and give examples of homeostasis.

8. Compare negative feedback and positive feedback.

EXERCISE 1-6.

INSTRUCTIONS

Fill in the blanks in the paragraph below using the following terms: negative feedback, positive feedback, homeostasis.

The maintenance of a constant internal body state, known as __________ (1), is critical for health. Different body parameters, such as body temperature and blood glucose concentration, are kept constant using ___________ (2). Conversely, ____________ (3) does not keep a body parameter constant. Instead, it reinforces the stimulus. This form of feedback facilitates dramatic events, such as childbirth.

9. List and define the main directional terms for the body.

EXERCISE 1-7: Directional Terms (Text Fig. 1-7)

INSTRUCTIONS

1. Write the name of each directional term on the numbered lines in different colors.
2. Color the arrow corresponding to each directional term with the corresponding color.

1. ______________________
2. ______________________
3. ______________________
4. ______________________
5. ______________________
6. ______________________
7. ______________________
8. ______________________

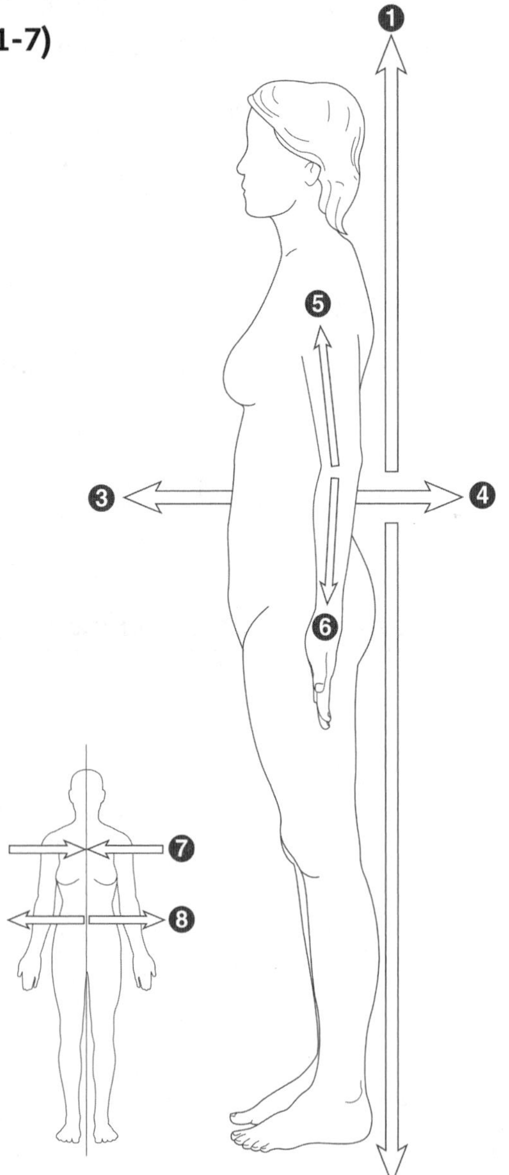

EXERCISE 1-8.

INSTRUCTIONS

Write the appropriate term in the blank.

posterior	anterior	medial	distal
caudal	lateral	horizontal	

1. A term that indicates a location toward the front ____________
2. A term that means farther from the origin of a part ____________
3. A directional term that means away from the midline (toward the side) ____________
4. A term that means nearer to the sacral (lowermost) region of the spinal cord ____________
5. A term that describes the position of the shoulder blades in relation to the collar bones ____________

10. List and define the three planes of division of the body.

EXERCISE 1-9: Planes of Division (Text Fig. 1-8)

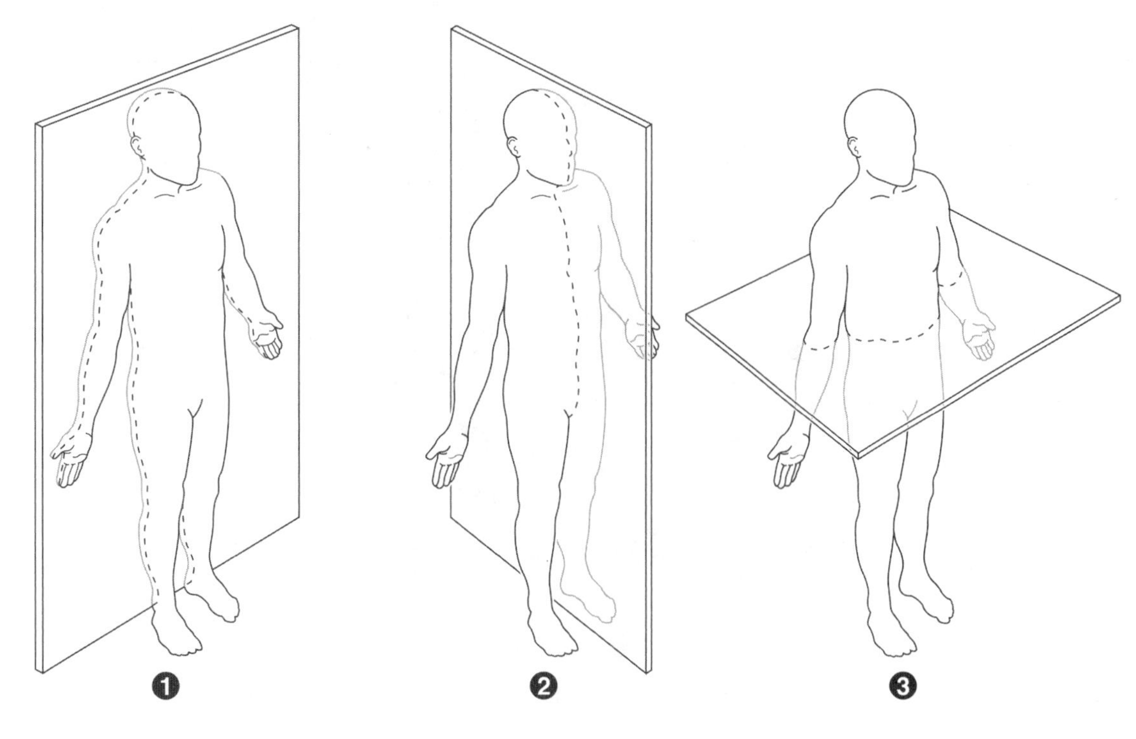

INSTRUCTIONS

1. Write the names of the three planes of division on the correct numbered lines in different colors.
2. Color each plane in the illustration with the corresponding color.

1. ____________________
2. ____________________
3. ____________________

11. Name the subdivisions of the dorsal and ventral cavities.

EXERCISE 1-10: Lateral View of Body Cavities (Text Fig. 1-11)

INSTRUCTIONS

1. Write the names of the different body cavities and other structures in the appropriate spaces in different colors. Try to choose related colors for the dorsal cavity subdivisions and ventral cavity subdivisions.
2. Color parts 2, 3, and 6 to 9 with the corresponding color.

1. ______________________
2. ______________________
3. ______________________
4. ______________________
5. ______________________
6. ______________________
7. ______________________
8. ______________________
9. ______________________

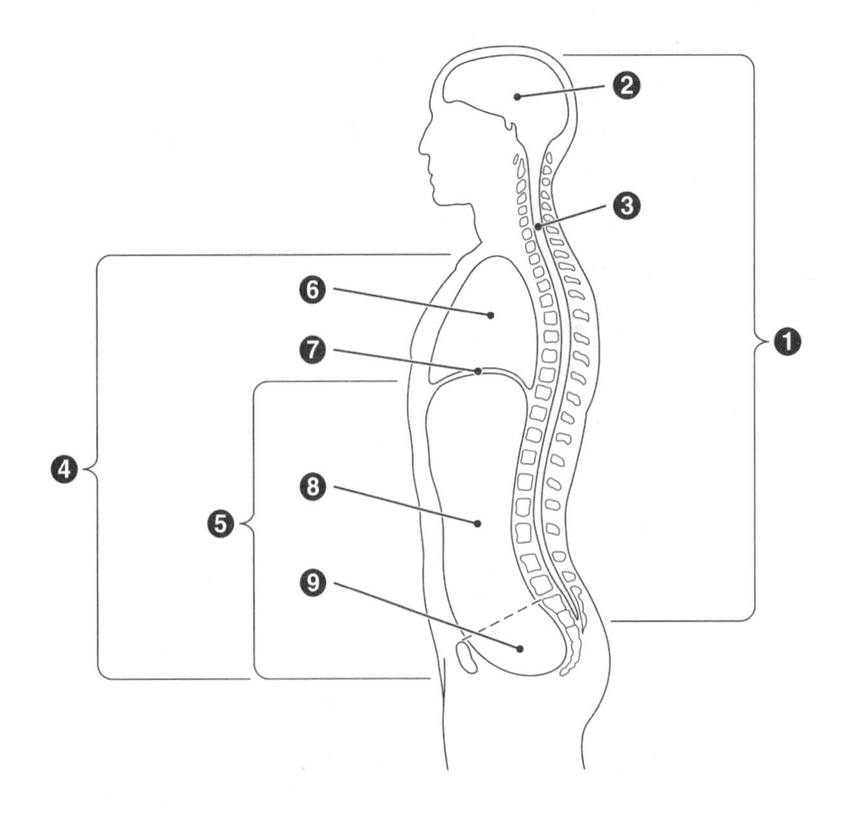

12. Name and locate the subdivisions of the abdomen.

EXERCISE 1-11: Regions of the Abdomen (Text Fig. 1-13)

INSTRUCTIONS

1. Write the names of the nine regions of the abdomen on the appropriate numbered lines in different colors.
2. Color each region with the corresponding color.

1. ______________________
2. ______________________
3. ______________________
4. ______________________
5. ______________________
6. ______________________
7. ______________________
8. ______________________
9. ______________________

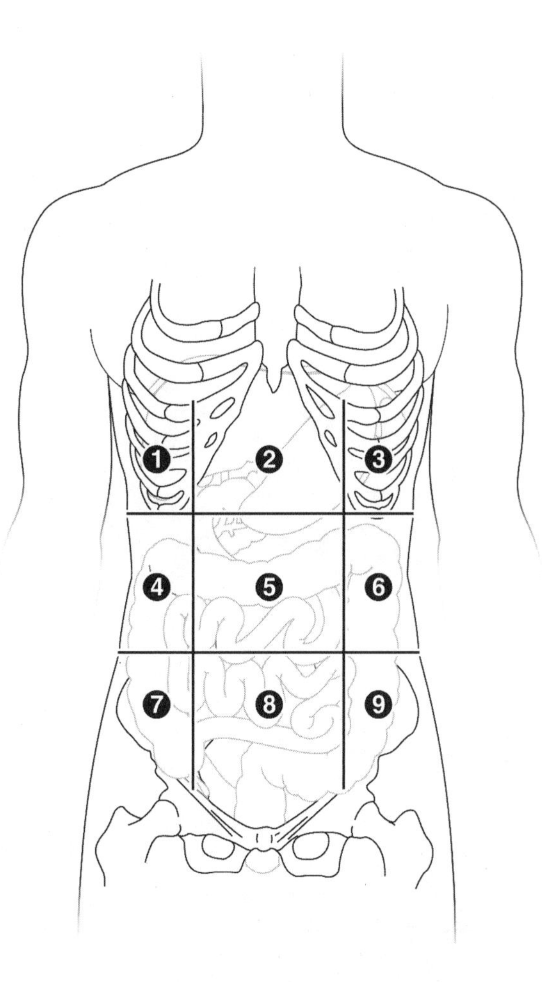

EXERCISE 1-12: Quadrants of the Abdomen (Text Fig. 1-14)

INSTRUCTIONS

1. Write the names of the four quadrants of the abdomen on the appropriate numbered lines in different colors.
2. Color the each quadrant with the corresponding color.

1. ________________________
2. ________________________
3. ________________________
4. ________________________

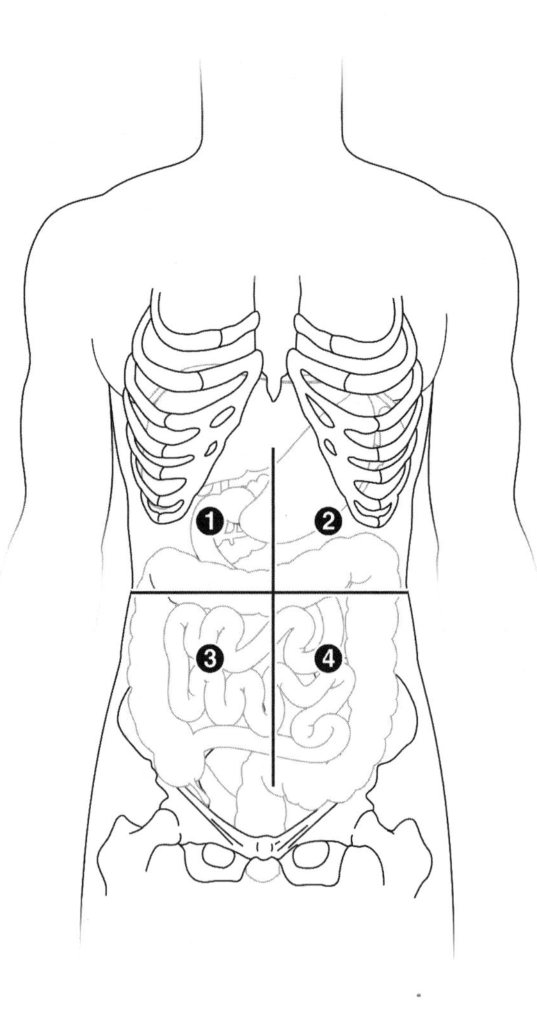

13. Name the basic units of length, weight, and volume in the metric system.

14. Define the metric prefixes *kilo-*, *centi-*, *milli-*, and *micro-*.

EXERCISE 1-13.

INSTRUCTIONS

Write the appropriate term in the space provided.

kilo-	milli-	centi-	micro-
gram	liter	meter	

1. A prefix meaning *one hundredth* ________
2. The basic unit of length ________
3. A prefix meaning *one millionth* ________
4. The basic unit of weight ________
5. A prefix meaning *one thousand* ________

15. Show how word parts are used to build words related to the body's organization.

EXERCISE 1-14.

INSTRUCTIONS

Complete the following table by writing the correct word part or meaning in the space provided. Write a word that contains each word part in the Example column.

Word Part	Meaning	Example
1. -tomy	________	________
2. -stasis	________	________
3. ________	nature, physical	________
4. homeo-	________	________
5. ________	apart, away from	________
6. ________	down	________
7. ________	upward	________
8. path-	________	________
9. -logy	________	________

Making the Connections

The following concept map deals with the body's cavities and their divisions. Complete the concept map by filling in the blanks with the appropriate word or phrase for the cavity, division, subdivision, or region.

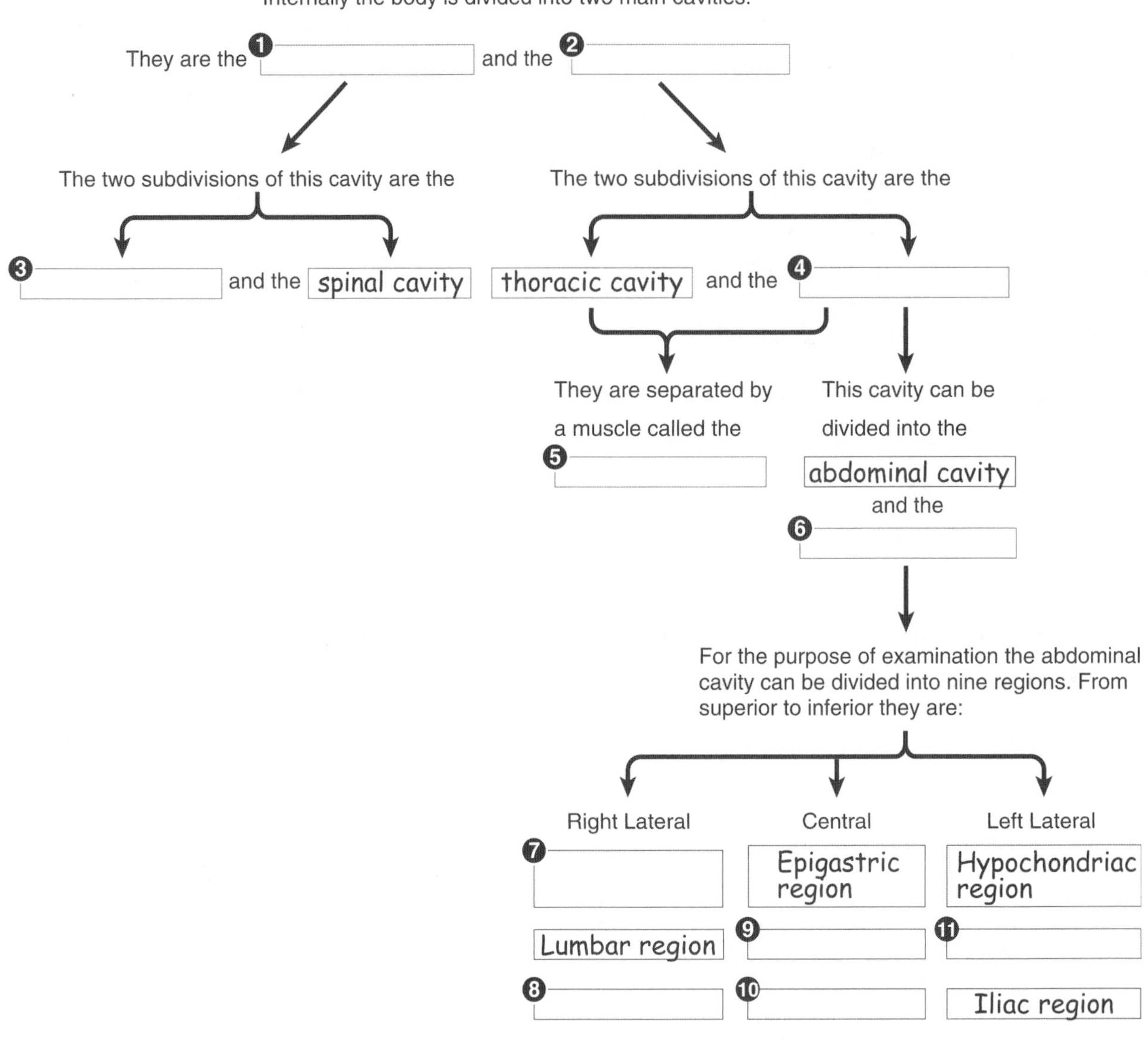

Testing Your Knowledge

Building Understanding

I. Multiple Choice

Select the best answer and write the letter of your choice in the blank.

1. The coronal plane is also called the: 1. _______
   a. frontal plane
   b. transverse plane
   c. cross-sectional plane
   d. sagittal plane
2. The diaphragm separates: 2. _______
   a. the cranial and spinal cavities
   b. the dorsal and ventral cavities
   c. the thoracic and abdominal cavities
   d. the abdominal and pelvic cavities
3. The breakdown of complex molecules into more simple ones is called: 3. _______
   a. anabolism
   b. synthesis
   c. negative feedback
   d. catabolism
4. Lymph is an example of which type of fluid? 4. _______
   a. extracellular
   b. intracellular
   c. superior
   d. extraneous
5. The heart and the blood vessels compose the: 5. _______
   a. circulatory system
   b. nervous system
   c. integumentary system
   d. digestive system
6. The navel is found in the: 6. _______
   a. lumbar region
   b. umbilical region
   c. iliac region
   d. hypogastric region
7. The study of normal body function is called: 7. _______
   a. physiology
   b. pathology
   c. anatomy
   d. chemistry
8. A penny-shaped slice of a banana is probably which type of section? 8. _______
   a. longitudinal section
   b. sagittal section
   c. cross section
   d. coronal section

II. Completion Exercise

➤ Group A: General Terminology

Write the word or phrase that correctly completes each sentence.

1. In the anatomic position, the body is upright and palms are facing ________
2. Fluid inside cells is called ________
3. Catabolism releases energy in the form of ________
4. Negative feedback is a mechanism for maintaining an internal state of balance known as ________
5. The sum of all catabolic and anabolic reactions in the body is called ________
6. The process of childbirth is an example of a type of feedback called ________

➤ Group B: Body Cavities, Directional Terms, and Planes of Division

1. The term that means nearer to the head is ________
2. The space enclosing the brain and spinal cord forms a continuous cavity called the ________
3. The abdomen may be divided into four regions, each of which is called a(n) ________
4. The cavity that houses the bladder is the ________
5. The plane that divides the body into anterior and posterior parts is the ________
6. The ventral body cavity that contains the stomach, most of the intestine, the liver, and the spleen is the ________
7. The abdomen may be subdivided into nine regions, including three along the midline. The region closest to the sternum (breastbone) is the ________
8. The space between the lungs is called the ________
9. The diaphragm separates the abdominopelvic cavity from the ________

➤ Group C: The Metric System

Write the word that correctly completes each sentence.

1. The number of milligrams in a gram is ________
2. The number of centimeters in an inch is ________
3. The number of millimeters in a centimeter is ________
4. Using the metric system, your height would probably be calculated in ________
5. A liter is a slightly greater volume than a ________

Understanding Concepts

I. True/False

For each question, write *T* for true or *F* for false in the blank to the left of each number. If a statement is false, correct it by replacing the underlined term and write the correct statement in the blank below the question.

_______ 1. Proteins are broken down into their component parts by the process of anabolism.

____

_______ 2. Your mouth is inferior to your nose.

____

_______ 3. The brain is caudal to the spine.

____

_______ 4. Your umbilicus is lateral to the left lumbar region.

____

_______ 5. The right iliac region is found in the right lower quadrant.

____

II. Practical Applications

Study each discussion. Then write the appropriate word or phrase in the space provided.

➤ Group A: Directional Terms

1. The gallbladder is located just above the colon. The directional term that describes the position of the gallbladder with regard to the colon is _______________.
2. The kidneys are closer to the sides of the body than is the stomach. The directional term that describes the kidneys with regard to the stomach is _______________.
3. The entrance to the stomach is nearest the point of origin or beginning of the stomach, so this part is said to be _______________.
4. The knee is located closer to the hip than is the ankle. The term that describes the position of the ankle with regard to the knee is _______________.
5. The ears are closer to the back of the head than is the nose. The term that describes the position of the ears with regard to the nose is _______________.
6. The stomach is below the esophagus; it may be described as _______________.
7. The head of the pancreas is nearer the midsagittal plane than is its tail portion, so the head part is more _______________.

➤ Group B: Body Cavities and the Metric System

Study the following cases and answer the questions based on the nine divisions of the abdomen and your knowledge of the metric system.

1. Mr. A bruised his ribs in a dirt buggy accident. He experienced tenderness in the upper left side of his abdomen. In which of the nine abdominal regions are the injured ribs located?

   ____
2. Ms. D had a history of gallstones. The operation to remove these stones involved the upper right part of the abdominal cavity. Which abdominal division is this? ________________
3. Following her operation, Ms. D was able to bring her stones home in a jar. She was told that her stones weighed 0.025 kg in total. How many milligrams do her stones weigh?

   ____
4. Ms. C is 8 weeks pregnant. Her uterus is still confined to the most inferior division of the abdomen. This region is called the ________________.
5. Ms. C is experiencing heartburn as a result of her pregnancy. The discomfort is found just below the breastbone, in the ________________.
6. Following the birth of her child, Ms. C opted for a tubal ligation. The doctor threaded a fiberoptic device through a small incision in her navel as part of the surgery. Ms. C will now have a very small incision in which portion of the abdomen? ________________
7. Ms. C's incision was 2 mm in length. What is the length of her incision in centimeters?

   ____

➤ Group C: Body Systems

The triage nurse in the emergency room was showing a group of students how she assessed patients with disorders in different body systems. Study each situation, and answer the following questions based on your knowledge of the 11 body systems.

1. One person was complaining of dizziness and blurred vision. Vision is controlled by the

   ________________.
2. One person had been injured in a snowboarding accident, spraining his wrist joint. The wrist joint is part of the ________________.
3. A woman had attempted a particularly onerous yoga pose and felt a sharp pain in her left thigh. Now she was limping. The nurse suspected a tear to structures belonging to the

   ________________.
4. An extremely tall individual entered the clinic, complaining of a headache. The nurse suspected that he had excess production of a particular hormone. The specialized glands that synthesize hormones make up the ________________.
5. A middle-aged woman was brought in with loss of ability to move the right side of her body. The nurse felt that a blood clot in a blood vessel of the brain was producing the symptoms. Blood vessels are part of the ________________.
6. A man complaining of pain in the abdomen and vomiting blood was brought in by his family. A problem was suspected in the system responsible for taking in food and converting it to usable products. This system is the ________________.
7. Each client was assessed for changes in the color of the outer covering of the body. The outer covering is called the skin, which is part of the ________________.
8. A young woman was experiencing pain in her pelvic region. The doctor suspected a problem with her ovaries. The ovaries are part of the ________________.
9. An older man was experiencing difficulty with urination. The production of urine is the function of the ________________.

III. Short Essays

1. Compare and contrast the terms **anabolism** and **catabolism**. List one similarity and one difference.

2. Which type of feedback, negative or positive, is used to maintain homeostasis? Defend your answer, referring to the definition of homeostasis.

3. Explain why specialized terms are needed to indicate different positions and directions within the body. Provide a concrete example.

Conceptual Thinking

1. Consider the role of negative feedback in the ability of a thermostat to keep your house at the same temperature. What would happen on a hot day if your thermostat worked according to the principles of positive feedback?

2. Consider the role of positive feedback in childbirth. Is childbirth a good example of homeostasis? Would the baby be delivered if stretching of the uterine muscles inhibited oxytocin secretion?

3. A disease at the chemical level can have an effect on the whole body. That is, a change in a chemical affects a cell, which alters the functioning of a tissue, which disrupts an organ, which disrupts a system, which results in body dysfunction. Illustrate this concept by rewriting the following description in your own words using the different levels of organization: chemical, cell, tissue, organ, system, and body. (Hint: blood is a tissue.) Which of the **bold** terms applies to each level of organization? Which level of organization is not explicitly stated?

   Mr. S. experiences pain throughout his **body**. The movement of **blood** through his vessels is impaired. His **blood cells** are misshapen. A chemical found in red blood cells called **hemoglobin** is abnormal.

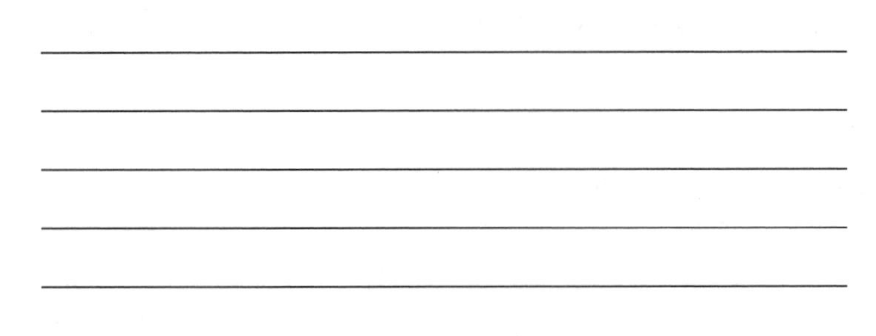

Expanding Your Horizons

As a student of anatomy and physiology, you have joined a community of scholars stretching back into prehistory. The history of biological thought is a fascinating one, full of murder and intrigue. We think that scientific knowledge is entirely objective. However, as the following books will show, theories of anatomy, physiology, and disease depend upon societal factors such as economic class, religion, and gender issues.

Resources

1. Serafini A. *The Epic History of Biology*. Cambridge, MA: Perseus; 1993.
2. Magner LN. *A History of the Life Sciences*. New York: Dekker; 1993.
3. Asimov I. *A Short History of Biology*. Westport, CT: Greenwood; 1980.

CHAPTER 2

Chemistry, Matter, and Life

Overview

Chemistry is the physical science that deals with the composition of matter. To appreciate the importance of chemistry in the field of health, it is necessary to know about elements, atoms, molecules, compounds, and mixtures. An **element** is a substance consisting of just one type of atom. Although exceedingly small particles, atoms possess a definite structure. The **nucleus** contains **protons** and **neutrons**, and the element's **atomic number** indicates the number of protons in its nucleus. The **electrons** surround the nucleus, where they are arranged in specific orbits called **energy levels**.

The union of two or more atoms produces a **molecule**; the atoms in the molecule may be alike (as in the oxygen molecule) or different (sodium chloride, for example). In the latter case, the substance is called a **compound**. A combination of compounds, each of which retains its separate properties, is a **mixture**. Mixtures include solutions, such as salt water, and suspensions. In the body, chemical compounds are constantly being formed, altered, broken down, and recombined into other substances.

Water is a vital substance composed of hydrogen and oxygen. It makes up more than half of the body and is needed as a solvent and a transport medium. Hydrogen, oxygen, carbon, and nitrogen are the elements that constitute about 96% of living matter, whereas calcium, sodium, potassium, phosphorus, sulfur, chlorine, and magnesium account for most of the remaining 4%.

When atoms combine with each other to form compounds they are held together by chemical bonds. Bonds that form as the result of the attraction between oppositely charged ions are called **ionic bonds**. Bonds that form when two atoms share electrons between them are called **covalent bonds**.

Inorganic compounds include acids, bases, or salts. **Acids** are compounds that can donate hydrogen ions (H^+) when in solution. **Bases** usually contain the hydroxide ion (OH^-) and can accept hydrogen ions. The **pH scale** is used to indicate the strength of an acid or base. **Salts** are formed when an acid reacts with a base.

Isotopes (forms of elements) that give off radiation are said to be **radioactive**. Because they can penetrate tissues and can be followed in the body, they are useful in diagnosis. Radioactive substances also have the ability to destroy tissues and can be used in the treatment of many types of cancer.

Proteins, carbohydrates, and lipids are the organic compounds characteristic of living organisms. **Enzymes**, an important group of proteins, function as catalysts in metabolism.

Addressing the Learning Outcomes

1. Define an element.

2. Describe the structure of an atom.

EXERCISE 2-1.

INSTRUCTIONS

Write the appropriate term in each blank.

nucleus	proton	element	electrons
neutrons	atom	ions	

1. A positively charged particle inside the atomic nucleus ____________
2. The smallest complete unit of matter ____________
3. An uncharged particle inside the atomic nucleus ____________
4. A substance composed of one type of atom ____________
5. The part of the atom containing protons and neutrons ____________
6. A negatively charged particle outside the atomic nucleus ____________

EXERCISE 2-2: Parts of the Atom, Molecule of Water (Text Figs. 2-1 and 2-2)

INSTRUCTIONS

1. Write the names of the types of atoms in this figure (2 hydrogen and 1 oxygen) on the numbered lines in two different light colors.
2. LIGHTLY color the hydrogen and oxygen atoms in the figure with the corresponding color.
3. Write the names of the parts of the atom (electron, proton, neutron) on the numbered lines in different, darker colors.
4. Color the electrons, protons, and neutrons on the figure in the appropriate colors. You should find 10 electrons, 10 protons, and 8 neutrons in total.

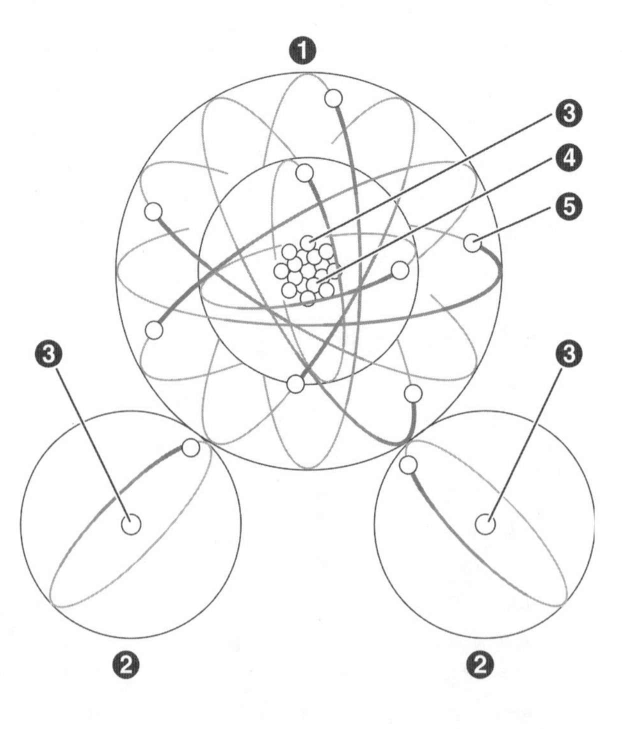

1. ____________________
2. ____________________
3. ____________________
4. ____________________
5. ____________________

3. Differentiate between molecules and compounds.

EXERCISE 2-3.

INSTRUCTIONS

Indicate which of the following statements apply to molecules (M), compounds (C), or both (B) by writing the appropriate letter in the blank.

1. contain two or more atoms ____________
2. contain two or more identical atoms ____________
3. contain two or more different atoms ____________

4. Explain why water is so important to the body.

EXERCISE 2-4.

Which of the following properties are NOT true of water? There is one correct answer.

A. all substances can dissolve in water
B. water participates in chemical reactions
C. water is a stable liquid at ordinary temperatures
D. water carries substances to and from cells

5. Define *mixture*; list the three types of mixtures and give two examples of each.

EXERCISE 2-5.

INSTRUCTIONS

Write the appropriate term in each blank.

solution	suspension	solute	colloid
solvent	aqueous	mixture	

1. A substance that is dissolved in another substance ____________
2. A mixture in which substances will settle out unless the mixture is shaken ____________
3. Term used to describe a solution mostly formed of water ____________
4. The substance in which another substance is dissolved ____________
5. Cytoplasm and blood plasma are examples of this type of mixture ____________
6. Any combination of two or more substances in which each constituent maintains its identity ____________

6. Differentiate between ionic and covalent bonds.

7. Define an electrolyte.

EXERCISE 2-6.

INSTRUCTIONS

Write the appropriate term in each blank.

cations ionic covalent electrolytes anions

1. Negatively charged ions ________________
2. A bond formed by the sharing of electrons between two atoms ________________
3. Compounds that form ions when in solution ________________
4. Positively charged ions ________________
5. A bond formed by the transfer of electron(s) from one atom to another ________________

8. Define the terms *acid, base,* and *salt.*

9. Explain how the numbers on the pH scale relate to acidity and basicity (alkalinity).

10. Define *buffer* and explain why buffers are important in the body.

EXERCISE 2-7.

INSTRUCTIONS

Fill in the blanks in the paragraph below using the following terms: pH scale, salt, acid, base, buffer, hydroxide, alkali, hydrogen, high, low.

Any substance that can donate a hydrogen ion to another substance is called a(n) __________(1). Any substance that can accept a hydrogen ion is called a(n) __________ (2) or a(n) __________ (3). Many of these contain a(n) ________ ion (4). A reaction between a hydrogen-accepting substance and a hydrogen-donating substance produces a __________ (5). The ______ (6) measures the concentration of hydrogen ions in a solution. A solution with a large concentration of hydrogen ions will have a _____ (7) pH; a solution with a large concentration of hydroxyl ions will have a ____ (8) pH. A substance that helps to maintain a stable hydrogen ion concentration in a solution is called a(n) ____________ (9); these substances are critical for health.

11. Define *radioactivity* and cite several examples of how radioactive substances are used in medicine.

EXERCISE 2-8.

Which of the following statements about radioactivity are TRUE? There are four right answers.

A. All isotopes are radioactive.
B. Isotopes have the same atomic weight.
C. Isotopes have the same number of protons.
D. Isotopes have the same number of electrons.
E. Isotopes have the same number of neutrons.
F. Radioactive isotopes disintegrate easily.
G. Radioactive isotopes can be used for diagnosis and treatment.

12. List three characteristics of organic compounds.

EXERCISE 2-9.

The three characteristics of organic compounds are:

1. ______________________________
2. ______________________________
3. ______________________________

13. Name the three main types of organic compounds and the building blocks of each.

EXERCISE 2-10.

INSTRUCTIONS

Write the appropriate term in each blank.

protein	amino acid	phospholipid	carbohydrate
monosaccharide	steroid	disaccharide	

1. Two simple sugars linked together ____________
2. A building block of proteins ____________
3. A lipid containing a ring of carbon atoms ____________
4. A lipid that contains phosphorus in addition to carbon, hydrogen, and oxygen ____________
5. A category of organic compounds that includes simple sugars and starches ____________

EXERCISE 2-11: Carbohydrates (Text Fig. 2-7)

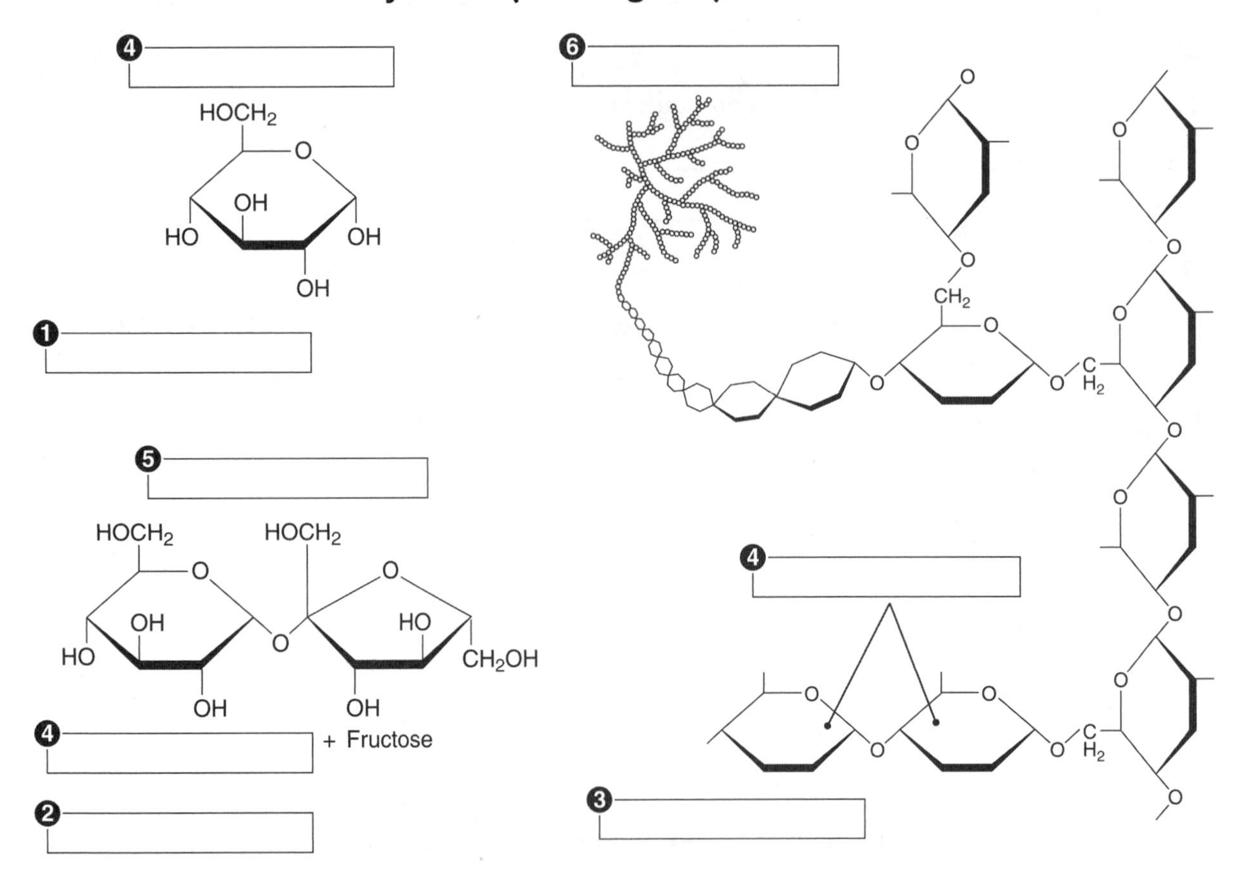

INSTRUCTIONS

1. Write the terms *monosaccharide*, *disaccharide*, and *polysaccharide* in the appropriate numbered boxes 1-3.
2. Write the terms *glucose*, *sucrose*, and *glycogen* in the appropriate numbered boxes 4-6 in different colors.
3. Color the glucose, sucrose, and glycogen molecules with the corresponding colors. To simplify your diagram, only use the glucose color to shade the glucose molecule in the monosaccharide.

EXERCISE 2-12: Lipids (Text Fig. 2-8)

INSTRUCTIONS

1. Write the terms *glycerol* and *fatty acids* on the numbered lines in two different colors.
2. Find the boxes surrounding these two components on the diagram and color them lightly in the corresponding colors.
3. Write the terms *triglyceride* and *cholesterol* in the boxes under the appropriate diagrams.

1. ___________________________
2. ___________________________

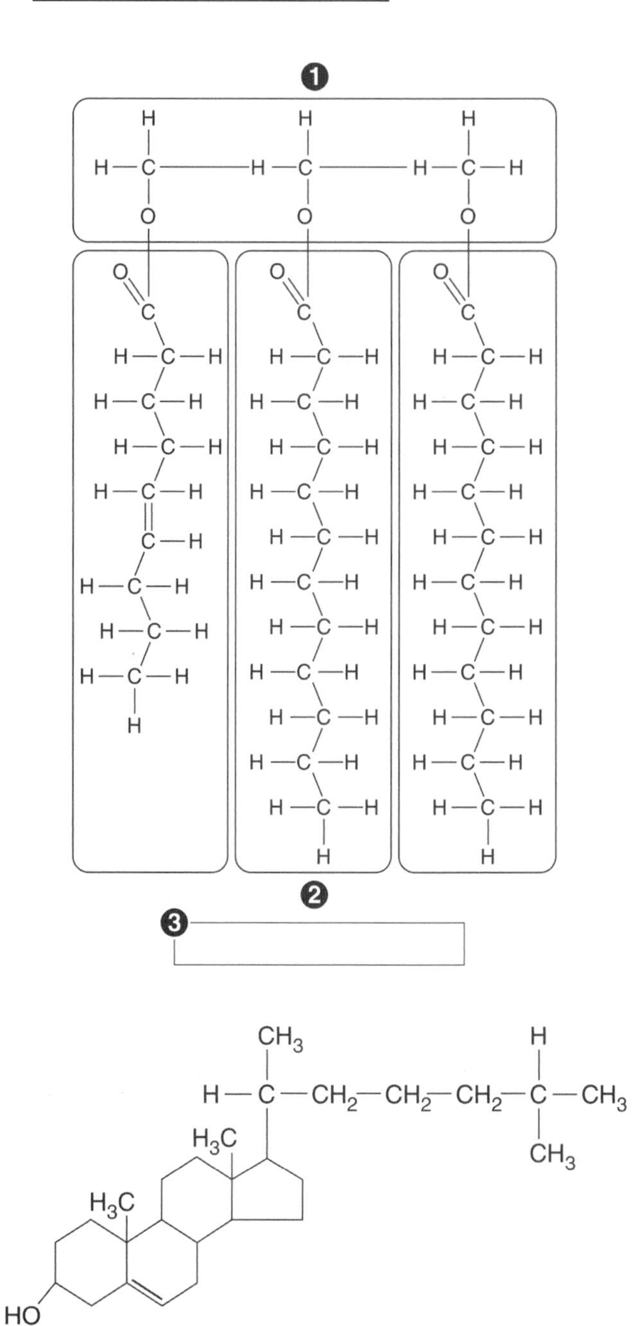

EXERCISE 2-13: Proteins (Text Fig. 2-9)

INSTRUCTIONS

1. Write the terms *amino group* and *acid group* on the numbered lines in different colors.
2. Find the shapes surrounding these two components on the diagram, and color them lightly in the corresponding colors.
3. Place the following terms in the appropriate numbered boxes: amino acid, coiled, pleated, folded.

1. ___________________________

2. ___________________________

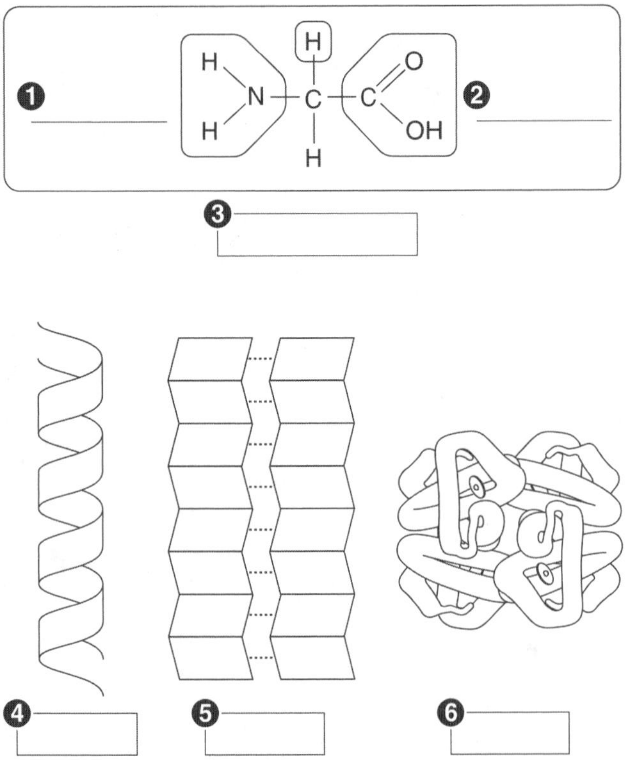

1 ________ H N H H C H C O OH 2 ________

3 ________

4 ________ 5 ________ 6 ________

14. Define *enzyme*; describe how enzymes work.

EXERCISE 2-14: Enzyme Action (Text Fig. 2-10)

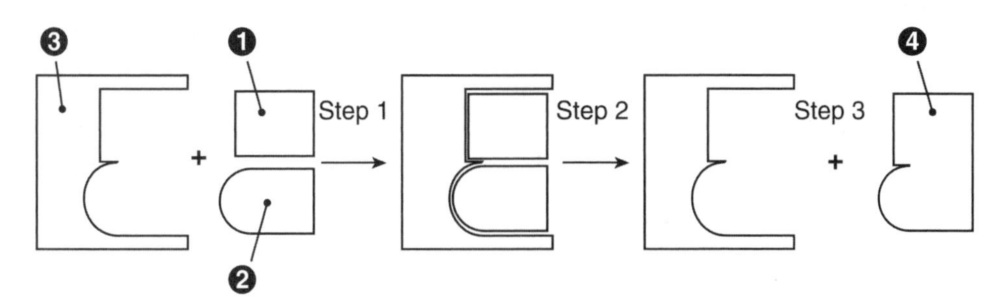

INSTRUCTIONS:

1. Write the terms *substrate 1, substrate 2,* and *enzyme* on the appropriate numbered lines in different colors (red and blue are recommended).
2. Color the structures on the diagram with the corresponding color.
3. What color will result from the combination of your two substrate colors? Write "product" in this color on the appropriate line, and then color the product.

1. ______________________________
2. ______________________________
3. ______________________________
4. ______________________________

15. Show how word parts are used to build words related to chemistry, matter, and life.

EXERCISE 2-15.

INSTRUCTIONS

Complete the following table by writing the correct word part or meaning in the space provided. Write a word that contains each word part in the Example column.

Word Part	Meaning	Example
1. ________	to fear	________________
2. ________	to like	________________
3. glyc/o	________	________________
4. ________	different	________________
5. hydr/o	________	________________
6. hom/o-	________	________________
7. ________	many	________________
8. aqu/e	________	________________
9. sacchar/o	________	________________
10. ________	together	________________

Making the Connections

The following concept map deals with the components of matter. Complete the concept map by filling in the blanks with the appropriate word or phrase.

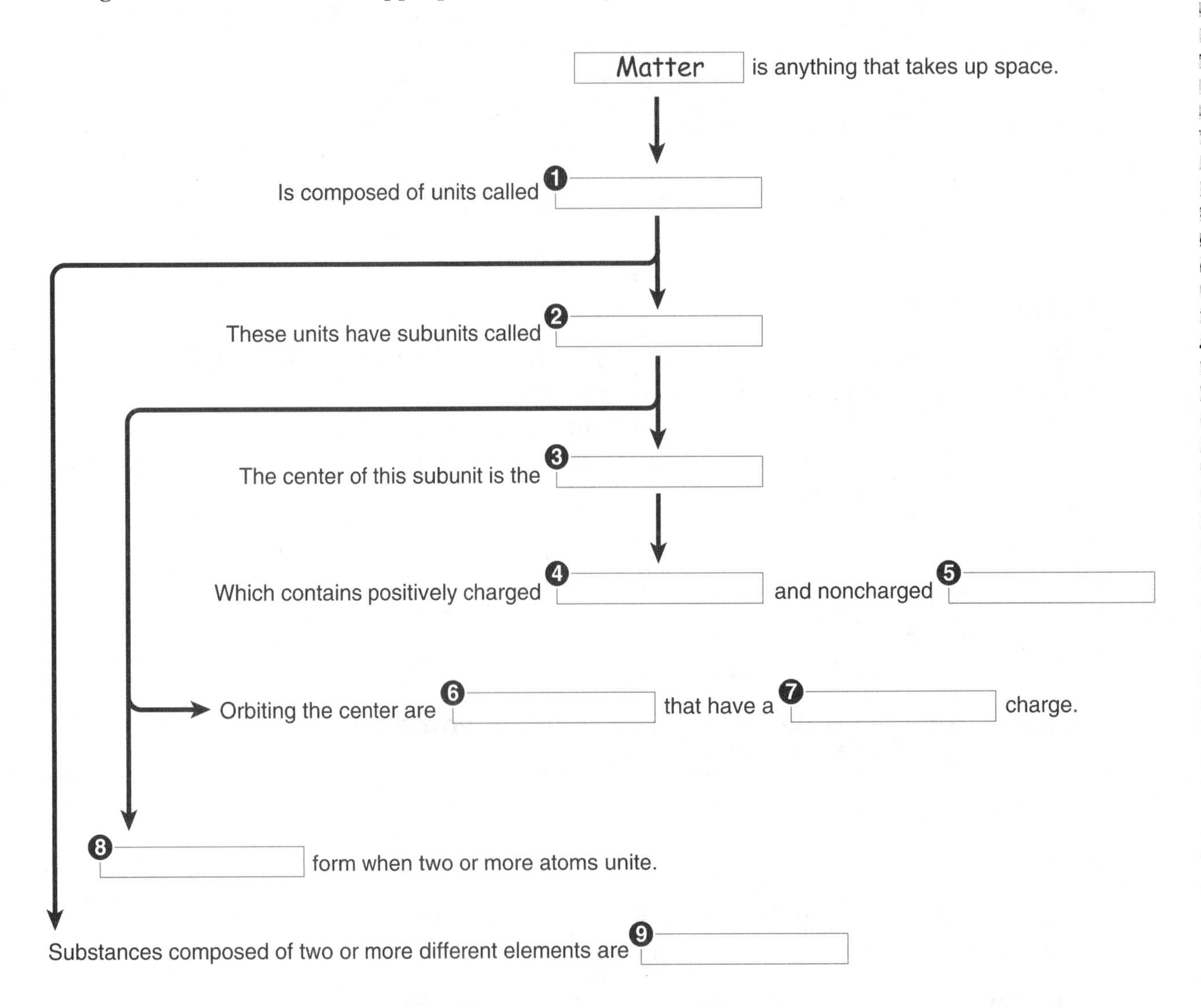

Optional Exercise: Construct your own concept map using the following terms: amino acid, protein, monosaccharide, enzyme, polysaccharide, glycogen, glucose, lipid, and triglyceride.

Testing Your Knowledge

Building Understanding

I. Multiple Choice

Select the best answer and write the letter of your choice in the blank.

1. The element that is the basis for organic chemistry is: 1. _______
   a. nitrogen
   b. carbon
   c. oxygen
   d. hydrogen
2. The smallest particle of an element that has all the properties of that element is a(n):
   a. proton 2. _______
   b. neutron
   c. electron
   d. atom
3. If an electron is added to an atom it becomes a type of charged particle known as a(n):
   a. proton 3. _______
   b. anion
   c. cation
   d. electrolyte
4. Electrons are arranged around the nucleus in specific orbits called: 4. _______
   a. energy levels
   b. ellipses
   c. pathways
   d. isotopes
5. A pH of seven is said to be: 5. _______
   a. basic
   b. neutral
   c. acidic
   d. radioactive
6. Building block for sugars are: 6. _______
   a. amino acids
   b. free fatty acids
   c. glycerol molecules
   d. monosaccharides
7. Covalent bonds are: 7. _______
   a. formed between two ions
   b. polar or nonpolar
   c. never used in organic molecules
   d. usually very unstable

8. Salt completely dissolved in water is an example of a(n): 8. _______
   a. colloid
   b. suspension
   c. aqueous solution
   d. isotope
9. A fat can also be called a(n): 9. _______
   a. polysaccharide
   b. lipid
   c. inorganic molecule
   d. isotope

II. Completion Exercise

Write the word or phrase that correctly completes each sentence.

1. The most abundant element by mass in the human body is ______________.
2. An isotope that disintegrates, giving off rays of atomic particles, is said to be ______________.
3. The number of electrons lost or gained by an atom in a chemical reaction is called its ____________.
4. Water can dissolve many different things. For this reason it is called the ________________.
5. Sugar water is a mixture in which one substance is dissolved in another and remains evenly distributed. This type of mixture is a(n)______________.
6. A mixture that is not a solution but does not separate because the particles in the mixture are so small is a(n) ______________.
7. Metabolic reactions require organic catalysts called ______________.
8. Many essential body activities depend on certain compounds that form ions when in solution. Such compounds are called _______________.
9. The name given to a chemical system that prevents changes in hydrogen ion concentration is ______________.
10. The study of the composition and properties of matter is called _______________.

Understanding Concepts

I. True-False

For each question, write *T* for true or *F* for false in the blank to the left of each number. If a statement is false, correct it by replacing the underlined term and write the correct statement in the blank below the question.

_______ 1. Sodium chloride (NaCl) is an example of an <u>element</u>.

____

_______ 2. If a neutral atom has 12 protons, it will have <u>11</u> electrons.

____

_______ 3. An atom with an atomic number of 15 will have <u>15</u> protons.

____

_______ 4. Butter dissolves poorly in water. Butter is thus an example of a hydrophobic substance.

____

_______ 5. When table salt is dissolved in water, the sodium ion donates one electron to the chloride ion. The chloride ion thus has 17 protons and 17 electrons.

____

_______ 6. You put some soil in water and shake well. After 10 minutes, you note that some of the dirt has settled to the bottom of the jar. With respect to the dirt at the bottom of the jar, your mixture is a colloid.

____

_______ 7. A pH of 10.0 is basic.

____

_______ 8. Carbon dioxide is composed of two oxygen atoms and one carbon atom bonded by the sharing of electrons. The shared electrons are usually closer to the oxygens than to the carbon. Carbon dioxide is formed by ionic bonds.

____

_______ 9. Maltose, a disaccharide, is composed of two glucose molecules.

____

_______ 10. Hemoglobin, a protein, is composed of long chains of phospholipids.

____

II. Practical Applications

Study each discussion. Then write the appropriate word or phrase in the space provided. The following medical tests are based on principles of chemistry and physics.

1. Young Ms. M was experiencing intense thirst and was urinating more than usual. Her doctor suspected that she might have diabetes mellitus. This diagnosis was confirmed by findings of glucose in her urine and high glucose levels in her blood. Glucose and other sugars belong to a group of organic compounds classified as _______________.
2. Young Mr. L was brought to the clinic because he had suffered two convulsions during the night. The doctor suspected that he might have epilepsy. He obtained a graphic record of his brain wave activity to aid in this diagnosis. This brain wave record is called a(n) _______________.
3. Ms. F. just ran a 50-kilometer ultramarathon in the Sahara desert. She was brought into the clinic with symptoms of decreased function of many systems. Her history revealed poor fluid intake for several weeks. Her symptoms were due to a shortage of the most abundant compound in the body, which is _______________.

4. Mr. Q has been experiencing diarrhea, intestinal gas, and bloating whenever he drinks milk. He was diagnosed with a deficiency in an enzyme called lactase, which digests the sugar in milk. Enzymes, like all proteins, are composed of building blocks called ________________.
5. Mr. V came to the clinic complaining of severe headaches. The doctor ordered a PET scan that measures the use of a monosaccharide by the brain. This monosaccharide, which can be assembled into a glycogen molecule in body tissues, is called ________________.
6. Daredevil Mr. L was riding his skateboard on a high wall when he fell. He had pain and swelling in his right wrist. His examination included a procedure in which rays penetrate body tissues to produce an image on a photographic plate. The rays used for this purpose are called ________________.
7. Ms. F. was given an intravenous solution containing sodium, potassium, and chloride ions. These elements come from salts that separate into ions in solution and are referred to as ________________.

III. Short Essays

1. Describe the structure of a protein. Make sure you include the following terms in your answer: pleated sheet, helix, protein, amino acid.

____

____

____

____

____

2. What is the difference between a solvent and a solute? Name the solvent and the solute in salt water.

____

____

____

3. What is the difference between inorganic and organic chemistry?

____

____

____

4. Why is the shape of an enzyme important in its function?

____

____

____

5. Compare and contrast solutions and suspensions. Name one similarity and one difference.

____

____

____

Conceptual Thinking

1. Using the periodic table of the elements in Appendix 3, answer the following questions:
   a. How many protons does silicon (Si) have? ____________
   b. How many electrons does copper (Cu) have? ______________
   c. Phosphorus (P) exists as many isotopes. One isotope is called P^{32}, based on its atomic weight. The atomic weight can be calculated by adding up the number of protons and the number of neutrons. How many neutrons does P^{32} have? _______________
   d. How many electrons does the magnesium ion Mg^{2+} have? (The 2+ indicates that the magnesium atom has lost two electrons). _____________________

2. There is much more variety in proteins than in carbohydrates. Explain why.

____

____

____

Expanding Your Horizons

Drink Your Orange Juice: The Benefits of Vitamin C

You may have read the term "antioxidant" in newpaper articles or on the Internet. But what is an antioxidant, and why do we want them? An understanding of atomic structure is required to answer this question. Radiation or chemical reactions can cause molecules to pick up or lose an electron, resulting in an unpaired electron. Molecules with unpaired electrons are called **free radicals.** For instance, an oxygen molecule can gain an electron, resulting in superoxide (O_2-), and a hydroxyl ion (OH^-) can lose an electron, resulting in a neutral OH free radical. Unpaired electrons are very unstable; thus, free radicals steal electrons from other substances, converting them into free radicals. This chain reaction disrupts normal cell metabolism and can result in cancer. Antioxidants, including vitamin C, give up electrons without converting into free radicals, thereby stopping the chain reaction. The two references listed below will tell you more about free radicals and how antioxidants may fight the aging process and protect against Parkinson disease.

Resources

1. Brown K. A radical proposal. *Scientific American Presents* 2000; 11:38–43.
2. Youdim MB, Riederer P. Understanding Parkinson's disease. *Sci Am* 1997; 276:52–59.

CHAPTER 3

Cells and Their Functions

Overview

The **cell** is the basic unit of life; all life activities result from the activities of cells. The study of cells began with the invention of the light microscope and has continued with the development of electron microscopes. Cell functions are carried out by specialized structures within the cell called **organelles**. These include the nucleus, ribosomes, mitochondria, Golgi apparatus, endoplasmic reticulum (ER), lysosomes, peroxisomes, and centrioles. Two specialized organelles, cilia and flagella, function in cell locomotion and the movement of materials across the cell surface.

An important cell function is the manufacture of proteins, including enzymes (organic catalysts). Protein manufacture is carried out by the ribosomes in the cytoplasm according to information coded in the deoxyribonucleic acid (DNA) of the nucleus. Specialized molecules of RNA called messenger RNA play a key role in the process by carrying copies of the information in DNA to the ribosomes. DNA also is involved in the process of cell division or mitosis. Before cell division can occur, the DNA must replicate so each daughter cell produced by mitosis will have exactly the same kind and amount of DNA as the parent cell.

The **plasma (cell) membrane** is important in regulating what enters and leaves the cell. Some substances can pass through the membrane by **diffusion**, which is simply the movement of molecules from an area where they are in higher concentration to an area where they are in lower concentration. The diffusion of water through the cell membrane is termed **osmosis**. Because water can diffuse very easily across the membrane, cells must be kept in solutions that have the same concentrations as the cell fluid. If the cell is placed in a solution of higher concentration, a **hypertonic solution**, it will shrink; in a solution of lower concentration, a **hypotonic solution**, it will swell and may burst. The cell membrane can also selectively move substances into or out of the cell by **active transport**, a process that requires energy (ATP) and transporters. Large particles and droplets of fluid are taken in by the processes of **phagocytosis** and **pinocytosis**.

When cells undergo a genetic change, or **mutation**, which causes them to multiply out of control, the result is a tumor. A tumor that spreads to other parts of the body is termed **cancer**. Risk factors that influence the development of cancer include heredity, chemicals (carcinogens), ionizing radiation, physical irritation, diet, and certain viruses.

Addressing the Learning Outcomes

1. List three types of microscopes used to study cells.

EXERCISE 3-1.

INSTRUCTIONS

Write the appropriate term in each blank.

compound light microscope
transmission electron microscope
scanning electron microscope
micrometer
centimeter

1. 1/1,000 of a millimeter ________________
2. Microscope that provides a three-dimensional view of an object ________________
3. The most common microscope, which magnifies an object up to 1,000 times ________________
4. A microscope that magnifies an object up to one million times ________________

2. Describe the function and composition of the plasma membrane.

EXERCISE 3-2: Structure of the Plasma Membrane (Text Fig. 3-3)

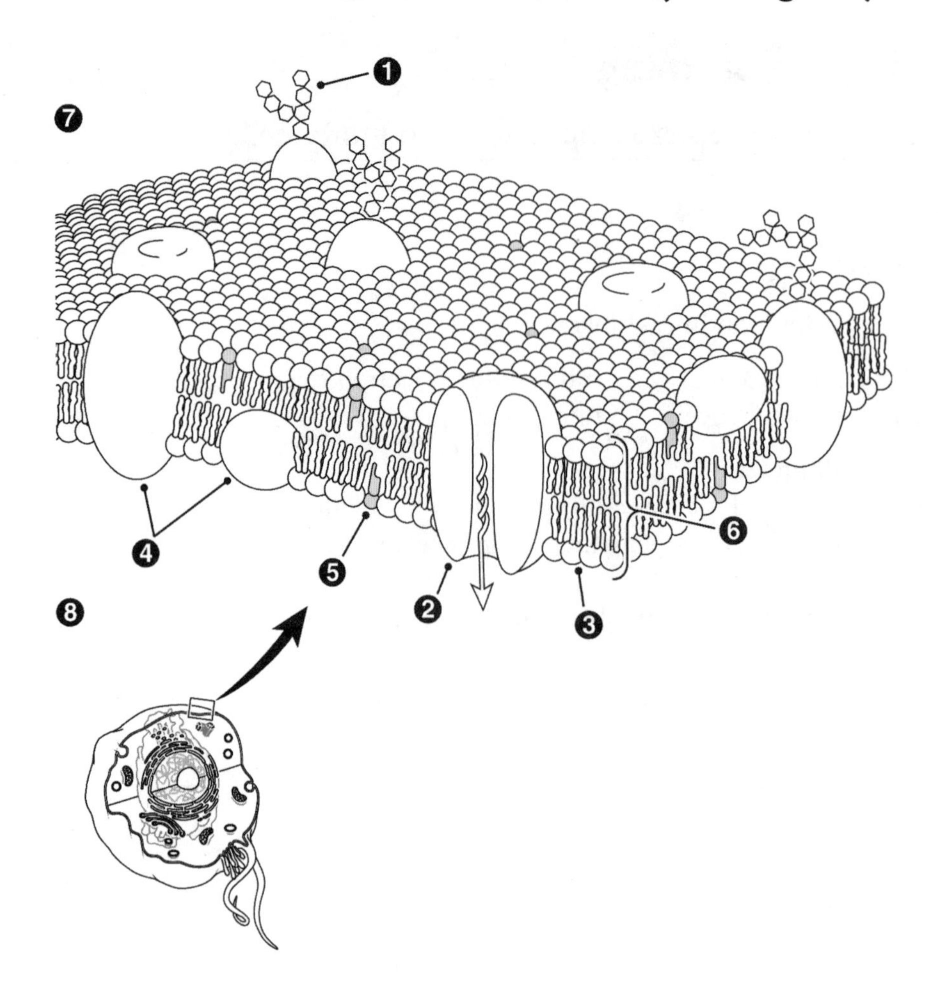

INSTRUCTIONS

1. Write the name of each labeled membrane component on the numbered lines in different colors. Choose a light color for component number 3.
2. Color the different structures on the diagram with the corresponding color (except for structures 6 to 8). Color ALL of components 1 and 3 to 5, not just those indicated by the leader lines. For instance, component 1 is found in three locations.

1. ____________________
2. ____________________
3. ____________________
4. ____________________
5. ____________________
6. ____________________
7. ____________________
8. ____________________

EXERCISE 3-3: Membrane Proteins (Text Table 3-2)

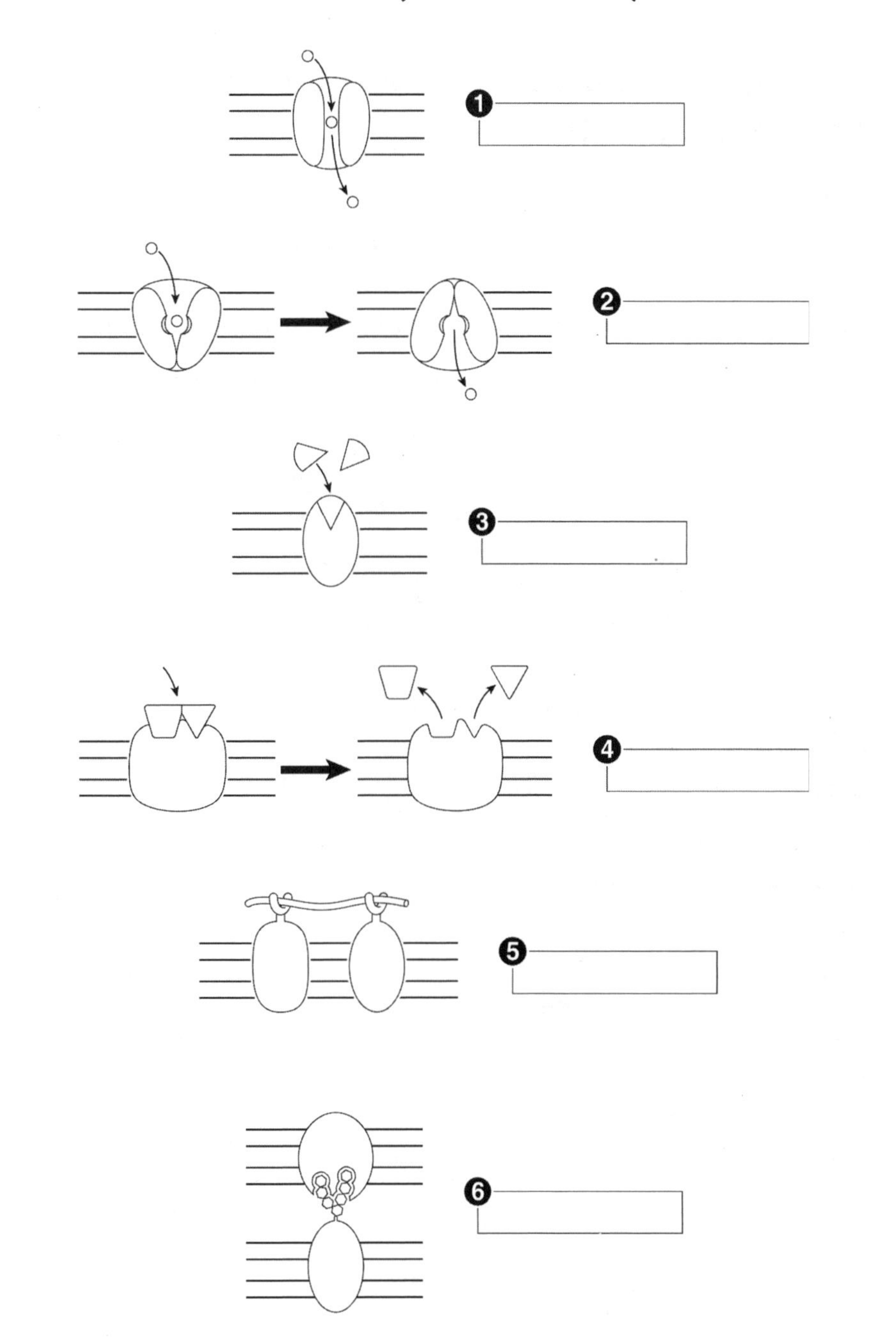

INSTRUCTIONS

1. Write the appropriate membrane protein function in boxes 1 to 6 in different colors.
2. Color the protein in each diagram with the corresponding color.

3. Describe the cytoplasm of the cell, including the name and function of the main organelles.

EXERCISE 3-4: Typical Animal Cell Showing the Main Organelles (Text Fig. 3-2)

INSTRUCTIONS

1. Write the name of each labeled part on the numbered lines in different colors. Make sure you use light colors for structures 1 and 7.
2. Color the different structures on the diagram with the corresponding color.

*Note: Parts 12 and 15 have the same appearance in this diagram; write the name of one of the possible parts in blank 12 and the other in blank 15.

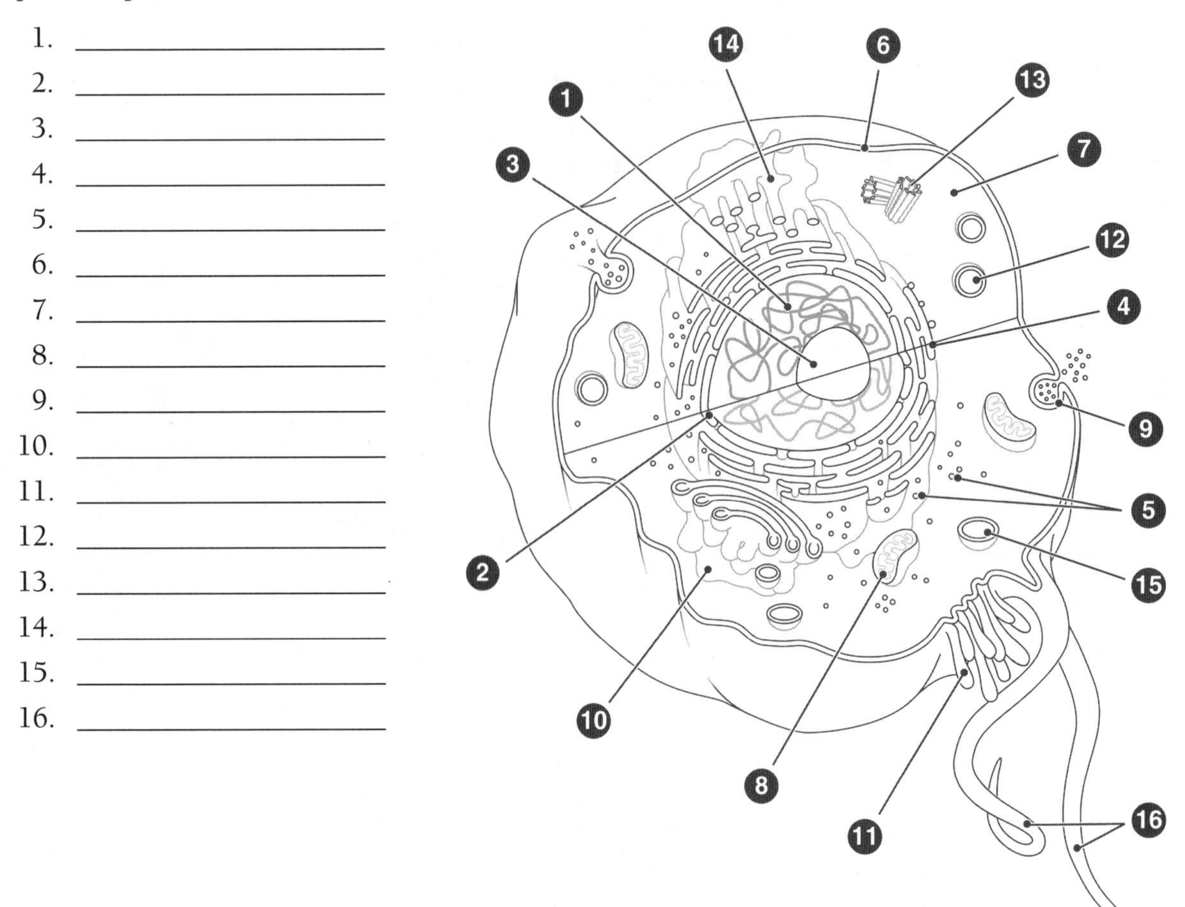

EXERCISE 3-5.

INSTRUCTIONS

Write the appropriate term in each blank.

lysosome	nucleolus	cilia	ribosome
Golgi apparatus	vesicle	mitochondrion	nucleus

1. A structure that assembles ribosomes ________________
2. A structure that assembles amino acids into proteins ________________
3. A set of membranes involved in packaging proteins for export ________________
4. A small saclike structure used to transport substances within the cell ________________
5. A membraneous organelle that generates ATP ________________
6. A small saclike structure that degrades waste products ________________
7. The site of DNA storage ________________

4. Describe the composition, location, and function of the DNA in the cell.

EXERCISE 3-6: Basic Structure of a DNA Molecule (Text Fig. 3-6)

INSTRUCTIONS

1. Write the name of each part of the DNA molecule on the numbered lines in contrasting colors.
2. Color the different parts on the diagram with the corresponding color. Try to color each part everywhere it occurs in the DNA molecule.

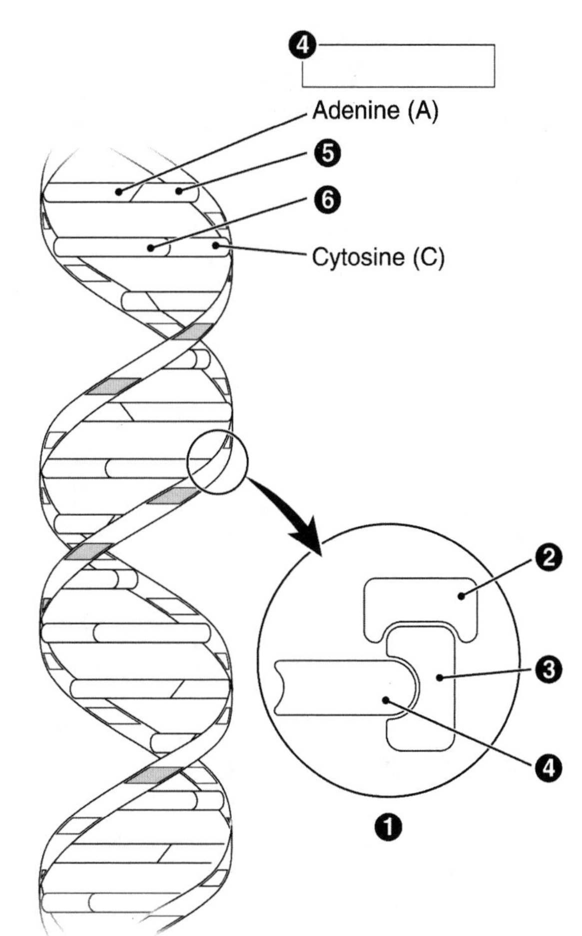

1. ________________________
2. ________________________
3. ________________________
4. ________________________
5. ________________________
6. ________________________

5. Compare the function of three types of RNA in the cells.

6. Explain briefly how cells make proteins.

EXERCISE 3-7.

INSTRUCTIONS

Write the appropriate term in each blank from the list below.

DNA	nucleotide	transcription	translation
transfer RNA (tRNA)	messenger RNA (mRNA)	ribosomal RNA (rRNA)	

1. The process by which RNA is synthesized from the DNA ________________
2. A building block of DNA and RNA ________________
3. An important component of ribosomes ________________
4. The structure that carries amino acids to the ribosome ________________
5. The nucleic acid that carries information from the nucleus to the ribosomes ________________
6. The process by which amino acids are assembled into a protein ________________

7. Name and briefly describe the stages in mitosis.

EXERCISE 3-8: Stages of Mitosis (Text Fig. 3-9)

INSTRUCTIONS

Identify interphase and the indicated stages of mitosis. Find the DNA in each stage and color it.

1. ______________________
2. ______________________
3. ______________________
4. ______________________
5. ______________________

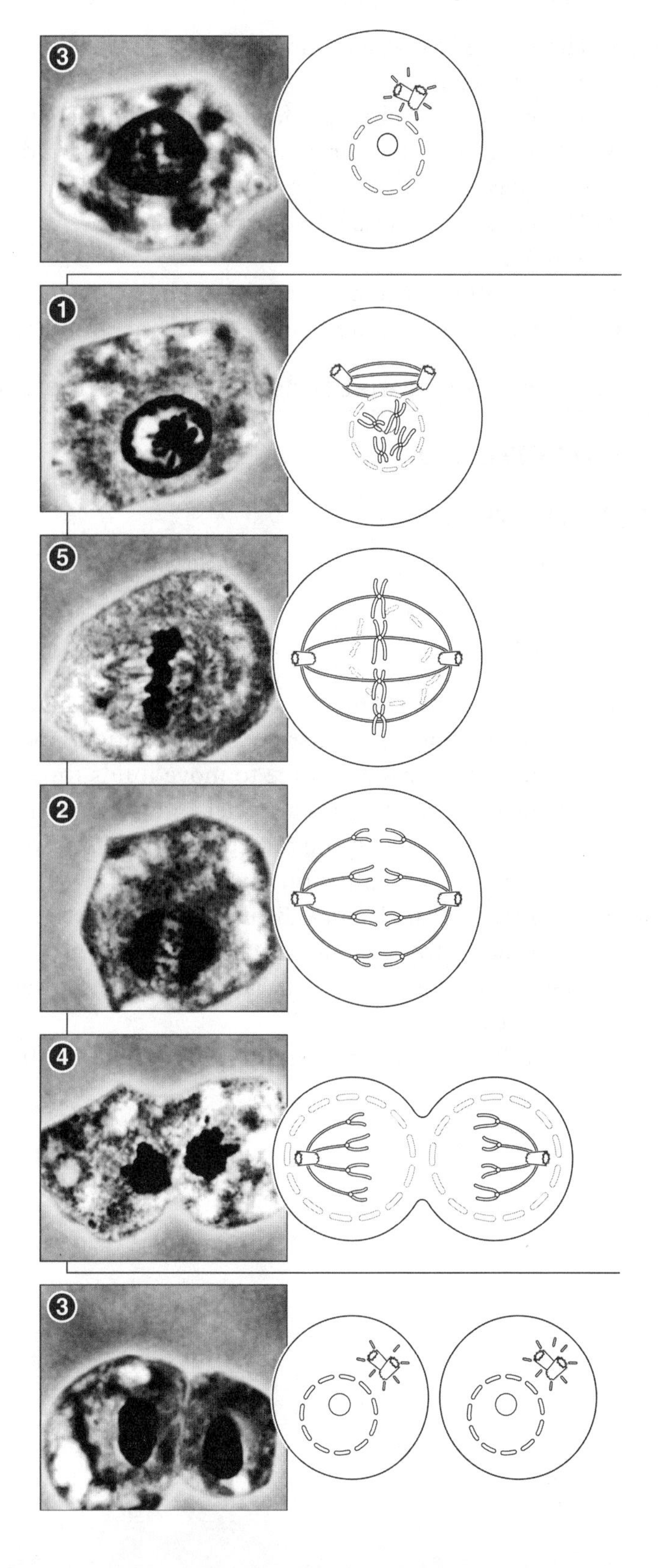

EXERCISE 3-9.

INSTRUCTIONS

Write the appropriate term in each blank from the list below.

mitosis	anaphase	telophase	meiosis
metaphase	prophase	interphase	

1. The process by which one cell divides into two identical daughter cells ________________
2. The nuclear membrane reforms during this phase ________________
3. Centrioles begin to form and chromosomes condense during this phase ________________
4. The phase of mitosis when chromosomes are aligned in the middle of the cell ________________
5. DNA synthesis occurs during this phase ________________
6. The chromosomes are being pulled apart in this phase ________________

8. Define eight methods by which substances enter and leave cells.

EXERCISE 3-10.

INSTRUCTIONS

Write the appropriate term in each blank using the terms listed below.

exocytosis	endocytosis	active transport	facilitated diffusion
osmosis	diffusion	filtration	pinocytosis

1. The process that utilizes a carrier to move materials across the plasma membrane against the concentration gradient using ATP ________________
2. The use of hydrostatic force to move fluids through a membrane ________________
3. The process that utilizes a carrier to move materials across the plasma membrane in the direction of the concentration gradient ________________
4. A special form of diffusion that applies only to water ________________
5. The spread of molecules throughout an area ________________
6. The process by which a cell takes in large particles ________________
7. The process by which materials are expelled from the cell using vesicles ________________
8. Small fluid droplets are brought into the cell using this method ________________

9. Explain what will happen if cells are placed in solutions with concentrations the same as or different from those of the cell fluids.

EXERCISE 3-11: Osmosis (Text Fig. 3-18)

INSTRUCTIONS

Label each of the following solutions using the term that indicates the solute concentration in the solution relative to the solute concentration in the cell.

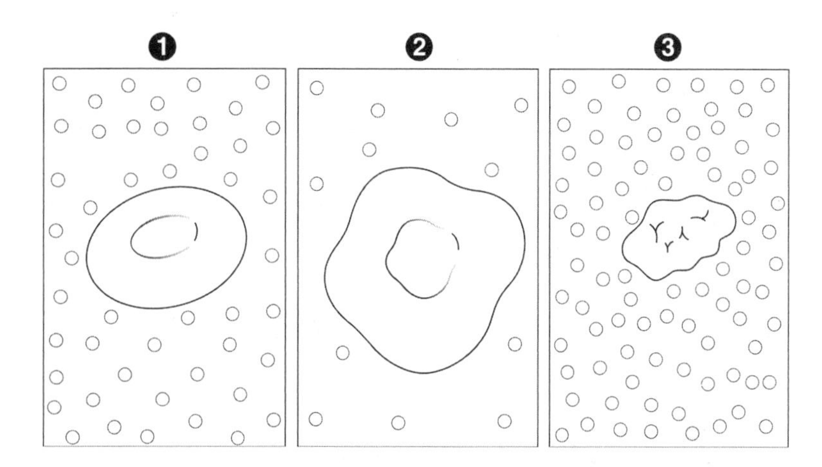

1. ___________________
2. ___________________
3. ___________________

10. Define cancer.

EXERCISE 3-12.

INSTRUCTIONS

For each of the following statements, state whether they are true (T) or false (F).

1. Cancers result from genetic mutations. ___________
2. Slower-growing cells are more likely to develop into cancers. ___________
3. The immune system often kills cancerous cells. ___________
4. Tumors that do not spread to other tissues are called cancers. ___________

11. List several risk factors for cancer.

EXERCISE 3-13.

INSTRUCTIONS

List six risk factors for cancer in the spaces below.

1. ___________________
2. ___________________
3. ___________________
4. ___________________
5. ___________________
6. ___________________

12. Show how word parts are used to build words related to cells and their functions.

EXERCISE 3-14.

INSTRUCTIONS

Complete the following table by writing the correct word part or meaning in the space provided. Write a word that contains each word part in the Example column.

Word Part	Meaning	Example
1. phag/o	________	________________
2. ________	to drink	________________
3. -some	________	________________
4. lys/o	________	________________
5. ________	cell	________________
6. ________	above, over, excessive	________________
7. hem/o	________	________________
8. ________	same, equal	________________
9. hypo-	________	________________
10. ________	in, within	________________

Making the Connections

The following concept map deals with the movement of materials through the plasma membrane. Complete the concept map by filling in the blanks with the appropriate word or phrase that describes the indicated process.

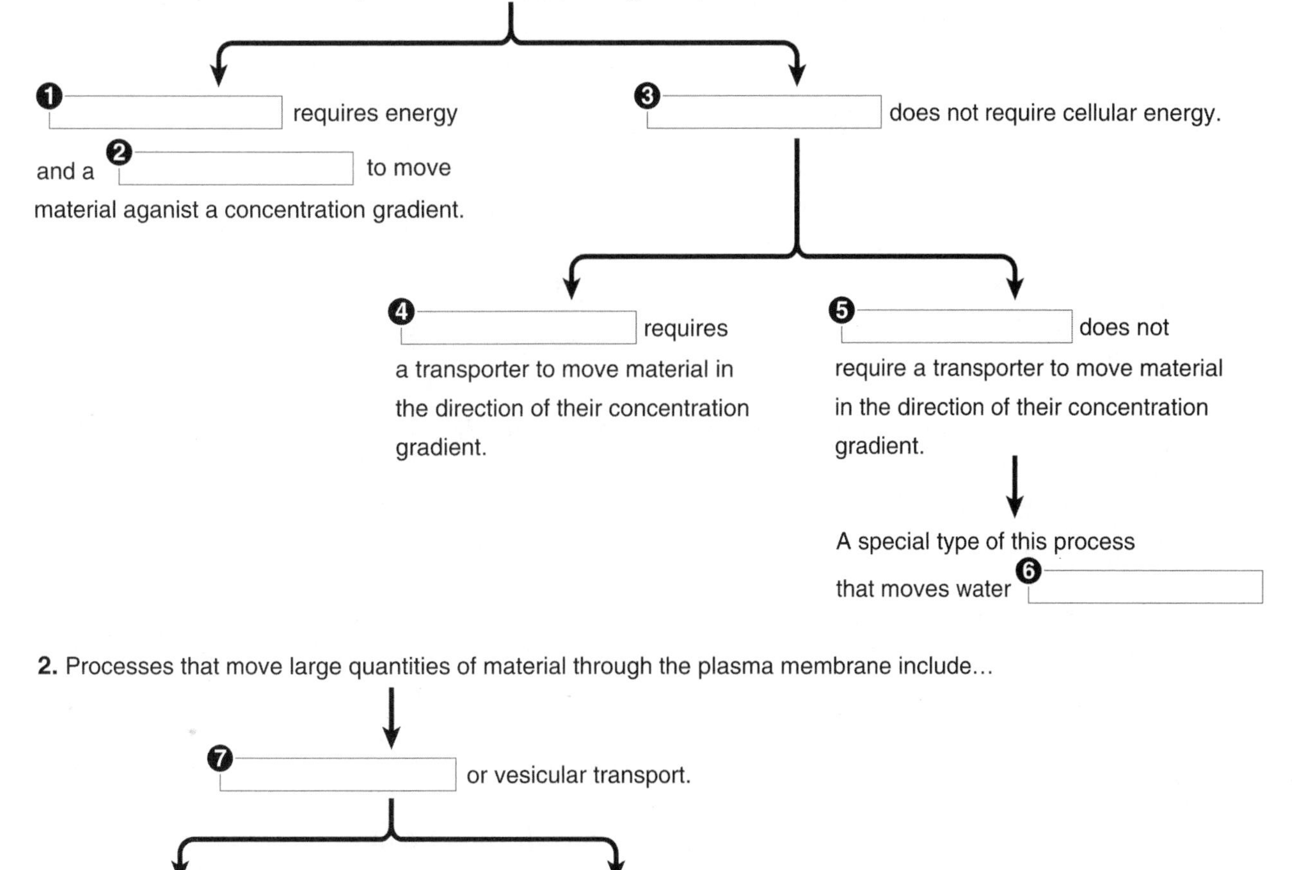

Bulk movement into a cell
8 ________

Bulk movement out of a cell
9 ________

Large particles
10 ________

Droplets
11 ________

Optional Exercise: Make your own concept map, based on the components of the cell. Choose your own terms to incorporate into your map, or use the following list: nucleus, mitochondria, cell membrane, protein, RNA, DNA, ATP, vesicle, ribosome, and endoplasmic reticulum.

Testing Your Knowledge

Building Understanding

I. Multiple Choice

Select the best answer and write the letter of your choice in the blank.

1. Which of the following organelles consists of a series of membranes studded with ribosomes? 1. _______
   a. mitochondrion
   b. rough endoplasmic reticulum
   c. smooth endoplasmic reticulum
   d. Golgi apparatus
2. A natural part of growth and remodeling involves the process of programmed cell death known as 2. _______
   a. mitosis
   b. mutation
   c. apoptosis
   d. phagocytosis
3. Which of the following are required for active transport? 3. _______
   a. vesicles and cilia
   b. transporters and ATP
   c. osmotic pressure and centrioles
   d. osmosis and lysosomes
4. The stage of mitosis during which the DNA condenses into visible chromosomes is called: 4. _______
   a. metaphase
   b. anaphase
   c. prophase
   d. telophase
5. Large proteins can be secreted from the cell using the process of: 5. _______
   a. pinocytosis
   b. osmosis
   c. endocytosis
   d. exocytosis
6. Which of the following substances is NOT a constituent of the plasma membrane? 6. _______
   a. DNA
   b. proteins
   c. carbohydrates
   d. phospholipids
7. Which of the following tools has the greatest magnification? 7. _______
   a. scanning electron microscope
   b. transmission electron microscope
   c. light microscope
   d. magnifying glass

8. A membrane protein that permits the passage of specific substances is called a(n): 8. ______
   a. channel
   b. receptor
   c. linker
   d. enzyme

II. Completion Exercise

Write the word or phrase that correctly completes each sentence.

1. The type of light microscope in use today is the ________________.
2. The plasma membrane contains two kinds of lipids: cholesterol and ________________.
3. In some cells the plasma membrane is folded outward into multiple small projections called ________________.
4. The four nitrogen bases found in DNA are A, C, G, and ________________.
5. The four nitrogen bases found in RNA are A, C, G, and ________________.
6. The assembly of amino acids into proteins is called ________________.
7. When a red blood cell draws in water and bursts it is said to undergo ________________.
8. The chromosomes duplicate during the period between mitoses, which is called ________________.
9. A cell has four chromosomes before entering the process of mitosis. After mitosis, the number of chromosomes in each daughter cell will be ________________.
10. Transporters are used for the processes of active transport and ________________.
11. Droplets of water and dissolved substances are brought ino the cell by the process of ________________.
12. The number of daughter cells formed when a cell undergoes mitosis is ________________.
13. Bacteria are brought into the cell by the process of ________________.

Understanding Concepts

I. True/False

For each question, write *T* for true or *F* for false in the blank to the left of each number. If a statement is false, correct it by replacing the underlined term and write the correct statement in the blank below the question.

_______ 1. The nucleotide sequence "ACCUG" would be found in DNA.

____

_______ 2. A living cell (with a tonicity equivalent to 0.9% NaCl) is placed in a solution containing 2% NaCl. This solution is hypertonic.

____

_______ 3. Glucose is moving into a cell, down its concentration gradient, using a carrier protein. Glucose is travelling by active transport.

____

_______ 4. A toxin has entered a cell. The cell is no longer capable of generating ATP. The most likely explanation for this effect is that the toxin has destroyed the <u>mitochondria</u>.

____

_______ 5. The best microscope to view a ribosome would be a <u>scanning electron microscope</u>.

____

_______ 6. It is impossible to count individual chromosomes during <u>interphase</u>.

____

II. Practical Applications

Study each discussion. Then write the appropriate word or phrase in the space provided. The following are observations you might make while working for the summer in a hospital laboratory.

1. A sample of breast tissue that was thought to be cancerous arrived at the pathology lab. Breast cancer is an example of a disease that occurs more often within certain families than others, suggesting that one cancer risk factor is ___________________.
2. The tissue was in a liquid called normal saline so that the cells would neither shrink nor swell. Normal saline contains 0.9% salt and is thus considered to be ___________________.
3. The pathologist (Dr. C) sliced the tissue very thinly and placed it on a microscope slide, but he was unable to see anything. Unfortunately, he had forgotten to add a dye to the tissue. These special dyes are called ___________________.
4. Dr. C went back to his bench in order to get the necessary dye. He noticed that there were many impurities floating in the dye, and he decided to screen them out. He separated the solid particles from the liquid by forcing the liquid through a membrane, a process called ___________________.
5. Dr. C was clumsy and accidentally spilled the dye into a sink full of dishes. The water in the sink rapidly turned pink. The dye molecules had moved through the water by the process of ___________________.
6. Dr. C made up some new dye and treated the tissue. Finally, the tissue was ready for examination. Which type of microscope uses light to view stained tissues? ___________________.
7. The pathologist looked at the tissue and noticed that the nuclei of many cells were in the process of dividing. This division process is called ___________________.
8. Some cells were in the stage of cell division called prophase. The DNA was condensed into structures called ___________________.

III. Short Essays

1. Compare and contrast active transport and facilitated diffusion. List at least one similarity and at least one difference.

___

___

___

2. You are in the hospital for a minor operation, and the technician is hooking you up to an IV. He knows you are a nursing student and jokingly asks if you would like a hypertonic, hypotonic, or isotonic solution to be placed in your IV. Which solution would you pick? Explain your answer.

___

___

___

3. Explain what is meant by the term *risk factors* in cancer and list several of these factors.

___

___

___

4. Compare the transmission electron microscope with the scanning electron microscope. List one similarity and one difference.

___

___

___

Conceptual Thinking

1. Compare the structure of a cell to a factory or a city. Try to find cell structures that accomplish all of the different functions of the city or factory.

___

___

___

___

___

2. Your great-aunt M is 96 years old and loves to hear about what you are learning in class. She recently attended an Elderhostel camp, where people were talking about this new-fangled notion called DNA.
   a. She asks you to explain why DNA is so important. Explain the role of DNA in protein synthesis, using clear, uncomplicated language. You must define any term that your great-aunt might not know. You can use an analogy if you like.

   ____

   ____

   ____

   ____

   ____

   b. Next, your great-aunt wonders how the proteins get out of the cell. Explain the pathway a protein takes from the ribosome to the blood. You can use an illustration if you like.

   ____

   ____

   ____

   ____

   ____

3. You are a xenobiologist studying an alien cell isolated on Mars. Surprisingly, you notice that the cell contains some of the same substances as our cells. You quantify the concentration of these substances and determine that the cell contains 10% glucose and 0.3% calcium. The cell is placed in a solution containing 20% glucose and 0.1% calcium. The plasma membrane of this cell is is very different from ours. It is permeable to glucose but not to calcium. That is, only glucose can cross the plasma membrane without using transporters. Use this information to answer the following questions:
   a. Will glucose move into the cell or out of the cell? Which transport mechanism will be involved?

   ____

   ____

   b. Carrier proteins are present in the membrane that can transport calcium. If calcium moves down its concentration gradient, will calcium move into the cell or out of the cell? Which transport mechanism will be involved?

   ____

   ____

c. You place the cell in a new solution to study the process of osmosis. You know that sodium does not move across the alien cell membrane. You also know that the concentration of the intracellular fluid is equivalent to 1% sodium. The new solution contains 2% sodium.
   (i) Is the 2% sodium solution hypertonic, isotonic, or hypotonic? ___________________
   (ii) Will water flow into the cell or out of the cell? ____________________
   (iii) What will be the effect of the water movement on the cell? ___________________

Expanding Your Horizons

Carcinogens are a hot topic. Newspapers often carry large headlines proclaiming the newest carcinogen (milk, BBQ meat, mosquito repellent). You can find out the truth about these claims in the National Toxicology Program yearly report on carcinogens. The most recent report can be read online. The *Scientific American* article listed below can provide you with more information about the causes of cancer.

Trees can be dated by counting the rings. Is it possible to tell the age of a cell? The answer is a qualified "yes." It is usually possible to estimate the number of divisions a cell has undergone. The DNA region at the end of chromosomes (the telomere) shortens every time a cell undergoes mitosis. Older cells thus have shorter telomeres. When telomeres become very short, the cell stops dividing or dies. Abnormally short telomeres result in premature aging of skin and other tissues. Some cells have a special enzyme (telomerase) that restores the telomeres, making the cells essentially immortal. Perhaps this enzyme can be inserted into human heart and skin cells—a true fountain of youth. You can read about telomeres and their importance in cell aging and cancer in the following article:

Resources

1. Gibbs WW. Untangling the roots of cancer. *Sci Am* 2003; 289:56–65.
2. Strauss E. Counting the lives of a cell. *Sci Am* 2000; 11:50–55.

CHAPTER 4

Tissues, Glands, and Membranes

Overview

The cell is the basic unit of life. Individual cells are grouped according to function into **tissues**. The four main groups of tissues include **epithelial tissue**, which forms glands, covers surfaces, and lines cavities; **connective tissue**, which gives support and form to the body; **muscle tissue**, which produces movement; and **nervous tissue**, which conducts nerve impulses.

Glands produce substances used by other cells and tissues. **Exocrine glands** produce secretions that are released through ducts to nearby parts of the body. **Endocrine glands** produce hormones that are carried by the blood to all parts of the body.

The simplest combination of tissues is a **membrane**. Membranes serve several purposes, a few of which are mentioned here: they may serve as dividing partitions, line hollow organs and cavities, and anchor various organs. Membranes that have epithelial cells on the surface are referred to as **epithelial membranes**. Two types of epithelial membranes are **serous membranes**, which line body cavities and cover the internal organs, and **mucous membranes**, which line passageways leading to the outside.

Connective tissue membranes cover or enclose organs, providing protection and support. These membranes include the fascia around muscles, the meninges around the brain and spinal cord, and the tissues around the heart, bones, and cartilage.

If the normal pattern of cell growth is disrupted by the formation of cells that multiply out of control, the result is a **tumor**. A tumor that is confined locally and does not spread is called a benign tumor; a tumor that spreads from its original site to other parts of the body, a process termed **metastasis**, is called a malignant tumor. The general term for any type of malignant tumor is **cancer**. Tissue biopsy, radiography, ultrasound, computed tomography, and magnetic resonance imaging are the techniques most frequently used to diagnose cancer. Most benign tumors can be removed surgically; malignant tumors are usually treated by surgery, radiation, or chemotherapy, or by a combination of these methods.

The study of tissues—**histology**—requires much memorization. In particular, you may be challenged to learn the different types of epithelial and connective tissue as well as the classification scheme of epithelial and connective membranes. Learning the structure of these different tissues and membranes will help you understand the amazing properties of the body—how we can jump from great heights, swim without becoming waterlogged, and fold our ears over without breaking them. Also, the study of histology is necessary for understanding tissue diseases such as cancer.

Addressing the Learning Outcomes

1. Name the four main groups of tissues and give the location and general characteristics of each.

EXERCISE 4-1: Three Types of Epithelium (Text Fig. 4-1)

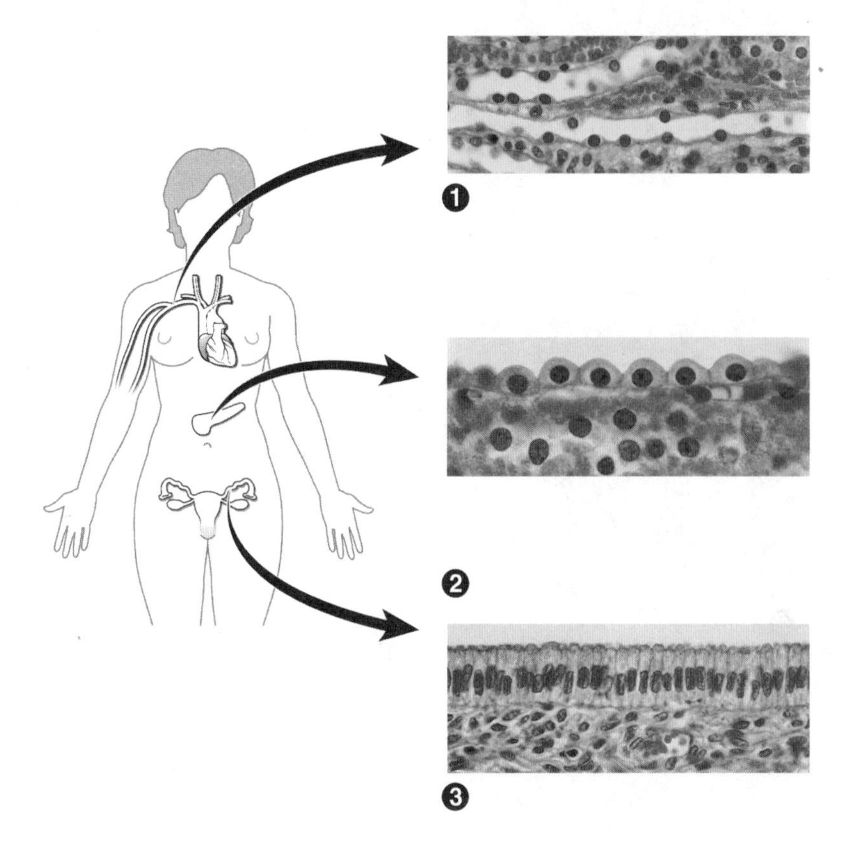

INSTRUCTIONS

Label each of the following types of epithelium.

1. ______________________
2. ______________________
3. ______________________

EXERCISE 4-2: Muscle Tissue (Text Fig. 4-7)

INSTRUCTIONS

Write the names of the three types of muscle tissue in the appropriate blanks in different colors. Color some of the muscle cells with the corresponding color. Look for the nuclei, and color them a different color.

1. ___________________________
2. ___________________________
3. ___________________________

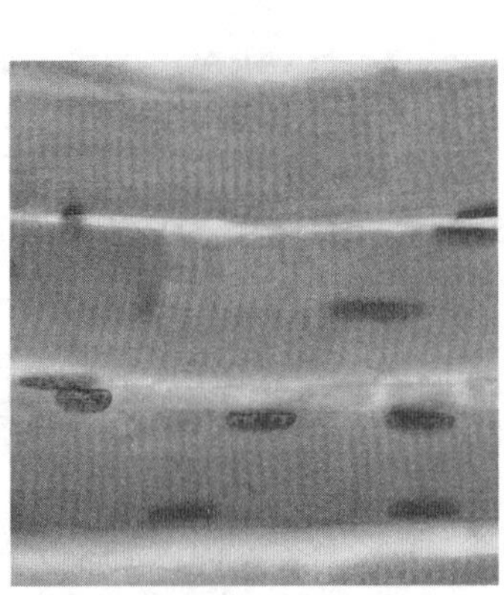

1

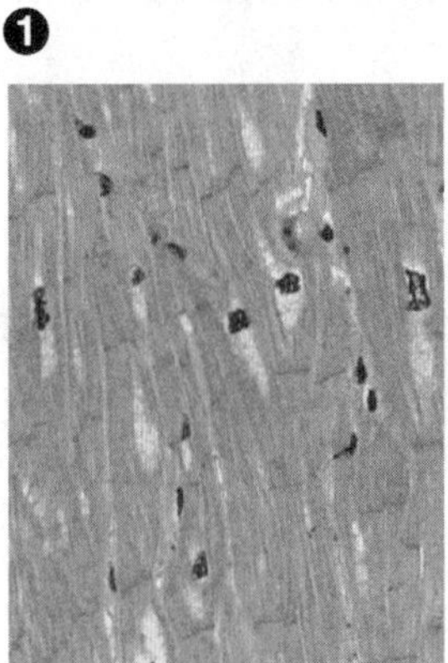

2

3

EXERCISE 4-3: Nervous Tissue (Text Fig. 4-8)

INSTRUCTIONS

1. Write the names of each tissue (indicating the plane of the section where appropriate) in boxes 7 to 9.
2. Label each of the following neural structures and tissues using different colors. Where possible, color each structure or tissue with the corresponding color.

1. ____________________
2. ____________________
3. ____________________
4. ____________________
5. ____________________
6. ____________________

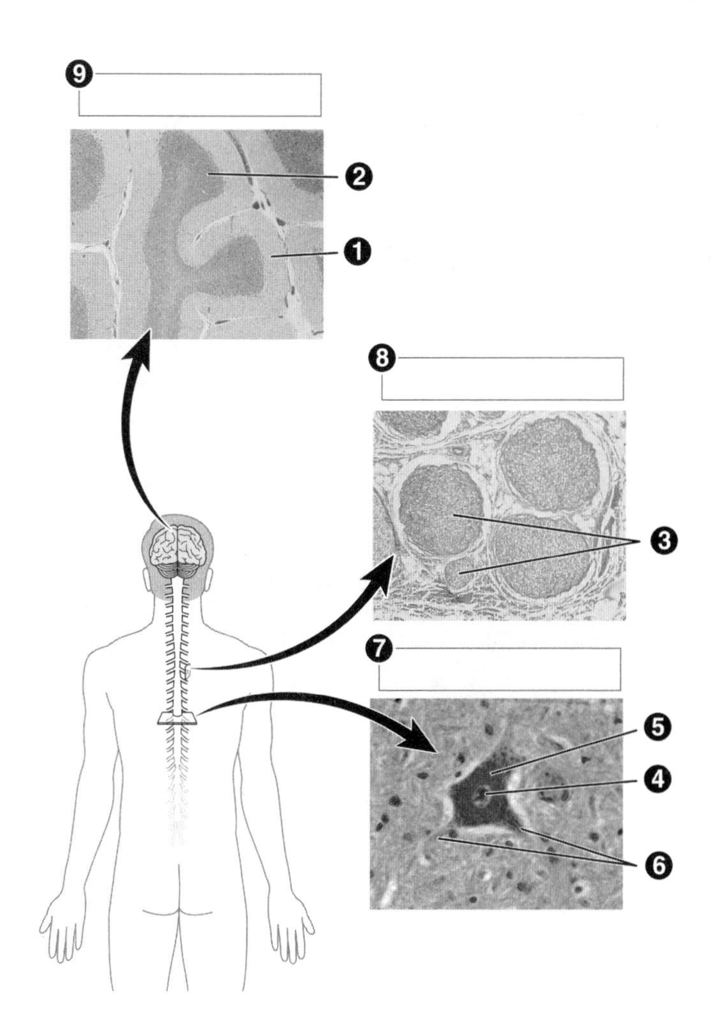

EXERCISE 4-4.

INSTRUCTIONS

Write the appropriate term in each blank.

tissue squamous stratified transitional
columnar simple cuboidal

1. A group of cells similar in structure and function ______________
2. Term that describes flat, irregular epithelial cells ______________
3. A term that means *in layers* ______________
4. Term that describes long and narrow epithelial cells ______________
5. Term that describes square epithelial cells ______________
6. Cells arranged in a single layer ______________

EXERCISE 4-5.

INSTRUCTIONS

Write the appropriate term in each blank.

bone myocardium voluntary muscle epithelial tissue
neuron smooth muscle neuroglia connective tissue

1. The thigh muscle is an example of ______________
2. Tissue that forms when cartilage gradually becomes impregnated with calcium salts ______________
3. The thick, muscular layer of the heart wall ______________
4. A type of tissue found in membranes and glands ______________
5. Visceral muscle is also known as ______________
6. A cell that carries nerve impulses is called a(n) ______________
7. A tissue in which cells are separated by large amounts of acellular material called matrix ______________

2. Describe the difference between exocrine and endocrine glands and give examples of each.

EXERCISE 4-6.

INSTRUCTIONS

Do each of the following characteristics apply to exocrine glands (EX), endocrine glands (EN), or both (B)? Write the appropriate abbreviation in each blank.

1. A gland that secretes into the blood ______________
2. A gland that secretes through ducts ______________
3. A gland that secretes onto the body surface ______________
4. The pituitary gland, for example ______________
5. A group of cells that produces substances for use by other parts of the body ______________
6. Salivary glands, for example ______________

3. Give examples of circulating, generalized, and structural connective tissues.

EXERCISE 4-7: Connective Tissue (Text Figs. 4-5 and 4-6)

INSTRUCTIONS

Write the names of the six examples of connective tissue in the appropriate blanks. Use red for the circulating connective tissue (1), three different shades of green for the generalized tissues (2 to 4), and two shades of blue for the structural tissues (5 to 6). Color some of the **cells** of each tissue type with the corresponding color.

1. ____________________
2. ____________________
3. ____________________
4. ____________________
5. ____________________
6. ____________________

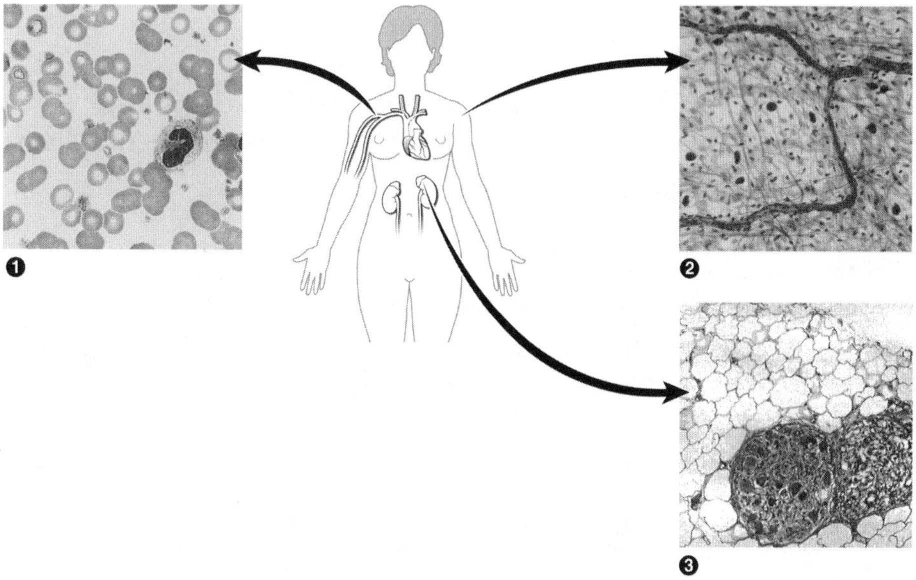

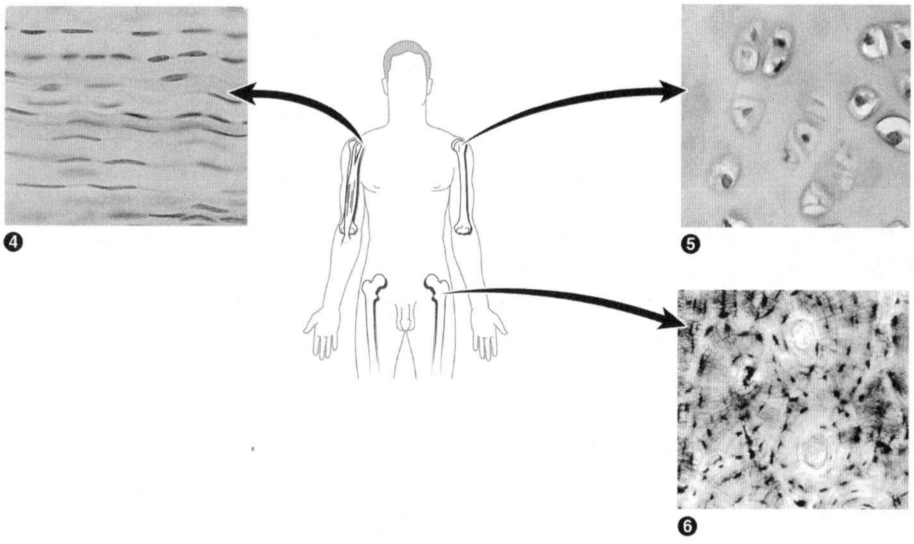

➊ ➋ ➌ ➍ ➎ ➏

EXERCISE 4-8.

INSTRUCTIONS

Write the appropriate term in each blank.

ligament	tendon	collagen	chondrocyte
capsule	fibrocartilage	hyaline cartilage	elastic cartilage

1. A cord of connective tissue that connects a muscle to a bone ________________
2. A tough membranous connective tissue that encloses an organ ________________
3. The cartilage found between the bones of the spine ________________
4. A fiber found in most connective tissues ________________
5. A cell that synthesizes cartilage ________________
6. A strong, gristly cartilage that makes up the trachea ________________

4. Describe three types of epithelial membranes.

EXERCISE 4-9.

INSTRUCTIONS

Write the appropriate term in each blank.

mesothelium	serous membrane	cutaneous membrane	parietal layer
visceral layer	mucous membrane	peritoneum	serous pericardium

1. An epithelial membrane that lines a body cavity or covers an internal organ ________________
2. The epithelial membrane also known as the skin ________________
3. A membrane that lines a space open to the outside of the body ________________
4. The portion of a serous membrane attached to an organ ________________
5. The portion of a serous membrane attached to the body wall ________________
6. The epithelial portion of serous membranes ________________
7. The serous membrane covering the heart ________________

5. List several types of connective tissue membranes.

Exercise 4-10.

INSTRUCTIONS

Write the appropriate term in each blank.

periosteum	fibrous pericardium	perichondrium	epithelial membrane
synovial membrane	superficial fascia	deep fascia	

1. The sheet of tissue that underlies the skin ________________
2. The connective tissue membrane that lines joint cavities ________________
3. A tough membrane composed entirely of connective tissue that serves to anchor and support an organ or to cover a muscle ________________
4. A layer of fibrous connective tissue around a bone ________________
5. The membrane that covers cartilage ________________

6. A general term describing a membrane composed of epithelial and connective tissue ________
7. A general term describing a membrane composed exclusively of connective tissue ________

6. Explain the difference between benign and malignant tumors and give several examples of each type.

EXERCISE 4-11.

INSTRUCTIONS

Write the appropriate term in each blank.

adenoma sarcoma osteoma nevus angioma
nevus malignant benign chondroma

1. Term for a tumor that does not spread ________
2. Term for a tumor that spreads to other tissues ________
3. The general term for a tumor composed of blood or lymphatic vessels ________
4. A benign connective tissue tumor that originates in a bone ________
5. A benign epithelial tumor that originates in a gland ________
6. A malignant tumor originating in connective tissue ________
7. A tumor that originates in cartilage ________
8. A small skin tumor that may become malignant ________

7. List some signs of cancer.

EXERCISE 4-12.

INSTRUCTIONS

In the blanks below, list some symptoms of cancer.

1. ________
2. ________
3. ________
4. ________
5. ________
6. ________

8. List six methods of diagnosing cancer.

EXERCISE 4-13.

INSTRUCTIONS

Write the appropriate term in each blank.

biopsy	ultrasound	positron emission tomography
computed tomography	magnetic resonance imaging	radiography

1. Mammography is an example of this technique ________
2. The use of magnetic fields and radio waves to diagnose soft tissue cancers ________
3. Technique using high-frequency sound waves ________
4. The removal of living tissue for microscopic examination ________
5. Technique that diagnoses tumors based on their abnormally high activity levels ________
6. Use of many x-rays to produce a cross-sectional image of body parts ________

9. Describe three traditional methods of treating cancer.

EXERCISE 4-14.

INSTRUCTIONS

List the three traditional cancer treatment methods in the blanks below. Briefly describe why each treatment works.

1. ____
____
____

2. ____
____
____

3. ____
____
____

10. Show how word parts are used to build words related to tissues, glands, and membranes.

EXERCISE 4-15.

INSTRUCTIONS

Complete the following table by writing the correct word part or meaning in the space provided. Write a word that contains each word part in the Example column.

Word Part	Meaning	Example
1. ________	cartilage	________________
2. ________	inflammation	________________
3. arthr/o	________	________________
4. ________	tumor, swelling	________________
5. osse/o	________	________________
6. -blast	________	________________
7. peri-	________	________________
8. ________	muscle	________________
9. ________	false	________________
10. hist/o	________	________________
11. neo-	________	________________
12. ________	vessel	________________
13. ________	side, rib	________________
14. leuk/o-	________	________________
15. neur/o	________	________________

Making the Connections

The following concept map deals with the classification of tissues. Complete the concept map by filling in the appropriate word or phrase that classifies or describes the tissue.

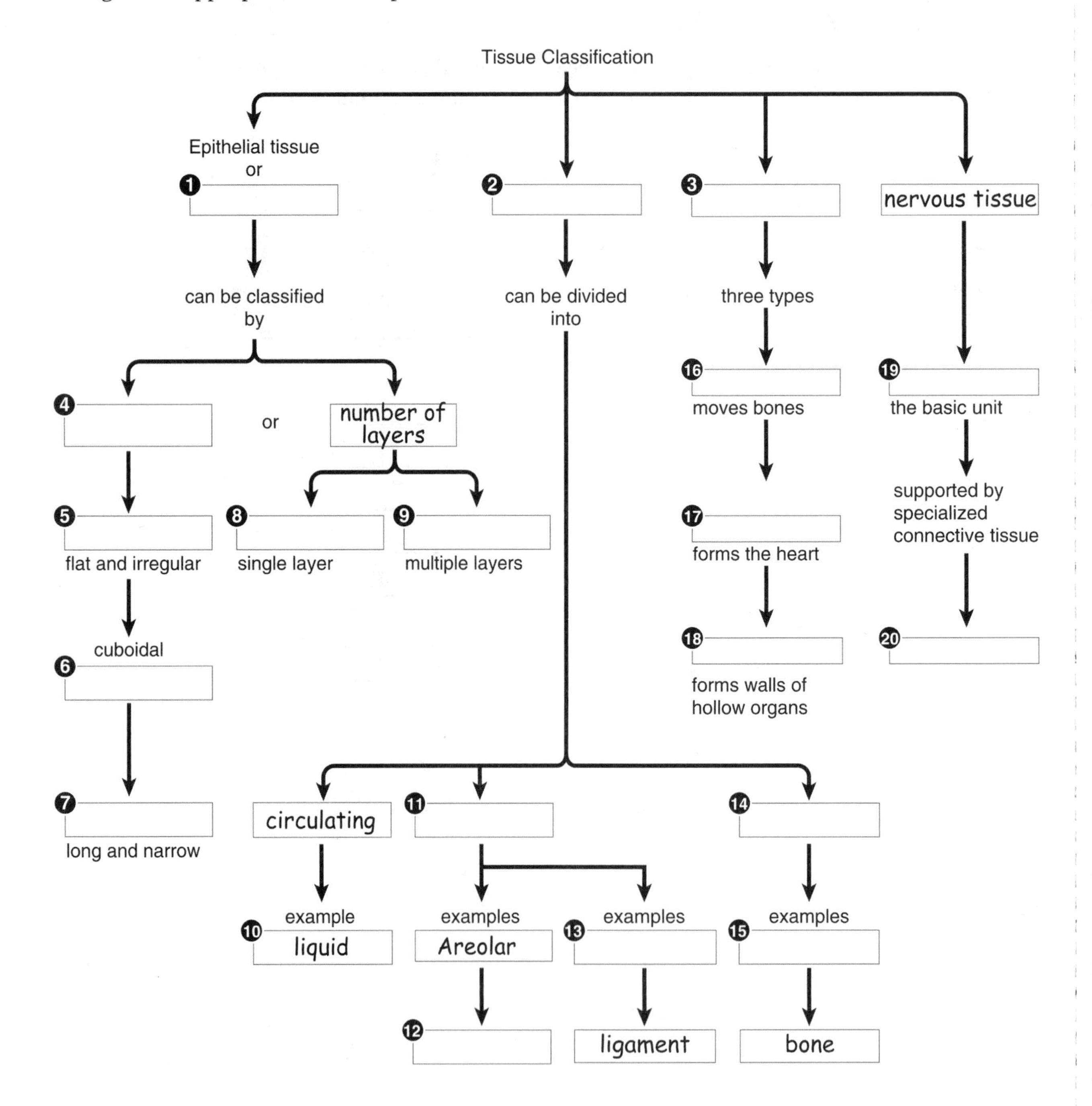

Optional Exercise: Assemble your own concept map summarizing the classification of membranes. Use the following terms: membrane, epithelial tissue membrane, connective tissue membrane, serous, mucous, cutaneous, peritoneum, pleurae, serous pericardium, parietal layer, visceral layer, meninges, fascia, deep, superficial, fibrous pericardium, periosteum, and perichondrium. You could also provide some examples of the different membranes, as shown in the concept map provided.

Testing Your Knowledge

Building Understanding

I. Multiple Choice

Select the best answer and write the letter of your choice in the blank.

1. Which of the following is a type of connective tissue? 1. _______
   a. transitional
   b. squamous
   c. cuboidal
   d. areolar
2. The phrase *stratified cuboidal epithelium* describes: 2. _______
   a. flat, irregular epithelial cells in a single layer
   b. square epithelial cells in many layers
   c. long, narrow epithelial cells in a single layer
   d. flat, irregular epithelial cells in many layers
3. The only type of muscle that is under voluntary control is: 3. _______
   a. smooth
   b. skeletal
   c. cardiac
   d. visceral
4. The proper scientific name for a nerve cell is: 4. _______
   a. neuroglia
   b. nevus
   c. neuron
   d. axon
5. Mucus is secreted from: 5. _______
   a. endocrine glands
   b. goblet cells
   c. areolar tissue
   d. tendons
6. Cartilage is produced by: 6. _______
   a. chondrocytes
   b. fibroblasts
   c. osteoblasts
   d. osteocytes
7. An example of structural connective tissue is: 7. _______
   a. an organ capsule
   b. adipose
   c. tendon
   d. bone
8. The tough connective tissue membrane that covers most parts of all bones is the: 8. _______
   a. perichondrium
   b. periosteum
   c. fascia
   d. ligament

9. A tumor that originates from the support tissue of the central nervous system is a(n): 9.______
   a. myoma
   b. angioma
   c. carcinoma
   d. glioma
10. A tumor that originates in the epithelium is a(n): 10.______
   a. myoma
   b. angioma
   c. carcinoma
   d. glioma

II. Completion Exercise

Write the word or phrase that correctly completes each sentence.

1. The secretion that traps dust and other inhaled foreign particles is ____________
2. Cells that form bone are called ____________
3. The cartilage found at the end of long bones is called ____________
4. The axons of some neurons have an insulating coating called ____________
5. The study of tissues is called ____________
6. The epithelial membrane that lines the walls of the abdominal cavity is called the ____________
7. A mucous membrane can also be called the ____________
8. A lubricant that reduces friction between the ends of bones is produced by a(n) ____________
9. The microscopic, hairlike projections found in the cells lining most of the respiratory tract are called ____________
10. The process of tumor cell spread is called ____________
11. A malignant tumor that originates in connective tissue is called a(n) ____________
12. The extent of tumor spread is determined by a process called ____________

Understanding Concepts

I. True/False

For each question, write *T* for true or *F* for false in the blank to the left of each number. If a statement is false, correct it by replacing the underlined term and write the correct statement in the blank below the question.

______ 1. The mouth is lined by a type of epithelial membrane called a <u>serous</u> membrane.

____

______ 2. The heart is an example of <u>skeletal muscle</u>.

____

_______ 3. Tumors that undergo metastasis are called <u>benign</u> tumors.

____

_______ 4. The strip of tissue connecting the kneecap to the thigh muscle is an example of a <u>tendon</u>.

____

_______ 5. Inflammation of the large abdominal serous membrane is <u>pleuritis</u>.

____

_______ 6. The pituitary gland releases prolactin into the bloodstream. The pituitary gland is thus an <u>exocrine</u> gland.

____

_______ 7. The <u>visceral</u> layer of the peritoneum is in contact with the stomach.

____

_______ 8. You have identified a new gland in the neck. This gland is connected to the mouth by a duct. Your new gland is an example of an <u>exocrine</u> gland.

____

II. Practical Applications

Study each discussion. Then write the appropriate word or phrase in the space provided.

➤ Group A: Cancer

You are spending some time in an oncology clinic as a volunteer. The nurse has asked you to perform an initial interview of a patient.

1. Ms. M was your first patient. You noted that she had a large wart on her chin. The scientific name for a wart is _______________.
2. Ms. M complained of headaches and blurred vision. You suspect a tumor of the neurons or the cells that that support the neurons. The support cells are called _______________.
3. Ms. M was subjected to a test that uses magnetic fields to show changes in soft tissues. This test is called _______________.
4. The test confirmed your diagnosis, showing a small tumor just beneath the membranes that surround the brain. These membranes are called the _______________.
5. The tumor appeared to be encapsulated, and other studies did not reveal any secondary tumors. Tumors that are confined to a single area are called _______________.
6. The neurosurgeon was able to successfully remove the tumor and sent it to be studied in the histology laboratory. To his surprise, the histologist determined that the tumor consisted of cells that resembled neurons. This type of tumor is called a(n) _______________.

7. The oncologist did not see the need for any further treatment. Ms. M was relieved, because she had always feared chemotherapy. Chemotherapy, in the context of cancer, involves treatment with a particular class of drugs called ______________.

➤ Group B: Tissues and Membranes

You are working as a sports therapist for a wrestling team. At a particularly brutal competition, you are asked to evaluate a number of injuries.

1. Mr. K suffered a crushing injury to the lower leg yesterday in a sumo wrestling match when his opponent fell on him. Initially, he had little pain. Now, he complains of numbness and pain in the foot and leg. This type of injury is made worse by the tight, fibrous covering of the muscles, known as the ______________.
2. Mr. K is also complaining of pain in the knee. You suspect an injury to the membrane that lines the joint cavity, a membrane called the ______________.
3. You note that Mr. K has a significant amount of fat. Fat is contained in a type of connective tissue called ______________.
4. Ms. J suffered a bloody nose in her wrestling match. Blood is a liquid form of the tissue classified as ______________.
5. Ms. J also suffered a painful bump on her ankle in the same match. The swelling involved the superficial tissues and the fibrous covering of the bone, or the ______________.
6. Mr. S was involved in a closely fought match when his opponent bent his ear back. Thankfully, the cartilage in his ear was able to spring back into shape. This kind of cartilage is called ______________.
7. Later, Mr. S suffered a penetrating wound to his abdomen when his opponent accidentally threw him into the seating area. You fear that the wound may have penetrated the membrane that lines his abdomen, called the ______________.
8. The wrestling coach comes over to talk to you during a break in the match. He has a question about his favorite shampoo. The advertisement stated that it contained collagen. He asks you which cells in the body synthesize collagen. These cells are ______________.

III. Short Essays

1. What are antineoplastic agents, and what is the rationale of using them to treat cancer?

____

____

____

2. Compare and contrast benign and malignant tumors. List at least one similarity and two differences.

____

____

____

3. Name two methods of cancer treatment other than chemotherapy, radiation, and surgery and briefly describe how they work.

____

____

____

Conceptual Thinking

1. Why is bone considered to be connective tissue? Define connective tissue in your answer.

____

____

____

2. Which tissue, epithelial or connective, would be best suited to the following functions?
   a. cushioning the kidneys against a blow
   b. creating a virtually waterproof barrier between the body and the environment
   c. preventing toxins from entering the blood from the gastrointestinal tract

____

____

____

Expanding Your Horizons

We are constantly reading headlines in the news about "New Treatments for Cancer!" Many of these claims are based on successful trials of drugs in mice that do not necessarily prove to be effective in humans. Investigate ongoing trials of different cancer treatments at the National Cancer Institute website (http://www.cancer.gov). The website of the National Center for Complementary and Alternative Medicine (http://nccam.nih.gov) discusses the efficacy of alternative treatments such as massage, acupuncture, and herbal remedies.

Your mother may have told you to "eat your broccoli so you won't get cancer!" We now know that lifestyle changes such as exercise and diet can reduce the incidence of cancer. Read about some of these lifestyle changes in the following article from *Scientific American*.

Resources

1. Howard K. Stopping cancer before it starts. *Sci Am* 2000; 11:80–87.

UNIT II Diseases and the First Line of Defense

CHAPTER 5

Disease and Disease-Producing Organisms

Overview

Disease may be defined as an abnormality of the structure or function of a body part, organ, or system. The categories of disease are many and varied. Among them are infections, degenerative diseases, nutritional disorders, metabolic disorders, immune disorders, neoplasms, and psychiatric disorders. Predisposing causes are factors that play a part in the development of disease, such as age, gender, heredity, living conditions and habits, emotional disturbance, physical and chemical damage, and pre-existing illness. An understanding of disease relies on the study of the anatomy (structure) and physiology (functions) of the body under normal and pathologic (abnormal) conditions. This understanding is aided by use of disease terminology. Healthcare workers diagnose diseases by observing a patient's symptoms and looking for signs of diseases. In recent years there has been an increased emphasis in healthcare professions to prevent diseases as well as treat them.

Infection, or invasion of the body by disease-producing microorganisms such as bacteria, fungi, viruses, and protozoa, is a major cause of disease in humans. Another major cause of human disease is **infestation**, a type of infection due to parasitic worms. Infection control depends on understanding how organisms are transmitted and how they enter and leave the body. Public health has been vastly improved through laboratory identification of pathogens and the application of **aseptic** and **chemotherapeutic** methods to prevent or control their spread.

This chapter does not contain any difficult concepts, but it is rich in detail. You may find it useful to construct a chart that summarizes the physical characteristics, diseases, and treatment methods for the different microorganisms and worms. The concept maps will also help you to master the content.

Addressing the Learning Outcomes

1. Define *disease* and list seven categories of disease.

EXERCISE 5-1.

INSTRUCTIONS

Write the appropriate terms in each blank.

neoplasm	infectious disease	psychiatric disorder	metabolic disorder
nutritional disorder	immune disorder	degenerative disorder	

1. A disease that can be transmitted between individuals ✓ infectious disease
2. A disorder involving tissue breakdown, such as that associated with aging ✓ degenerative
3. A cancer or other tumor ✓ neoplasm
4. Disorders involving a disruption in the chemical reactions of the body ✓ metabolic
5. Obesity is an example of this type of disorder ✓ nutritional disorder
6. Allergy is an example of this type of disorder ✓ immune
7. Another term for mental disorder ✓ psychiatric disorder

2. List seven predisposing causes of disease.

EXERCISE 5-2.

INSTRUCTIONS

List the seven predisposing causes of disease in the spaces below.

1. age ✓
2. gender ✓
3. heredity ✓
4. living conditions + habits ✓
5. emotional disturbance ✓
6. physical + chemical damage ✓
7. pre-existing illness ✓

3. Define terminology used in describing and treating disease.

EXERCISE 5-3.

INSTRUCTIONS

Write the appropriate term in each blank.

chronic	acute	endemic	idiopathic	epidemiology
iatrogenic	communicable	pandemic	epidemic	

1. Term for a disease that is the result of adverse effects of medical treatment ___iatrogenic___
2. Describing a relatively severe disorder of short duration ___acute___
3. A disease that is found continuously, but to a low extent, in a population ___endemic___
4. Term for a disease that persists over a long period ___chronic___
5. A disease that can be transmitted between individuals ___communicable___
6. A disease prevalent throughout an entire country or the world ___pandemic___
7. A disease of unknown cause ___idiopathic___
8. The study of the geographical distribution of a disease ___epidemiology___

EXERCISE 5-4.

INSTRUCTIONS

Write the appropriate term in each blank.

therapy	syndrome	etiology	symptom	prevalence
sign	prognosis	diagnosis	incidence	

1. A group of signs or symptoms that occur together
2. The study of the cause of a disorder
3. A condition of disease that is experienced by the patient
4. The process of determining the nature of an illness
5. Evidence of disease noted by a healthcare worker
6. A prediction of the probable outcome of a disease
7. A course of treatment
8. The overall frequency of a disease in a given group

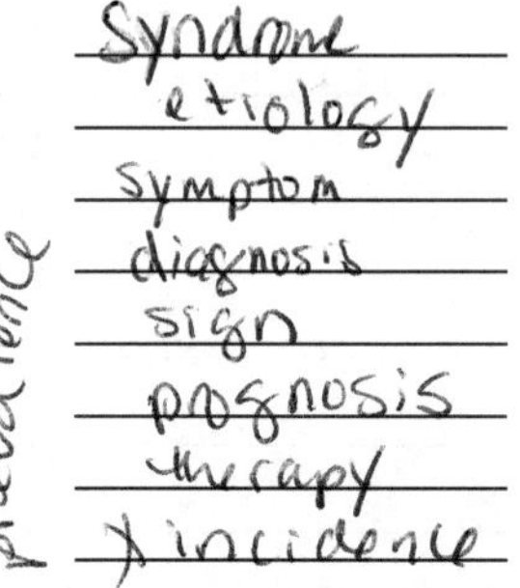

4. Define *complementary and alternative medicine*; cite several alternative or complementary fields of practice.

EXERCISE 5-5.

INSTRUCTIONS

For each of the following statements, write if they are true (T) or false (F).

1. Complementary and alternative medicine refers to practices that cannot be combined with traditional modern medicine practices. ___F___
2. Naturopathy uses manipulation to correct misalignment for musculoskeletal disorders. ___F___

3. Yoga and meditation are examples of complementary medical practices. T ✓
4. Complementary and alternative medical practices are evaluated by the same institute that evaluates traditional modern medical practices, the NIH. T ✓

5. Explain methods by which microorganisms can be transmitted from one host to another.

EXERCISE 5-6.

INSTRUCTIONS

Write the appropriate term in each blank.

systemic	local	opportunistic	pathogen
parasite	host	vector	portal of entry

1. Term for an infection that affects the whole body systemic ✓
2. Term for an infection that takes hold in a weakened host opportunistic ✓
3. Any organism that lives on or within another organism at that organism's expense parasite ✓
4. The avenue by which a microorganism enters the body portal of entry ✓
5. An animal that transmits a disease-causing organism from one host to another vector ✓
6. An organism that causes disease pathogen ✓

6. List four types of organisms studied in microbiology and give the characteristics of each.

EXERCISE 5-7.

Do each of the following characteristics apply to bacteria (B), viruses (V), protozoa (P), or fungi (F)?

1. Single-celled organisms with a cell membrane but without a nucleus ~~protozoa~~ Bacteria B
2. Group of microbes that includes the ciliates protozoa ✓
3. Obligate intracellular parasites resistant to antibiotics viruses ✓
4. Microbes composed of nucleic acid and a protein coat viruses ✓
5. Single-celled forms are called yeasts fungi ✓
6. Filamentous forms are called molds fungi ✓
7. Single-celled, animal-like microbes ~~bacteria~~ Protozoa
8. Some of these microbes produce endospores bacteria ✓

EXERCISE 5-8: Pathogenic Protozoa (Text Fig. 5-9)

INSTRUCTIONS

1. Write the name of each type of protozoon on the appropriate numbered lines 1 to 3 in a particular color. Use light colors; do not use red.
2. Color the protozoa on the diagram with the corresponding color.
3. Write the names of the different forms of protozoa in lines 4 to 7.
4. Color the red blood cells red.

1. ____________________
2. ____________________
3. ____________________
4. ____________________
5. ____________________
6. ____________________
7. ____________________

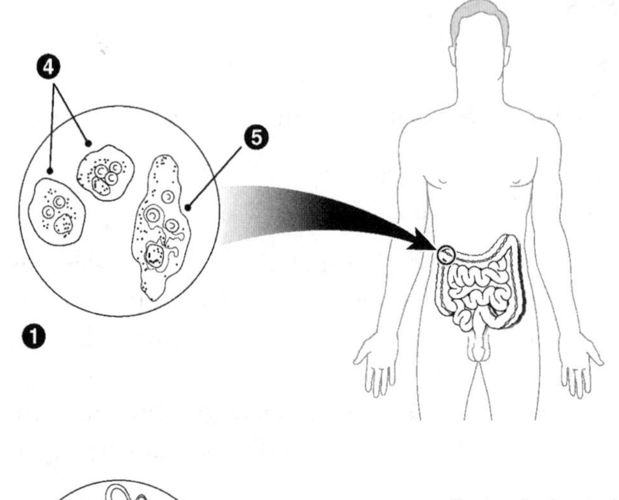

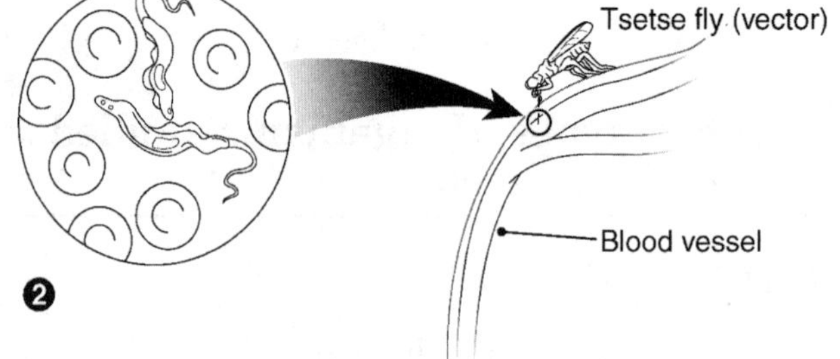

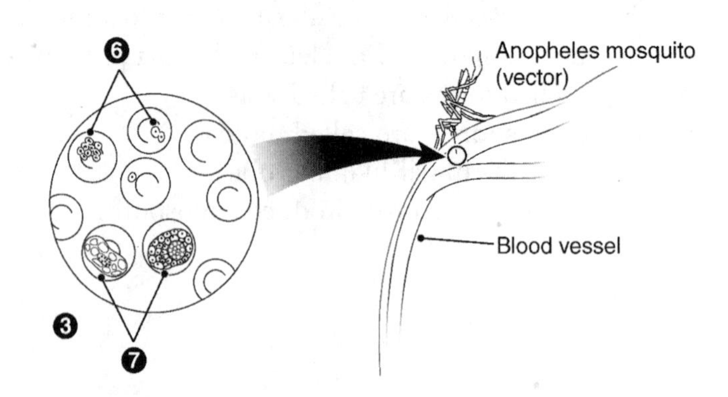

7. List some diseases caused by each type of microorganism.

EXERCISE 5-9.

INSTRUCTIONS

For each of the following diseases, state if they are bacterial (B), viral (V), protozoan (P), or fungal (F).

1. African sleeping sickness protozoan ✓
2. AIDS viral ✓
3. *Pneumocystis* pneumonia (PCP) fungi ✓
4. malaria protozoan ✓
5. thrush fungi ✓
6. Rocky Mountain spotted fever bacterial
7. diphtheria bacteria ✓
8. influenza viral ✓
9. chickenpox viral ✓
10. cholera bacteria ✓

8. Define *normal flora* and explain the value of the normal flora.

EXERCISE 5-10.

INSTRUCTIONS

Address this objective by writing an answer in the lines below. Use your own words—avoid copying from the textbook!

Normal flora - grow on or in your body -
we live in balance with these organisms
Normal flora - crowd out and prevent the growth of harmful varities of organisms.
without harm

9. Describe the three types of bacteria according to shape.

EXERCISE 5-11: Bacteria (Text Figs. 5-3, 5-4, and 5-5)

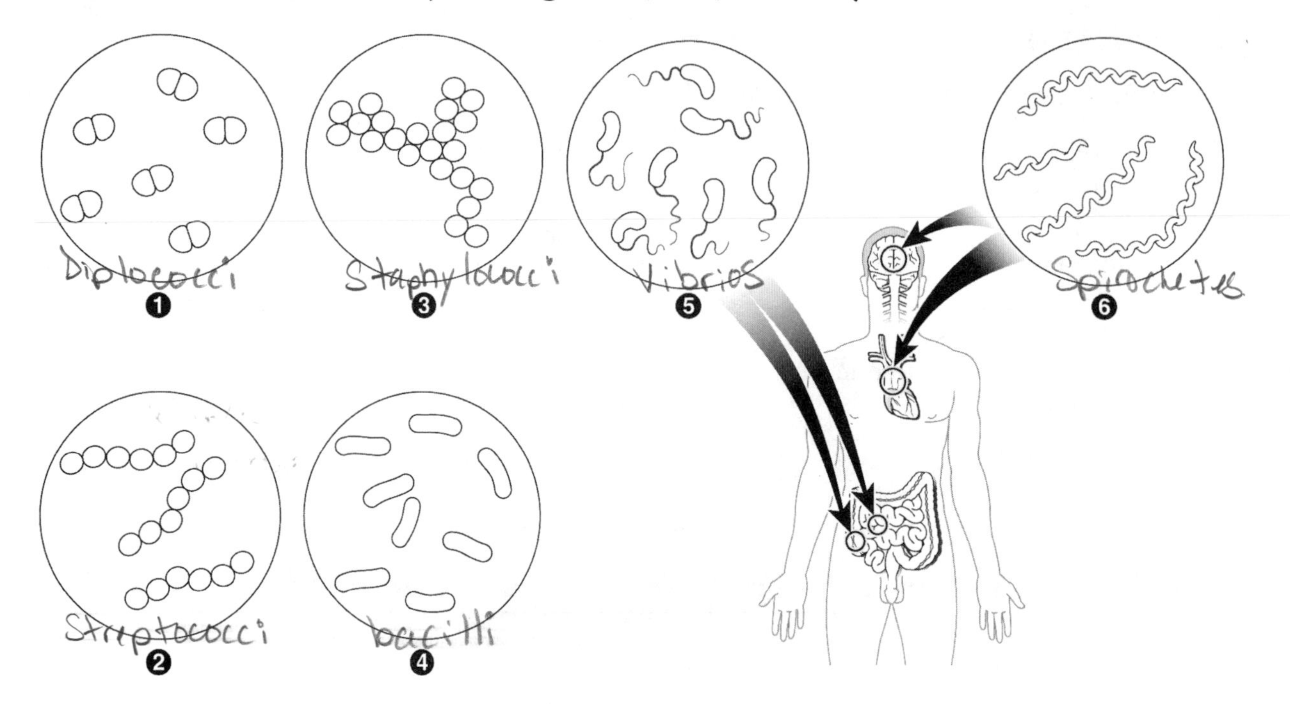

INSTRUCTIONS

1. Write the name of each type of bacteria on the appropriate numbered line in different colors.
2. Color the bacteria on the diagram with the corresponding color.

1. Diplococci
2. Streptococci
3. Staphylococci
4. bacilli
5. Vibrios
6. Spirochetes

✓

EXERCISE 5-12.

INSTRUCTIONS

Write the appropriate term in each blank.

diplococci	staphylococci	streptococci	vibrios	chlamydiae
bacilli	spirilla	spirochetes	rickettsiae	

✓

1. Spherical bacteria organized in clusters Staphylococci
2. Spherical bacteria arranged in chains streptococci
3. Long, wavelike cells resembling corkscrews Spirilla
4. Bacteria shaped like straight rods that may form endospores bacilli
5. Short rods with a small curvature vibrios
6. A group of bacteria smaller than the rickettsiae chlamydiae
7. Spherical bacteria occurring in pairs diplococci

10. List several diseases in humans caused by worms.

EXERCISE 5-13.

INSTRUCTIONS

Write the appropriate term in each blank.

infestation	pinworm	ascaris	trichina
filaria	vermifuge	hookworm	

1. A roundworm transmitted by eating undercooked pork trichma
2. The most common form of roundworm, which infests the intestines or lungs ascaris ✓
3. A general term describing an invasion by a parasitic worm infestation
4. An anthelmintic agent vermifuge
5. The worm that can cause elephantiasis filaria
6. A small worm that can cause anemia hookworm

EXERCISE 5-14: Parasitic Worms (Text Fig. 5-10)

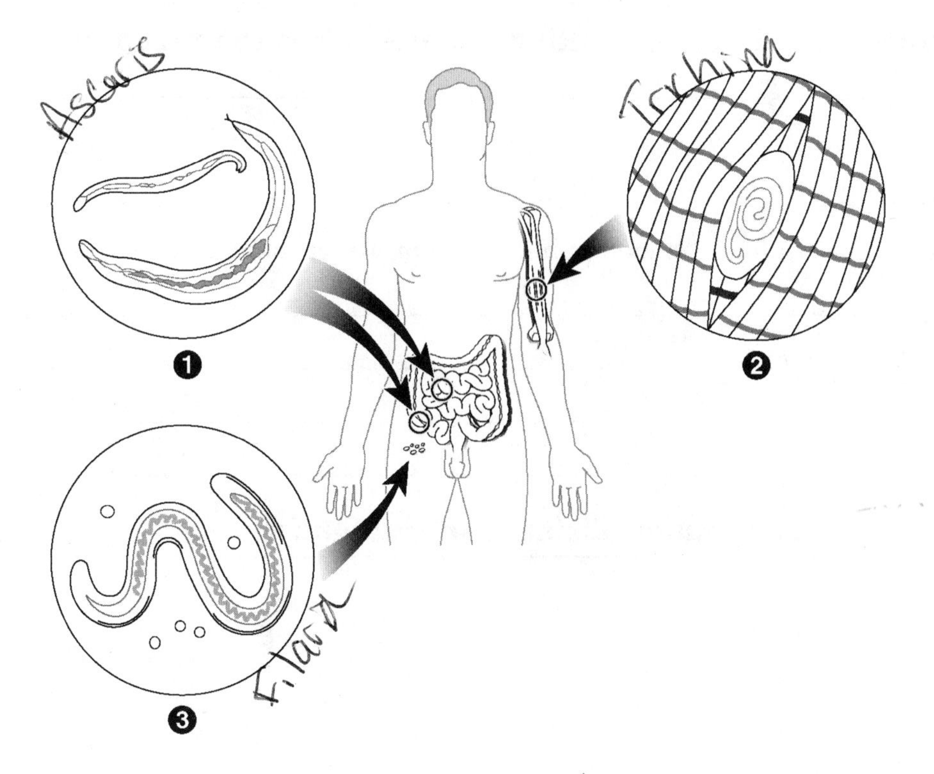

INSTRUCTIONS

1. Write the name of each type of worm on the appropriate numbered line in different colors.
2. Color the worm on the diagram with the corresponding color.

1. ____________________
2. ____________________
3. ____________________

11. Give some reasons for the emergence and spread of microorganisms today.

EXERCISE 5-15.

INSTRUCTIONS

List four reasons why infectious diseases are on the rise.

1. increased crowding
2.
3. global travel
   lifespan
4. poor food handling ↑ illness

12. Describe several public health measures taken to prevent the spread of disease.

EXERCISE 5-16.

INSTRUCTIONS

In the space below, list four common public health measures that improve human health.

1. Sewage + garbage disposal
2. water purification
3. Prevention of food contamination   proper food handling
4. Milk pasteurization

13. Differentiate *sterilization, disinfection,* and *antisepsis.*

EXERCISE 5-17: Aseptic Methods (Text Fig. 5-13)

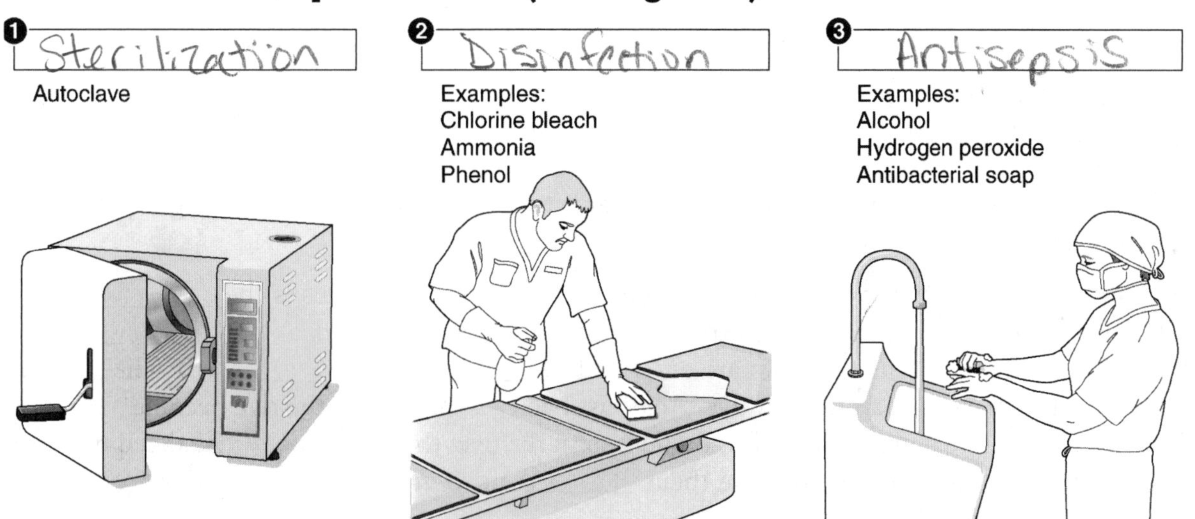

INSTRUCTIONS

Label each of the aseptic methods pictured here in the appropriate box.

14. Describe techniques included as part of body substance precautions.

EXERCISE 5-18.

INSTRUCTIONS

Identify which techniques are appropriate (G: good) and which might contribute to the spread of disease (B: bad).

1. changing gloves every two or three patients to avoid excess garbage ____
2. recapping needles to avoid accidental injections ____
3. using barriers (such as a mask) for every patient, not just potentially infective patients ____
4. washing hands after gloves are removed ____

15. List some antimicrobial agents and describe how they work.

EXERCISE 5-19.

INSTRUCTIONS

Fill in the blanks, using the following terms: reverse transcriptase inhibitor, antibiotic, protease inhibitor, nosocomial.

Bacterial infections are commonly treated with an ____________ (1), but overuse and abuse of these agents have led to an increase in drug-resistant ____________ (2) infections in hospitals. Effective treatments for viral infections are sparse. A ____________ (3) such as indinavir blocks a specific enzyme that viruses need to spread to other cells. A ____________ (4) blocks an enzyme viruses need to produce copies of their nucleic acid.

16. Describe several methods used to identify microorganisms in the laboratory.

EXERCISE 5-20.

INSTRUCTIONS

Fill in the blanks using the following terms: Gram stain, acid-fast stain, red, blue, gram-negative, gram-positive.

The most common staining procedure used to identify microorganisms is the Gram stain (1). Bacteria that appear bluish purple after staining are described as gram positive. Bacteria that do not retain the stain are described as gram negative (3). The acid-fast staining procedure uses a different dye. Acid-fast bacteria that retain the dye become red (4) in color. Bacteria that are not acid-fast do not retain the dye; they are usually visualized by counterstaining with another dye, to make them blue (5) in color.

17. Show how word parts are used to build words related to disease.

EXERCISE 5-21.

INSTRUCTIONS

Complete the following table by writing the correct word part or meaning in the space provided. Write a word that contains each word part in the Example column.

Word Part	Meaning	Example
1. pan-	all	pandemic
2. py/o	pus	pyogenes
3. idio	self, separate, distinct	idiopathic
4. aer/o	air gas	aerobic
5. myc/o	fungus	mycology
6. syn	together	syndrome
7. strepto	chain	streptococci
8. an-	absent, deficient	anaerobic
9. staphylo	grapelike cluster	staphylococci
10. iatro	physician medicine	iatrogenic

Making the Connections

The following concept map deals with disease transmission. Complete the concept map by filling in the blanks with the appropriate word or phrase for the organism or process indicated.

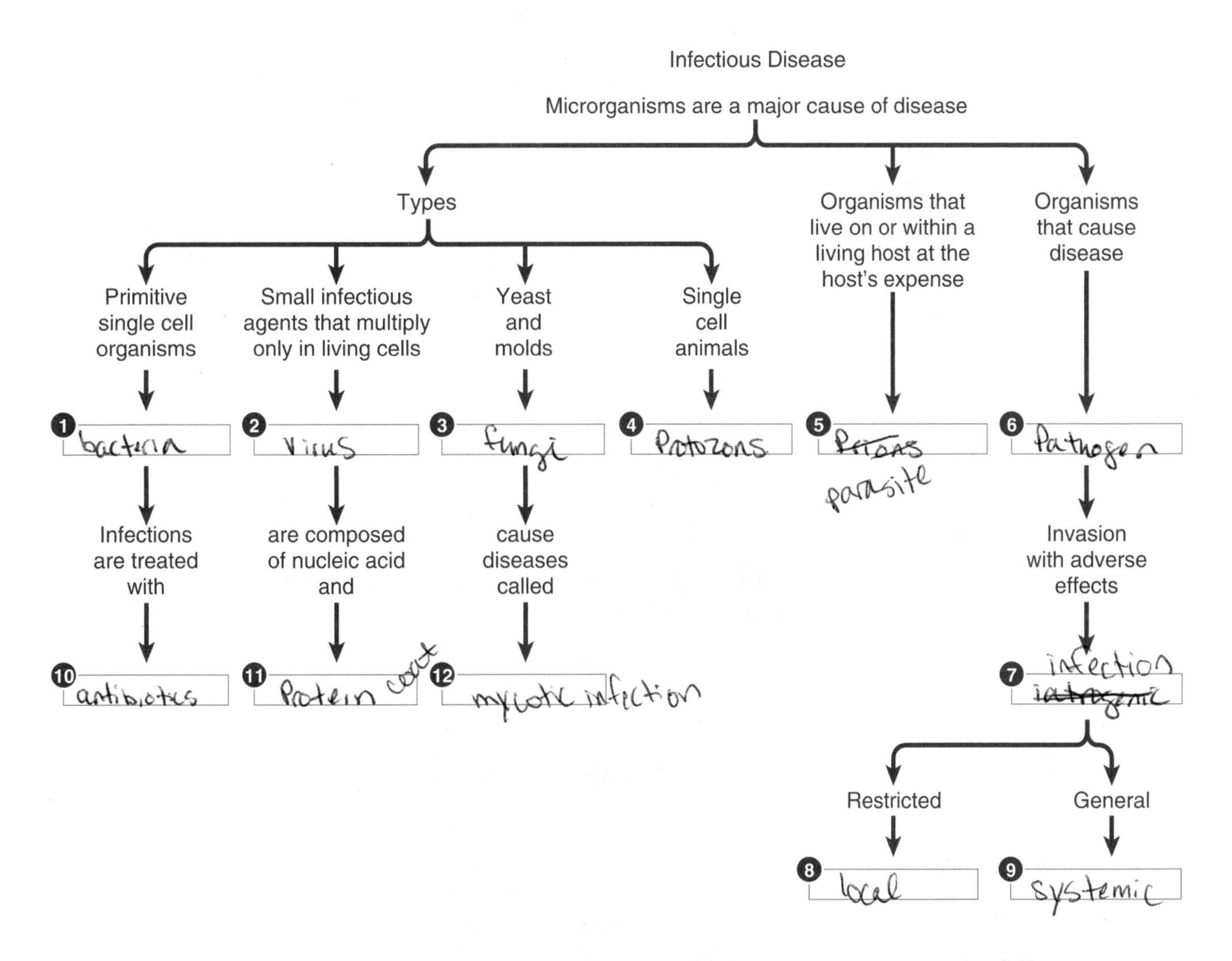

Optional Exercise: Construct a concept map of terms relating to worms using the following terms and any others you would like to include: worms, infestation, vermifuge, flatworms, roundworms, flukes, tapeworms, Ascaris, hookworms, pinworms, helminthology, and trichina.

Testing Your Knowledge

Building Understanding

I. Multiple Choice

Select the best answer and write the letter of your choice in the blank.

1. After Gram staining, a gram positive coccus would appear as a — 1. ______
   a. red rod
   b. purple rod
   c. red sphere
   d. purple sphere
2. Blood poisoning is a generalized infection described as — 2. ______
   a. aseptic
   b. nosocomial
   c. local
   d. systemic
3. The study of fungi is called — 3. ______
   a. mycology
   b. virology
   c. helminthology
   d. protozoology
4. The sexually transmitted disease syphilis is caused by long, wavy cells that perform twisting motions. These are — 4. ______
   a. streptococci
   b. spirochetes
   c. diplococci
   d. staphylococci
5. Which type of pathogen would NOT be affected by antibacterial agents? — 5. ______
   a. viruses
   b. spirochetes
   c. diplococci
   d. rickettsiae
6. Athlete's foot, tinea capitis, thrush, and Candida infections are all caused by — 6. ______
   a. viruses
   b. fungi
   c. rickettsias
   d. filariae
7. An example of an immune disorder is — 7. ______
   a. osteoporosis
   b. multiple sclerosis
   c. trichinosis
   d. scurvy

8. Pinworms, hookworms, trichina, and filaria are examples of 8. ______
   a. flatworms
   b. flagellates
   c. flukes
   d. roundworms
9. Creutzfeldt-Jacob disease is caused by infectious agents called 9. ______
   a. viruses
   b. bacteria
   c. rickettsiae
   d. prions

II. Completion Exercise

Write the word or phrase that correctly completes each sentence.

1. Short hairlike structures that help bacteria glide along solid surfaces and help them anchor to solid surfaces are ____________________.
2. An insect or other animal that transmits a disease-causing organism from one host to another is a(n) ____________________.
3. An obligate intracellular parasite composed of nucleic acid and a protein coat is a(n) ____________________.
4. Legionnaire disease, tuberculosis, and tetanus (lockjaw) are caused by rod-shaped microorganisms known as ____________________.
5. One type of flatworm, composed of many segments (proglottids), may grow to a length of 50 feet—hence the name ____________________.
6. Round bacteria found in pairs are called ____________________.
7. An alternative medical practice based on the philosophy of helping people develop healthy lifestyles is called ____________________.
8. The group of protozoa that consists of all obligate parasites is ____________________.
9. Bacteria that are pathogenic may cause injury and even death by the action of poisons referred to as ____________________.
10. Members of the protozoon (or sporozoon) genus *Plasmodium* cause a debilitating tropical disease called ____________________.
11. A device that sterilizes equipment using steam under pressure is a(n) ____________________.
12. Tetanus is caused by an organism that exists in two forms. One of these is the growing vegetative form. The other is a resting and resistant form, the ____________________.
13. Syphilis is a sexually transmitted disease that may eventually involve the brain and circulatory organs. It is caused by a corkscrew-shaped microorganism called a(n) ____________________.
14. Infections that are caused by antibiotic-resistant pathogens contracted in the hospital are termed ____________________.

Understanding Concepts

I. True/False

For each question, write *T* for true or *F* for false in the blank to the left of each number. If a statement is false, correct it by replacing the underlined term and write the correct statement in the blank below the question.

__F__ 1. Ms. J has severe anemia as a result of chemotherapy. Her anemia can be termed an idiopathic disease.

__iatrogenic__

__F__ 2. A large number of influenza infections were reported in cold regions of the United States last winter. During this period, influenza was pandemic.

__endemic__

__T__ 3. An aerobic organism requires oxygen for growth.

____

__T__ 4. Protozoa covered with tiny hairs are called ciliates.

____

__F__ 5. Gram-negative organisms appear bluish-purple under the microscope after staining with the Gram stain.

__Red.__

__F__ 6. Disinfection kills every living organism on or in an object.

__Sterilization__

II. Practical Applications

Ms. L, aged 10, has been brought to the clinic by her father. As a nursing student, you are responsible for her primary evaluation. Write the appropriate word or phrase in each space provided.

1. First, you evaluate Ms. L's signs and symptoms. Ms. L has been complaining of fatigue. Is fatigue a sign or a symptom? __Symptom__
2. You take a blood sample and determine that Ms. L has an abnormally low number of red blood cells (anemia). Is red blood cell number a sign or a symptom? __Sign__
3. You take a stool sample and observe large numbers of round worms in her feces. The presence of worms in the body is called a(n) __infestation__
4. Based on the worm's round appearance and the fact that Ms. L has anemia, you suspect that the worms responsible for Ms. L's condition are called __pinworms [illegible]__
5. The likely portal of entry for the organisms causing her illness is the __mouth [illegible]__
6. Ms. L will require a medication to kill the worms; the general term for this type of drug is __vermifuge.__

7. Ms. L's worm problem has rendered her more suceptible to other pathogens, and she has a *Candida* infection in her mouth. An infection that only affects weakened hosts is termed opportunistic
8. The *Candida* infection is caused by the group of microorganisms called fungi

III. Short Essays

1. Define the term *predisposing cause* in relation to disease, and give several examples of predisposing causes.

   makes you more suseptable
   age, gender, heredity, living conditions.

2. What is the science of pathophysiology, and why is it important in medicine?

   physical medicine

3. Describe some disadvantages of antibiotic therapy.

   Over used, ppl are becoming resistant.

4. Compare and contrast bacteria and protozoa.

   bacteria - single celled.
   protozoa - animal like

5. Compare antisepsis and bacteriostasis.

6. Why are opportunistic infections more prevalent now than in the past?

7. Have immunizations and antibiotics successfully conquered infectious disease? Explain.

Conceptual Thinking

1. Mr. S engaged in unsafe sexual practices. One week later, he experienced painful urination and an abnormal discharge from his penis. Microscopic analysis of the discharge revealed the presence of small round cells organized in pairs. A. Based on the microscopic findings, which type of microorganism is responsible for Mr. S's symptoms? Provide as much information as you can. B. What is the most probable diagnosis? C. What would be the most appropriate treatment?

   diplicocci

2. Young Ms. Y has been suffering from nausea, a stuffy nose, coughing, and a fever for several days. She has been diagnosed with influenza, and her mother requests that she be placed on antibiotics. Will the doctor fulfill her request? Why or why not?

   no - the flu is a virus and can not be kill w/ antibiotics.

3. Write a sentence using the words "etiology" and "idiopathic."

4. Based on your knowledge of word parts, what is a streptobacillus? Give as much information as you can.

   Strep - Chain, round, bacteria
   bacillus - rod shaped, could be ~~spore~~
   Spore form

Expanding Your Horizons

The SARS scare in the spring of 2003 was a reminder of how vulnerable we are to infectious diseases. Thankfully, the SARS epidemic was not as severe as many had feared. You can read about some historical epidemics in the articles listed below.

Alternative and complementary medicine practices are becoming increasingly mainstream. Investigate different practices using the website established by the National Institutes of Health (http://www.nccam.nih.gov/).

Resources

1. Foster KR, Jenkins MF, Toogood AC. The Philadelphia yellow fever epidemic of 1793. *Sci Am* 1998; 279:88–93.
2. McEvedy C. The bubonic plague. *Sci Am* 1988; 258:118–123.

CHAPTER 6

The Skin in Health and Disease

Overview

Because of its various properties, the skin can be classified as a membrane, an organ, or a system. The outermost layer of the skin is the **epidermis**. Beneath the epidermis is the **dermis** (the true skin), where glands and other accessory structures are mainly located. The **subcutaneous tissue** underlies the skin. It contains fat that serves as insulation. The accessory structures of the skin are the **sudoriferous** (sweat) **glands**, the oil-secreting **sebaceous glands**, **hair**, and **nails**.

The skin protects deeper tissues against drying and against invasion by harmful organisms. It regulates body temperature through evaporation of sweat and loss of heat at the surface. It collects information from the environment by means of sensory receptors.

The protein **keratin** in the epidermis thickens and protects the skin and makes up hair and nails. **Melanin** is the main pigment that gives the skin its color. It functions to filter out harmful ultraviolet radiation from the sun. Skin color is also influenced by the hemoglobin concentration and quantity of blood circulating in the surface blood vessels.

Much can be learned about the condition of the skin by observing for discoloration, injury, or lesions. Aging, exposure to sunlight, and the health of other body systems also have a bearing on the skin's condition and appearance. The skin is subject to numerous diseases, common forms of which are atopic dermatitis (eczema), acne, psoriasis, infections, and cancer.

This chapter does not contain any particularly difficult material. However, you must be familiar with the different tissue types discussed in Chapter 4 in order to understand the structure and function of the integument. Familiarity with the terminology used in describing and treating disease will help you learn about skin diseases.

Addressing the Learning Outcomes

1. Name and describe the layers of the skin.

EXERCISE 6-1.

INSTRUCTIONS

Write the appropriate term in each blank.

melanocyte	integument	keratin	dermis	epidermis
stratum corneum	dermal papillae	stratum basale	subcutaneous tissue	

1. A pigment-producing cell that becomes more active in the presence of ultraviolet light ________________
2. The protein in the epidermis that thickens and protects the skin ________________
3. The true skin, or corium ________________
4. The uppermost layer of skin, consisting of flat, keratin-filled cells ________________
5. Another name for the skin as a whole ________________
6. Portions of the dermis that extend into the epidermis ________________
7. The deepest layer of the epidermis, which contains living, dividing cells ________________

2. Describe the subcutaneous tissue.

EXERCISE 6-2.

INSTRUCTIONS

Match the structures in the list below with their functions.

adipose tissue elastic fibers blood vessels nerves

1. Connect the subcutaneous tissue with the dermis ________________
2. Insulates the body and acts as an energy reserve ________________
3. Carry sensory information from the skin to the brain ________________
4. Supply skin with nutrients and oxygen ________________

3. Give the location and function of the accessory structures of the skin.

EXERCISE 6-3.

INSTRUCTIONS

Write the appropriate term in each blank.

apocrine	eccrine	ceruminous	ciliary	sudoriferous
sebaceous	sebum	wax	vernix caseosa	

1. A general term for any gland that produces sweat ________________
2. Sweat glands found throughout the skin that help cool the body ________________
3. Glands that are found only in the ear canal ________________
4. Excess activity of these glands contributes to acne vulgaris ________________
5. The product of ceruminous glands ________________
6. Sweat glands in the armpits and groin that become active at puberty ________________
7. Glands that are only found on the eyelids ________________

EXERCISE 6-4: The Skin (Text Fig. 6-1)

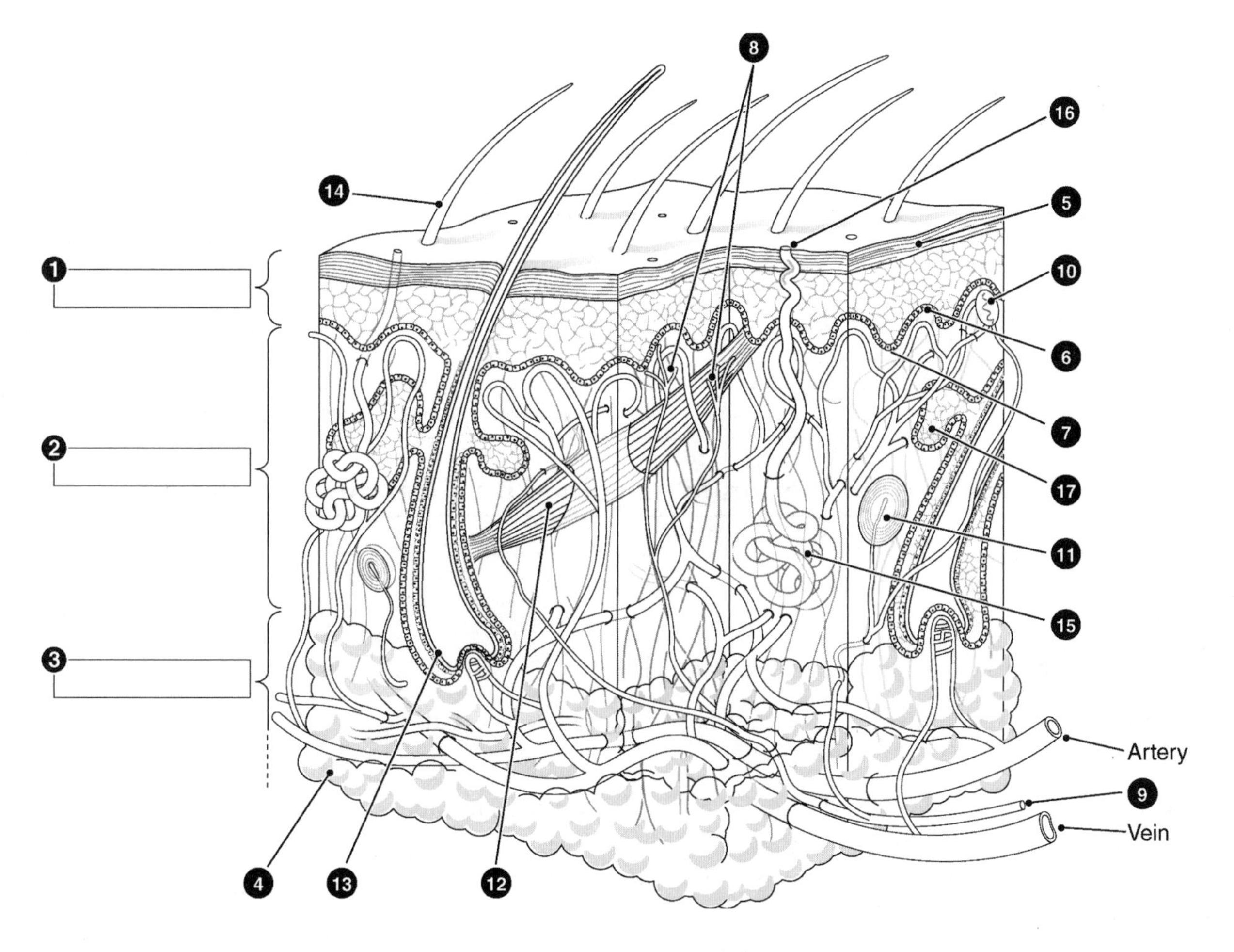

INSTRUCTIONS

1. Write the names of the three skin layers in the numbered boxes 1 to 3.
2. Write the name of each labeled part on the numbered lines in different colors. Use a light color for structures 4 and 12. Use the same color for structures 15 and 16, for structures 13 and 14, and for structures 8 and 9.
3. Color the different structures on the diagram with the corresponding color. Try to color every structure in the figure with the appropriate color. For instance, structure number 8 is found in three locations.

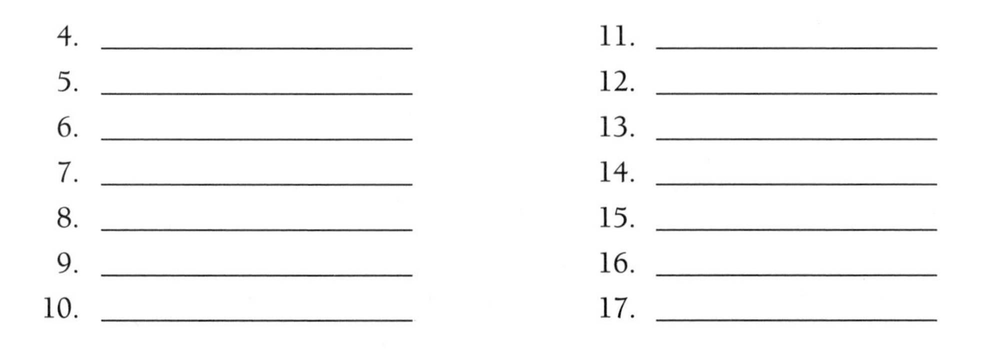

4. ____________________
5. ____________________
6. ____________________
7. ____________________
8. ____________________
9. ____________________
10. ____________________
11. ____________________
12. ____________________
13. ____________________
14. ____________________
15. ____________________
16. ____________________
17. ____________________

4. List the main functions of the skin.

EXERCISE 6-5.

INSTRUCTIONS

Fill in the numbered blanks of the table below.

Function	Structures Involved
(1)	Stratum corneum, shedding skin cells
Protection against dehydration	(2)
Regulation of body temperature	(3)
(4)	Nerve endings, specialized receptors

5. Summarize the information to be gained by observation of the skin.

EXERCISE 6-6.

INSTRUCTIONS

Write the appropriate term in each blank below.

carotenemia albinism cyanosis jaundice pallor flushing

1. A condition in which the skin takes on a bluish discoloration ____________
2. A condition in which the skin takes on a yellowish discoloration due to excess carrot consumption ____________
3. Paleness of the skin ____________
4. Redness of the skin, often related to fever ____________
5. A condition in which the skin takes on a yellowish discoloration due to excess bile pigments ____________

6. List the main disorders of the skin.

EXERCISE 6-7: Some Common Skin Lesions (Text Figs. 6-8 and 6-9)

INSTRUCTIONS

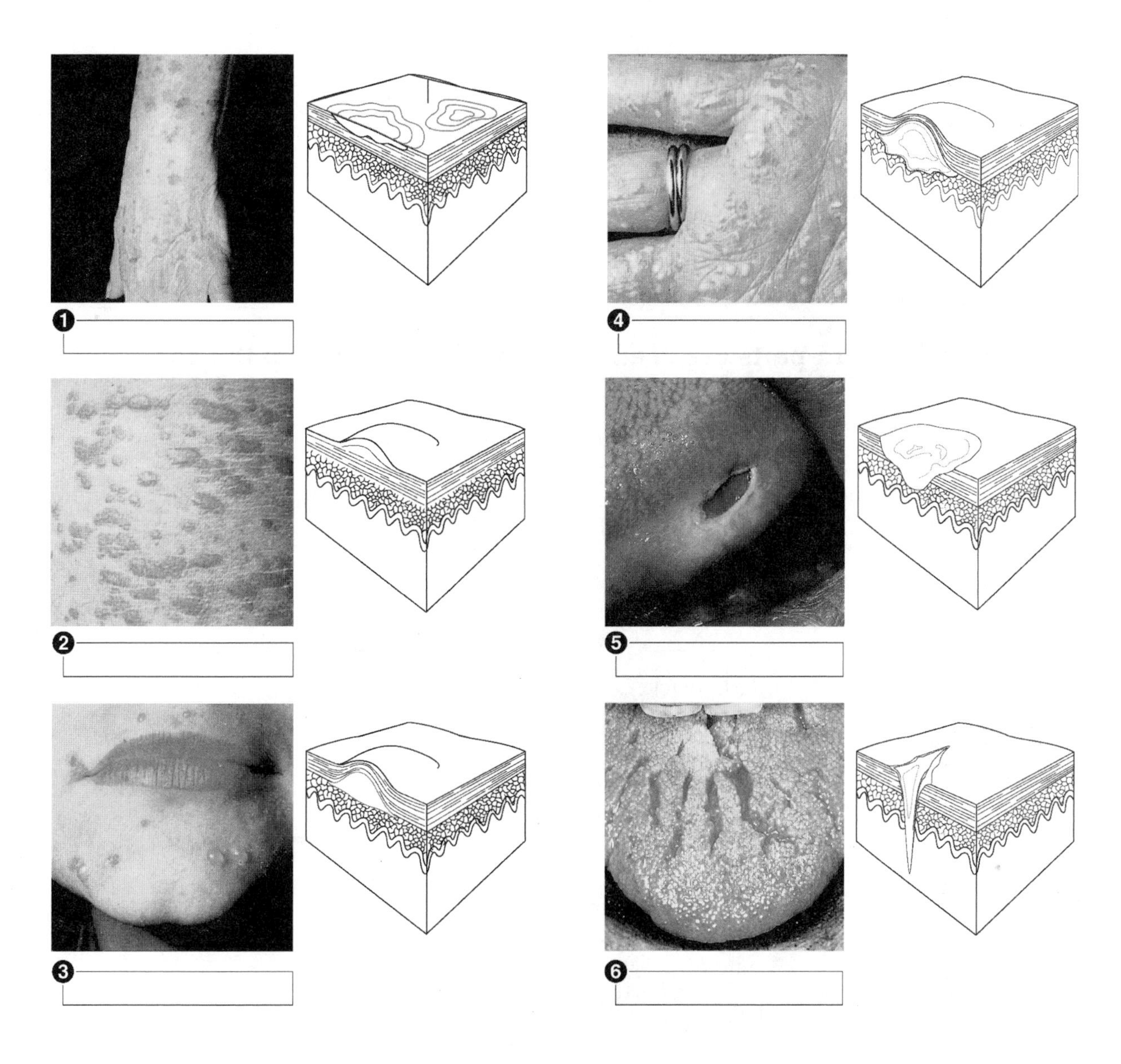

Identify each of the following types of skin lesions.

EXERCISE 6-8.

INSTRUCTIONS

Write the appropriate term in each blank.

pustule	macule	vesicle	
fissure	excoriation	laceration	ulcer

1. A spot that is neither raised nor depressed ____________
2. A scratch into the skin ____________
3. A crack in the skin ____________
4. A vesicle filled with pus ____________

5. A small sac that contains fluid; a blister ____________
6. A sore associated with disintegration and death of tissue ____________

EXERCISE 6-9.

INSTRUCTIONS

Write the appropriate term in each blank.

scleroderma erythema dermatosis dermatitis pruritus urticaria

1. Any skin disease ____________
2. Redness of the skin ____________
3. Any inflammation of the skin ____________
4. Another term for itching ____________
5. A disease resulting from excess collagen production ____________

7. Show how word parts are used to build words related to the skin.

EXERCISE 6-10.

INSTRUCTIONS

Complete the following table by writing the correct word part or meaning in the space provided. Write a word that contains each word part in the Example column.

Word Part	Meaning	Example
1. cyan/o-	________	____________
2. ________	dark, black	____________
3. ________	hard	____________
4. eryth-	________	____________
5. dermat/o-	________	____________
6. ________	horny	____________
7. ap/o-	________	____________
8. ________	state of	____________
9. ________	hair	____________
10. -emia	________	____________

Making the Connections

The following concept map deals with the structural features of the skin. Complete the concept map by filling in the appropriate word or phrase that describes the indicated skin structure.

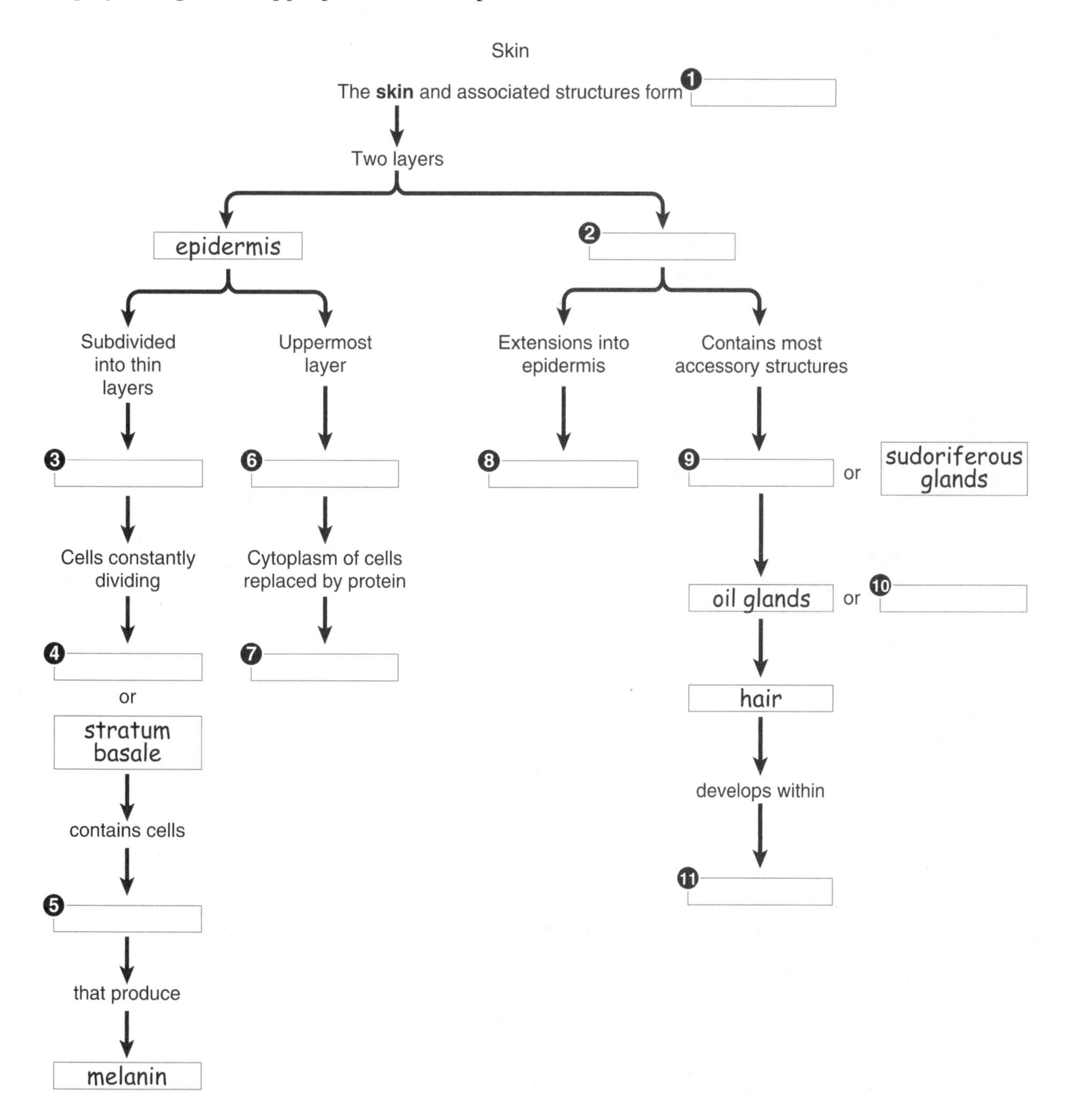

Optional Exercise: Make your own concept map based on the observation and pathology of skin. Choose your own terms to incorporate into your map or use the following list: surface lesion, macule, papule, nodule, vesicle, pustule, deeper lesion, excoriation, laceration, ulcer, fissure, dermatosis, dermatitis, psoriasis, impetigo, shingles, and acne vulgaris. Create links between the disease names and the types of associated skin lesions.

Testing Your Knowledge

Building Understanding

I. Multiple Choice

Select the best answer and write the letter of your choice in the blank.

1. New epidermal cells are produced by the 1. _______
   a. dermis
   b. stratum corneum
   c. stratum basale
   d. subcutaneous layer
2. A yellowish discoloration due to the presence of bile pigments in the blood is called 2. _______
   a. eczema
   b. impetigo
   c. alopecia
   d. jaundice
3. Which of the following glands is NOT a modified sweat gland? 3. _______
   a. mammary gland
   b. sebaceous gland
   c. ceruminous gland
   d. ciliary gland
4. Blood vessels are made smaller in order to decrease blood flow. This decrease in size is called 4. _______
   a. dilation
   b. constriction
   c. closure
   d. merger
5. A discoloration of the skin caused by diet is 5. _______
   a. pallor
   b. cyanosis
   c. liver spots
   d. carotenemia
6. Which of the following is NOT an accessory structure of the skin? 6. _______
   a. hair
   b. nails
   c. blood vessels
   d. sweat glands
7. The term *bedsore* is the common name for a 7. _______
   a. laceration
   b. decubitus ulcer
   c. psoriasis
   d. herpes
8. A gland that produces ear wax is a 8. _______
   a. ciliary gland
   b. ceruminous gland
   c. sudoriferous gland
   d. eccrine gland

9. Many babies are born with a cheesy covering known as 9. _______
   a. keratin
   b. melanin
   c. cerumen
   d. vernix caseosa
10. Which of the following skin disorders could be treated with antiviral agents? 10. _______
    a. ringworm
    b. acne vulgaris
    c. impetigo
    d. shingles

II. Completion Exercise

Write the word or phrase that correctly completes each sentence.

1. The outer layer of the epidermis, which contains flat, keratin-filled cells, is called ____________________.
2. Fingerprints are created by extensions of the dermis into the epidermis. These extensions are ____________________.
3. The main pigment of the skin is ____________________.
4. The light-colored, proximal end of a nail that overlies the thicker growing region is ____________________.
5. The muscle attached to a hair follicle that produces a "goose bump" when it contracts is the ____________________.
6. The subcutaneous layer is also called the hypodermis or the ____________________.
7. Baldness is very common in elderly men. Absence of hair from any areas where it is normally present is called ____________________.
8. Inflammation of the skin is called ____________________.
9. The ceruminous glands and the ciliary glands are modified forms of ____________________.
10. Hair and nails are composed mainly of a protein named ____________________.
11. A crack in the skin, such as that observed with athlete's foot, is called a(n) ____________________.
12. An acute contagious skin disease caused by staphylococci or streptococci may be extremely serious in infants and young children. This disease is ____________________.
13. Overactivity of the sebaceous glands during adolescence may play a part in the common skin disease called ____________________.
14. The body surface area involved in a burn may be estimated using the rule of ____________________.

Understanding Concepts

I. True/False

For each question, write *T* for true or *F* for false in the blank to the left of each number. If a statement is false, correct it by replacing the underlined term and write the correct statement in the blank below the question.

_______ 1. The nail cuticle, which seals the space between the nail plate and the skin above the nail root, is an extension of the stratum basale.

____

_______ 2. In cold weather, the blood vessels in the skin constrict in order to conserve heat.

____

_______ 3. A bluish discoloration of the skin due to insufficient oxygen is called pallor.

____

_______ 4. Sebum is produced by sudoriferous glands.

____

_______ 5. A firm, raised area of skin is called a macule.

____

_______ 6. The stratum corneum is the deepest layer of the epidermis.

____

II. Practical Applications

Study each discussion. Then write the appropriate word or phrase in the space provided.

➤ Group A

Mr. B has suffered a fall in a downhill mountain biking competition. You are a first-aid volunteer at the competition and are the first person to the scene of the accident.

1. Mr. B has light scratches on his left cheek. The medical term for a scratch is _________________.
2. A branch tore a long, jagged wound in his right arm. This wound is called a(n) _________________.
3. The skin of Mr. B's nose is very red, a symptom that is described as _________________.
4. The redness is accompanied by blisters. Blisters are also called bulla, or _________________.
5. You determine that the redness and blisters are due to sunburn and that the damage is confined to the epidermis. This type of burn is categorized as _________________.
6. When you examine Mr. B, you see numerous healed wounds. These wounds were healed through the actions of cells called _________________.
7. The new tissue that healed the wound is called a scar, or a _________________.
8. Some scars contain sharply raised areas resulting from excess collagen production. These areas are called _________________.

➤ Group B

As a volunteer with Doctors without Borders, you are working at a clinic in Malawi. Mr. L has brought his two children, ages 1 and 6, to the clinic.

1. You notice that the baby has small, pimple-like protrusions on her cheeks. These protrusions are called ________________.
2. The baby also has scaly, crusty areas in the folds of her elbows and knees and has been scratching incessantly. Severe itching is also ________________.
3. Based on her symptoms, you suspect that the baby has a noncontagious skin disorder called ________________.
4. The older child has a number of blisters on his hands that contain pus. Microscopic examination reveals the presence of staphylococci. This contagious skin disease is called ________________.
5. You also examined the father, who had small facial lesions and complained of pain, sensitivity, and itching. You noted that the lesions followed the path of nerves, leading to the diagnosis of ________________.
6. The father asks if antibiotics will help his condition. You reply that the best medication for his disorder is a(n) ________________.

III. Short Essays

1. Explain why the skin is valuable in diagnosis.

____

____

____

2. Compare and contrast eccrine and apocrine sweat (sudoriferous) glands.

____

____

____

3. Describe the role that the skin plays in the regulation of body temperature.

____

____

____

____

Conceptual Thinking

1. Describe the location and structure of the different tissue types (epithelial, muscle, nervous, connective) present in the integumentary system.

   ____

   ____

   ____

2. Chapter 5 discussed predisposing causes of disease. Briefly describe four predisposing causes of disease that are associated with alopecia.

   ____

   ____

   ____

Expanding Your Horizons

Why did different skin tones evolve? It is often thought that darker pigmentation (more melanin) has evolved to protect humans from skin cancer. However, since skin cancer occurs later in life (usually postreproduction), it cannot exert much evolutionary pressure. The advantages and disadvantages of darker skin tone are discussed in a *Scientific American* article.

Resources

1. Jablonski NG, Chaplin G. Skin deep. *Sci Am* 2002; 287:74–81.

UNIT III Movement and Support

➤ Chapter 7

The Skeleton: Bones and Joints

➤ Chapter 8

The Muscular System

CHAPTER 7

The Skeleton: Bones and Joints

Overview

The skeletal system protects and supports the body parts and serves as attachment points for the muscles, which furnish the power for movement. The bones also store calcium salts and are the site of blood cell production. The skeletal system includes some 206 bones; the number varies slightly according to age and the individual.

Although bone tissue contains a matrix of nonliving material, bones also contain living cells and have their own systems of blood vessels, lymphatic vessels, and nerves. Bone tissue may be either **spongy** or **compact**. Compact bone is found in the **diaphysis** (shaft) of long bones and in the outer layer of other bones. Spongy bone makes up the **epiphyses** (ends) of long bones and the center of other bones. **Red marrow**, present at the ends of long bones and the center of other bones, manufactures blood cells; **yellow marrow**, which is largely fat, is found in the central (medullary) cavities of the long bones.

Bone tissue is produced by cells called **osteoblasts**, which gradually convert cartilage to bone during development. The mature cells that maintain bone are called **osteocytes**, and the cells that break down (resorb) bone for remodeling and repair are the **osteoclasts**.

The skeleton is divided into two main groups of bones, the **axial skeleton** and the **appendicular skeleton**. The axial skeleton includes the skull, spinal column, ribs, and sternum. The appendicular skeleton consists of the bones of the upper and lower extremities, the shoulder girdle, and the pelvic girdle. Disorders of bones include metabolic disorders, tumors, infection, structural disorders, and fractures.

A **joint** is the region of union of two or more bones. Joints are classified into three main types on the basis of the material between the connecting bones. In **fibrous joints** the bones are held together by fibrous connective tissue, and in **cartilaginous joints** the bones are joined by cartilage. In **synovial joints**, the material between the bones is synovial fluid, which is secreted by the synovial membrane lining the joint cavity. The bones in synovial joints are connected by ligaments. Synovial joints show the greatest degree of movement, and the six types of synovial joints allow for a variety of movements in different directions.

Addressing the Learning Outcomes

1. List the functions of bones.

EXERCISE 7-1.

INSTRUCTIONS

List 5 functions of bones in the spaces provided.

1. ______
2. ______
3. ______
4. ______
5. ______

2. Describe the structure of a long bone.

EXERCISE 7-2: Structure of a Long Bone (Text Fig. 7-2)

INSTRUCTIONS

1. Write the names of the three parts of a long bone in the numbered boxes 1 to 3.
2. Write the name of each labeled part on the numbered lines in different colors. Use a dark color for structure 5.
3. Color the different structures on the diagram with the corresponding color.

4. ______
5. ______
6. ______
7. ______
8. ______
9. ______
10. ______
11. ______
12. ______

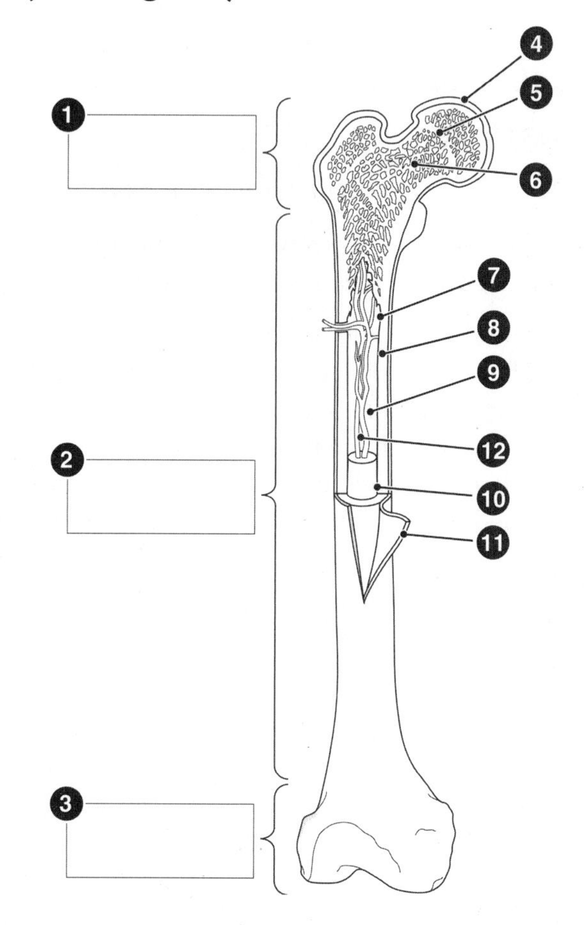

3. Differentiate between compact bone and spongy bone with respect to structure and location.

EXERCISE 7-3.

INSTRUCTIONS

Do the following statements apply to compact bone (C) or spongy bone (S)?

1. Makes up the interior of the epiphyses of long bones _______
2. Makes up the center of short bones_______
3. Makes up the shaft of a long bone _______
4. A meshwork of small, bony plates _____
5. Very hard bone with few spaces _____

4. Differentiate between red and yellow marrow with respect to function and location.

EXERCISE 7-4.

INSTRUCTIONS

Do the following statements apply to red marrow (R) or yellow marrow (Y)?

1. Found in the spaces of spongy bone _____
2. Composed largely of fat ____
3. Site of blood cell synthesis ____
4. Found in the shaft of a long bone ____

EXERCISE 7-5.

INSTRUCTIONS

Write the appropriate term in each blank.

diaphysis	epiphysis	medullary cavity	haversian canal
periosteum	endosteum	spongy bone	osteon

1. The shaft of a long bone _______________
2. The tough connective tissue membrane that covers bones _______________
3. The end of a long bone _______________
4. The type of bone tissue found at the end of long bones _______________
5. The thin membrane that lines the central cavity of long bones _______________
6. The hollow portion of a long bone containing yellow marrow _______________

5. Name the three different types of cells in bone and describe the functions of each.

EXERCISE 7-6.

INSTRUCTIONS

Write the name of the appropriate bone cell in each blank.

osteoblast osteocyte osteoclast

1. A cell that resorbs bone matrix ______________
2. A mature bone cell that is completely surrounded by hard bone tissue ______________
3. A cell that builds bone tissue ______________

6. Explain how a long bone grows.

EXERCISE 7-7.

INSTRUCTIONS

Label each of the following statements as true (T) or false (F).

1. Long bones grow in length after birth by producing new bone tissue in the middle of the diaphysis. ____
2. Once bone growth is complete, the epiphyseal plate turns into the epiphyseal line. ___
3. Long bones elongate by converting cartilage in the bone ends into bone tissue. ____
4. Osteoclasts and osteoblasts stop working once bone growth is complete. ____
5. As a bone lengthens, the medullary cavity becomes larger. ____

7. Name and describe various markings found on bones.

EXERCISE 7-8.

INSTRUCTIONS

Write the appropriate term in each blank.

crest	condyle	head	process
foramen	fossa	sinus	meatus

1. A short channel or passageway ______________
2. An air space found in some skull bones ______________
3. A rounded knoblike end separated by a slender region from the rest of the bone ______________
4. A rounded projection ______________
5. A distinct border or ridge ______________
6. A depression on a bone surface ______________
7. A hole that permits the passage of a vessel or nerve ______________

8. List the bones in the axial skeleton.

EXERCISE 7-9: The Skull (Text Fig. 7-5A)

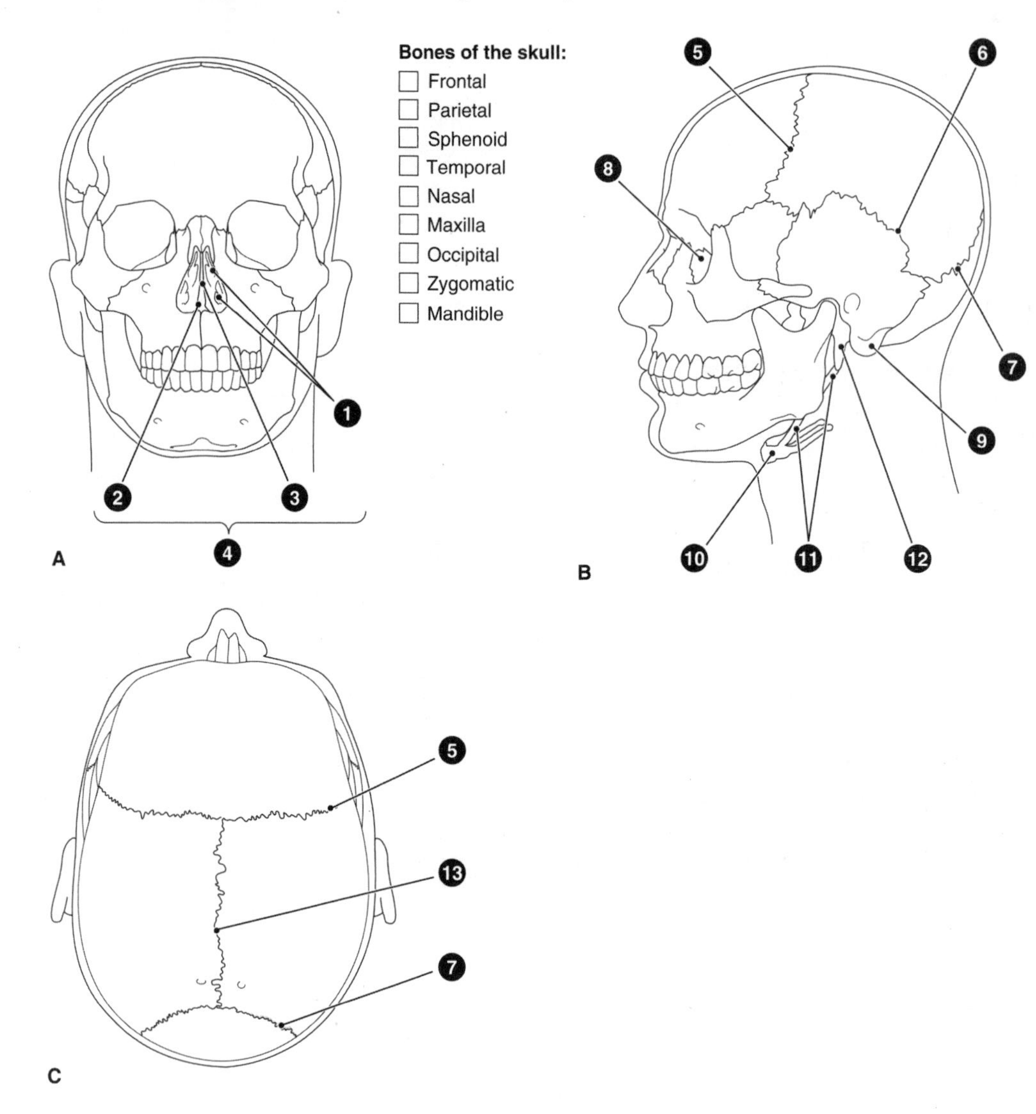

INSTRUCTIONS

1. Color the boxes next to the names of the skull bones in different, light colors.
2. Color the skull bones in parts A, B, and C of the diagram with the corresponding color.
3. Label each of the following numbered bones and bone features. If you wish, you can use different dark colors to write the names and color or outline the corresponding part for all structures except structures 4, 9, and 12.

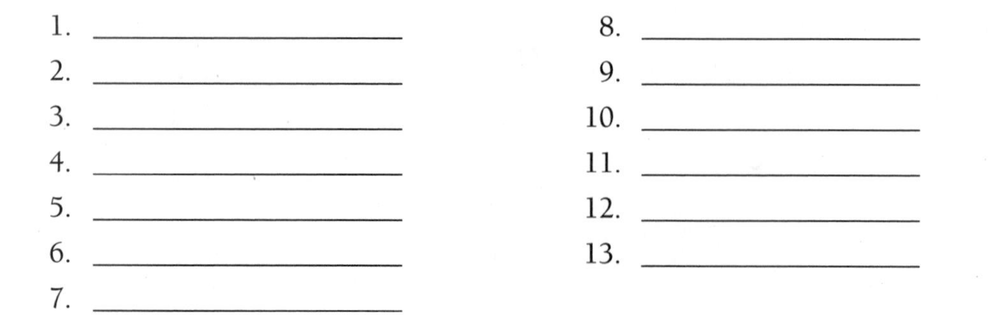

1. ______________________
2. ______________________
3. ______________________
4. ______________________
5. ______________________
6. ______________________
7. ______________________
8. ______________________
9. ______________________
10. ______________________
11. ______________________
12. ______________________
13. ______________________

EXERCISE 7-10: The Skull: Inferior, Superior, and Sagittal Views (Text Figs. 7-6, 7-7, and 7-8)

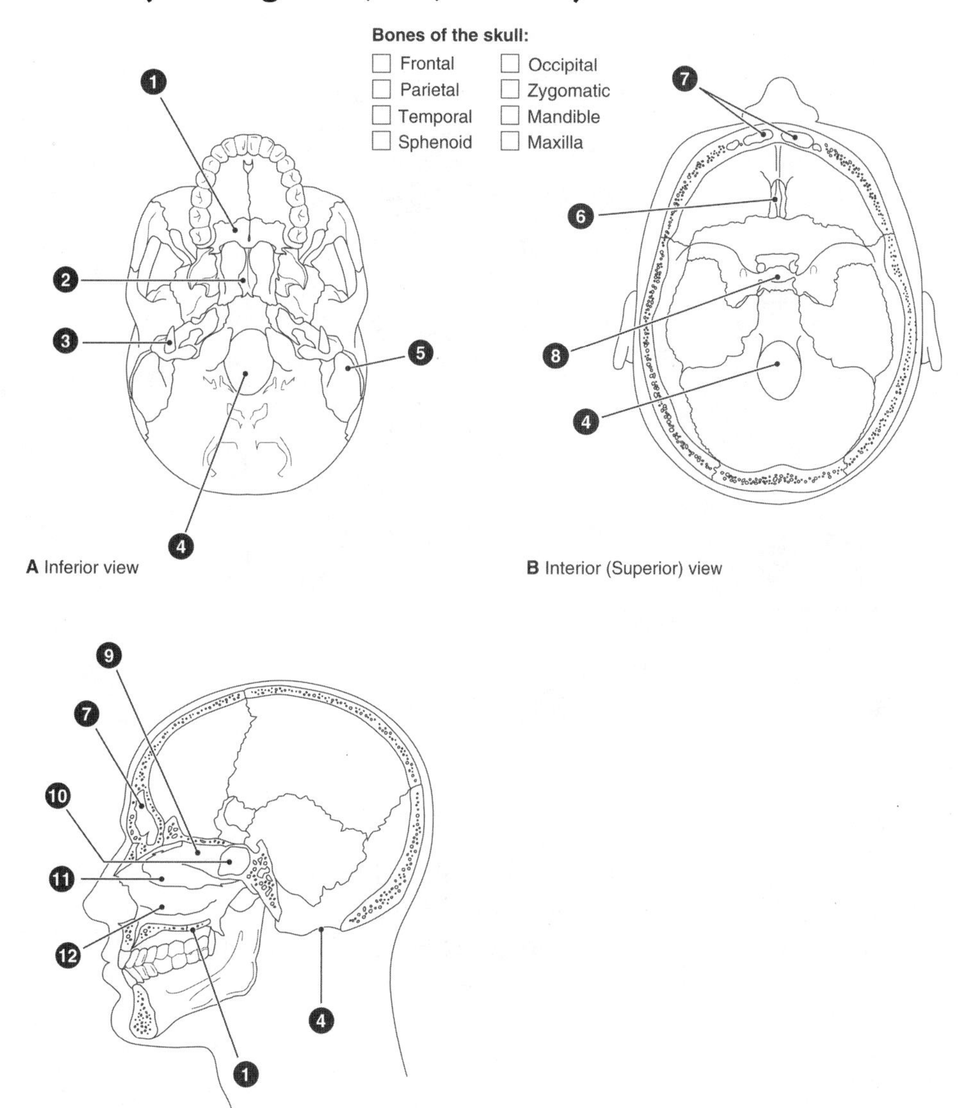

INSTRUCTIONS

1. Use the same colors you used in Exercise 7-9 to color the boxes next to the skull bone names.
2. Color the skull bones in parts A, B, and C of the diagram with the corresponding color.
3. Label each of the following numbered bones and bone features. If you wish, you can use different colors to write the name and color the corresponding part for all structures except structures 3, 5, and 8.

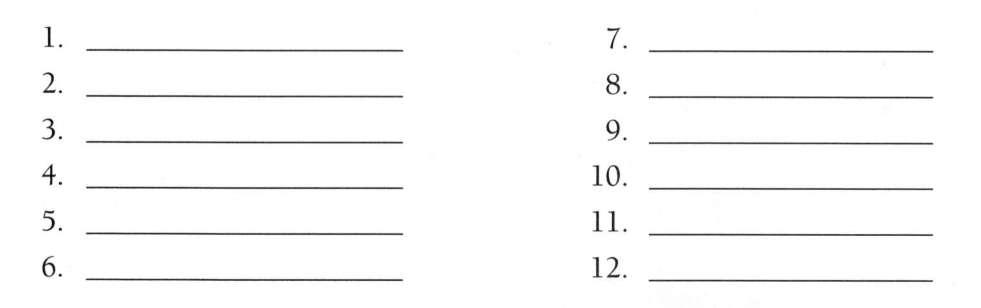

1. ____________________
2. ____________________
3. ____________________
4. ____________________
5. ____________________
6. ____________________
7. ____________________
8. ____________________
9. ____________________
10. ____________________
11. ____________________
12. ____________________

EXERCISE 7-11: Bones of the Thorax Anterior View (Text Fig. 7-14)

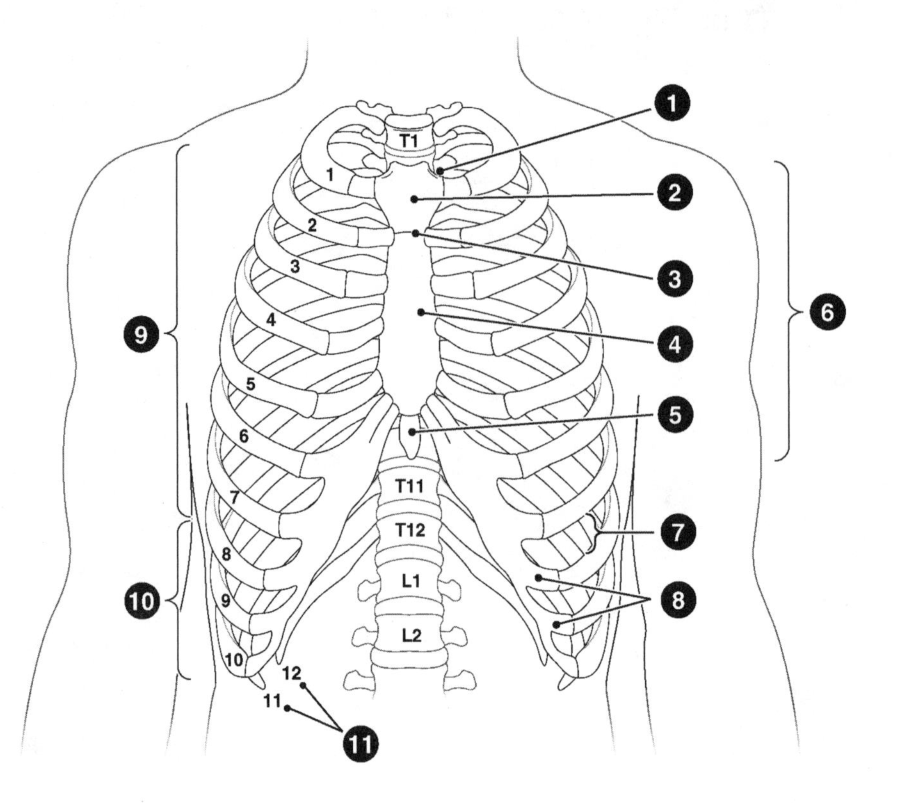

INSTRUCTIONS

1. Write the name of each labeled part on the numbered lines in different colors. Structures 1, 3, 6, 7 will not be colored, so write their names in black.
2. Color the different structures on the diagram with the corresponding color.

1. ___________________
2. ___________________
3. ___________________
4. ___________________
5. ___________________
6. ___________________
7. ___________________
8. ___________________
9. ___________________
10. ___________________
11. ___________________

EXERCISE 7-12: Vertebral Column Lateral View (Text Fig. 7-10)

INSTRUCTIONS

1. Color each of the following bones the indicated color.
   a. cervical vertebrae—blue
   b. thoracic vertebrae—red
   c. lumbar vertebrae—green
   d. sacrum—yellow
   e. coccyx—violet
2. Label each of the indicated bones and bone parts.

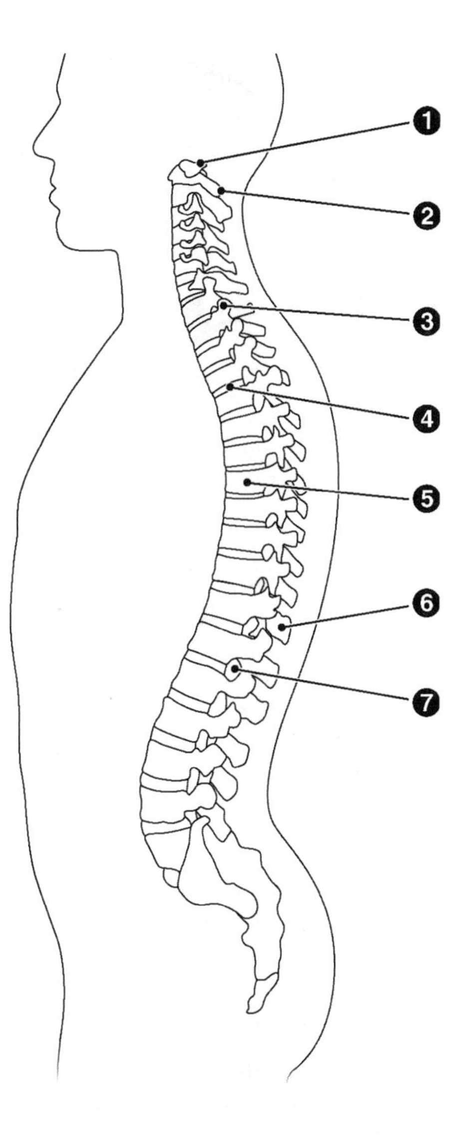

1. ____________________
2. ____________________
3. ____________________
4. ____________________
5. ____________________
6. ____________________
7. ____________________

EXERCISE 7-13: Vertebral Column Anterior View (Text Fig. 7-11)

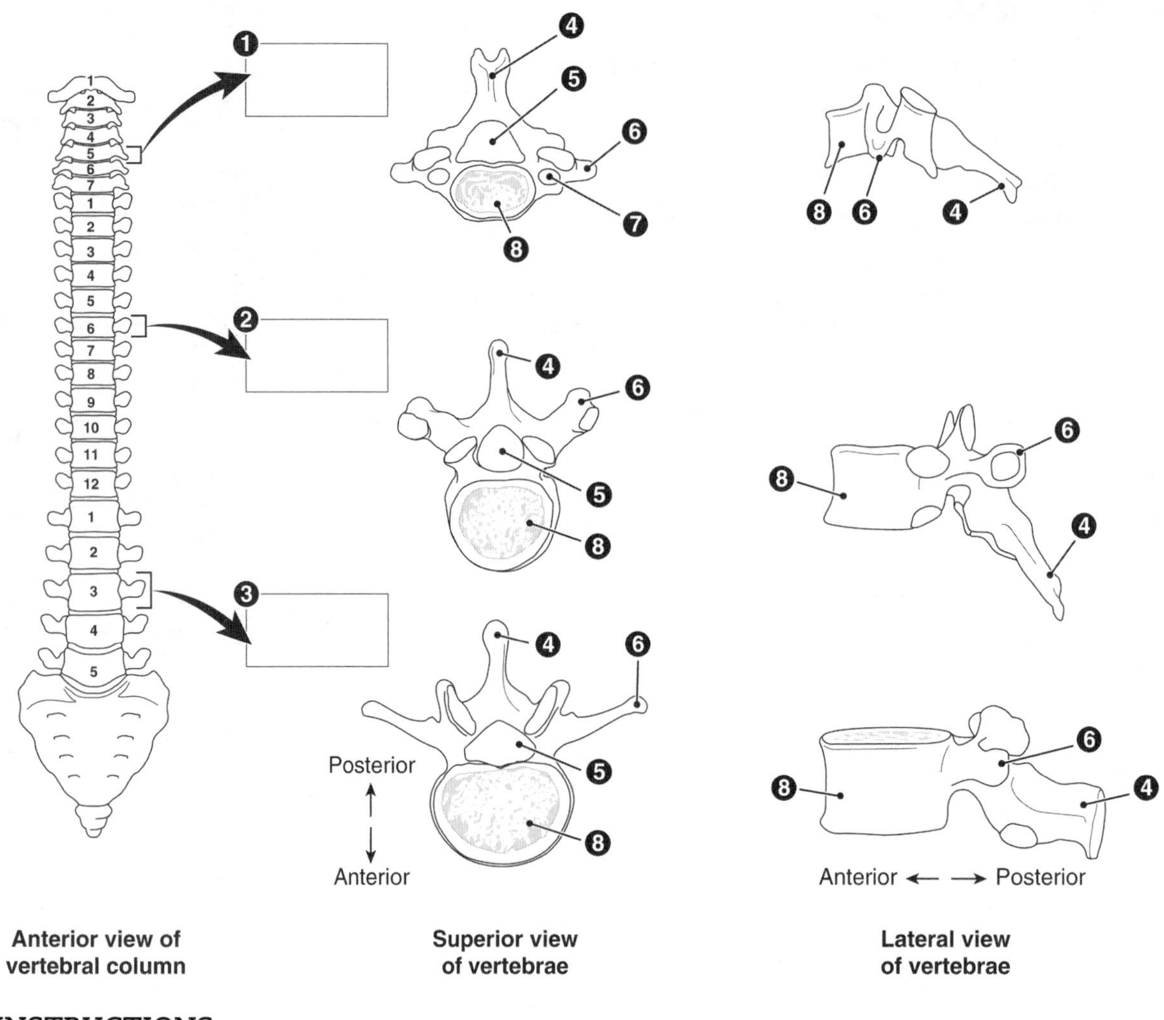

INSTRUCTIONS

1. Write the names of the three vertebral divisions in the numbered boxes 1 to 3.
2. Write the name of each labeled part on the numbered lines in different colors.
3. Color the different structures on the diagram with the corresponding color.

4. ____________________
5. ____________________
6. ____________________
7. ____________________
8. ____________________

EXERCISE 7-14.

INSTRUCTIONS

Write the appropriate term in each blank.

parietal	temporal	frontal	hyoid	nasal
maxilla	mandible	sphenoid bone	zygomatic bone	occipital

1. The only movable bone of the skull ________________
2. A bone of the upper jaw ________________
3. The U-shaped bone lying just below the mandible ________________
4. The bone that articulates with the parietal and temporal bones and forms the posterior inferior part of the cranium ________________
5. The bone that forms the forehead ________________
6. One of two slender bones that form the bridge of the nose ________________
7. One of two large bones that articulate with the frontal bone and form the superior lateral portions of the cranium ________________
8. The anatomical name for the cheekbone ________________

9. Explain the purpose of the infant fontanels.

EXERCISE 7-15.

INSTRUCTIONS

Write the appropriate term in each blank.

floating ribs	true ribs	fontanel	costal
xiphoid process	manubrium	clavicular notch	foramina

1. The T-shaped, superior portion of the sternum ________________
2. The portion of the sternum that is made of cartilage in children ________________
3. An adjective that refers to the ribs ________________
4. A soft spot in the infant skull that later closes ________________
5. The last two pairs of ribs, which are very short and do not extend to the front of the body ________________
6. The point of articulation between the sternum and the collarbone ________________
7. Ribs that attach to the sternum by individual cartilagenous extensions ________________

10. Describe the normal curves of the spine and explain their purpose.

EXERCISE 7-16.

INSTRUCTIONS

Write the appropriate term in each blank.

cervical region	thoracic region	lumbar region	coccyx
thoracic curve	lumbar curve	cervical curve	

1. A primary curve of the spine ________________
2. The second part of the vertebral column, made up of 12 vertebrae ________________
3. The spinal curve that appears when the infant holds her head up ________________
4. The spinal curve that appears when the infant begins to walk ________________
5. The most caudal part of the vertebral column ________________
6. The region of the spine that contains the largest, strongest vertebrae ________________
7. The region of the vertebral column made up of the first seven vertebrae ________________

11. List the bones in the appendicular skeleton.

EXERCISE 7-17: The Skeleton (Text Fig. 7-1)

INSTRUCTIONS

1. Write the name of each labeled part on the numbered lines in different colors. Use the same color for structures 24 and 25 and for structures 19 and 20.
2. Color the different structures on the diagram with the corresponding color. Try to color every structure in the figure with the appropriate color. For instance, structure number 3 is found in two locations.

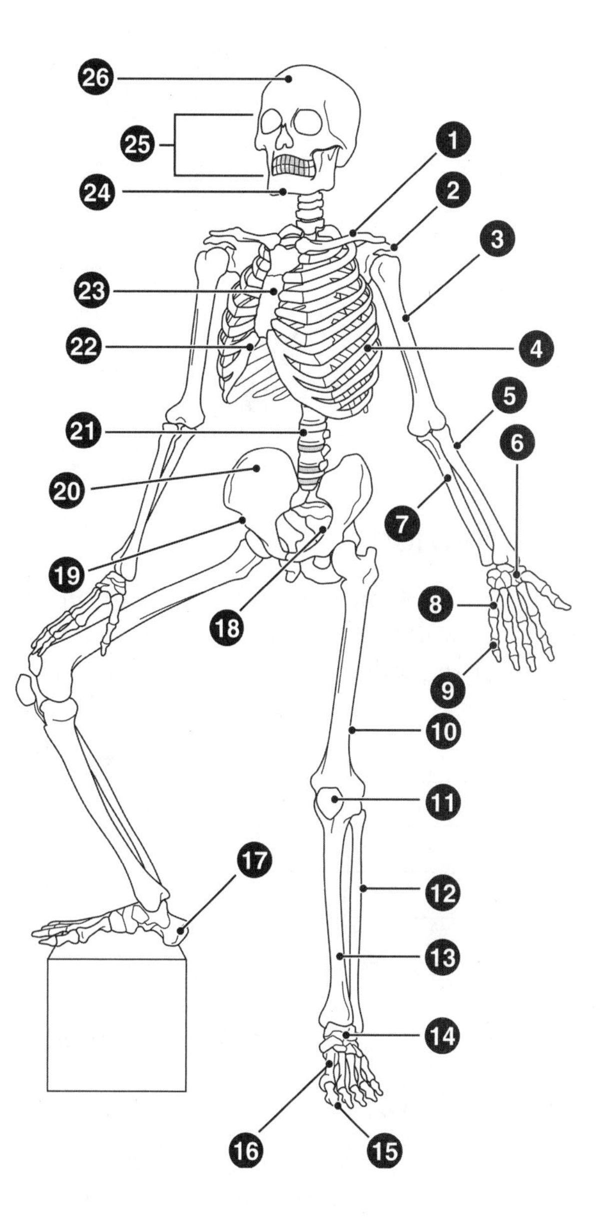

1. ____________________
2. ____________________
3. ____________________
4. ____________________
5. ____________________
6. ____________________
7. ____________________
8. ____________________
9. ____________________
10. ____________________
11. ____________________
12. ____________________
13. ____________________
14. ____________________
15. ____________________
16. ____________________
17. ____________________
18. ____________________
19. ____________________
20. ____________________
21. ____________________
22. ____________________
23. ____________________
24. ____________________
25. ____________________
26. ____________________

EXERCISE 7-18: Bones of the Shoulder Girdle (Text Fig. 7-15)

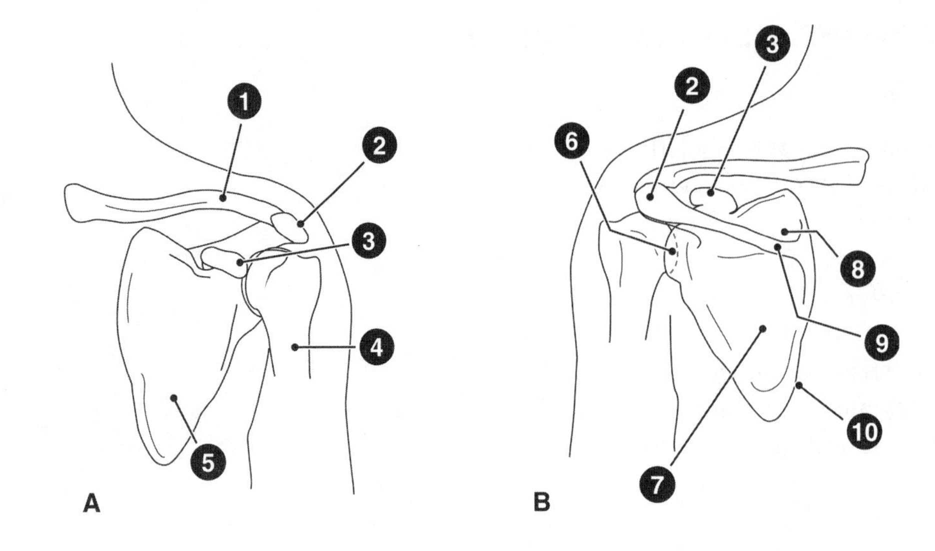

INSTRUCTIONS

1. Write the name of each labeled part on the numbered lines in different colors. Structures 1, 3, 6, 7 will not be colored, so write their names in black.
2. Color the different structures on the diagram with the corresponding color.

1. ____________________
2. ____________________
3. ____________________
4. ____________________
5. ____________________
6. ____________________
7. ____________________
8. ____________________
9. ____________________
10. ____________________

EXERCISE 7-19: Left Elbow Lateral View (Text Fig. 7-19)

INSTRUCTIONS

Label each of the indicated parts. The identity of parts 7 and 8 can be found in Fig. 7-17.

1. ____________________
2. ____________________
3. ____________________
4. ____________________
5. ____________________
6. ____________________
7. ____________________
8. 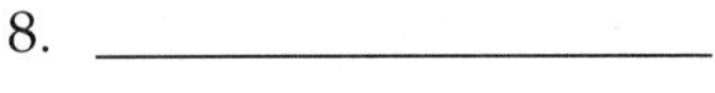

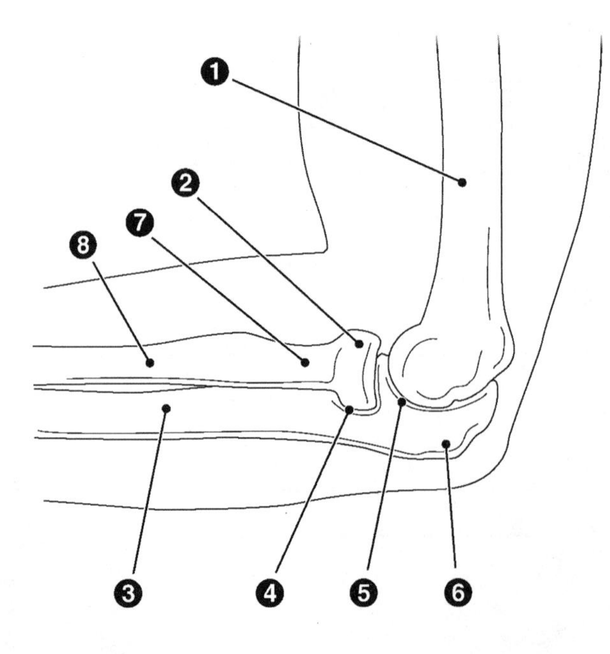

EXERCISE 7-20: Pelvic Bones (Text Fig. 7-21)

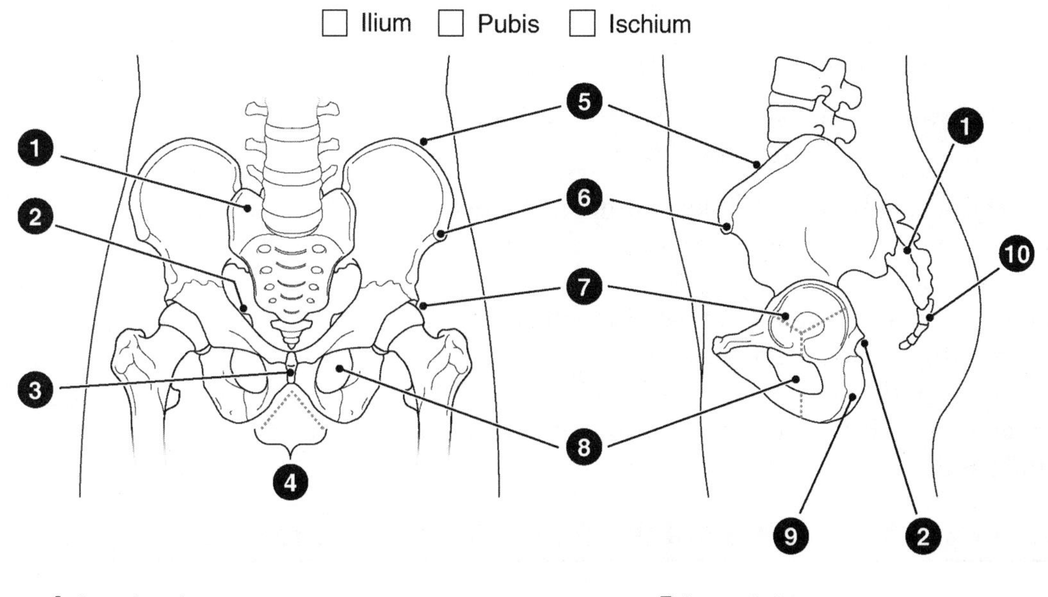

INSTRUCTIONS

1. Color the boxes next to the names of the pelvic bones in different, light colors.
2. Color the pelvic bones in parts A and B of the diagram with the corresponding color.
3. Label each of the following numbered bones and bone features. If you wish, you can use different dark colors to write the names and color or outline the corresponding part for structures 1, 3, 8, and 10.

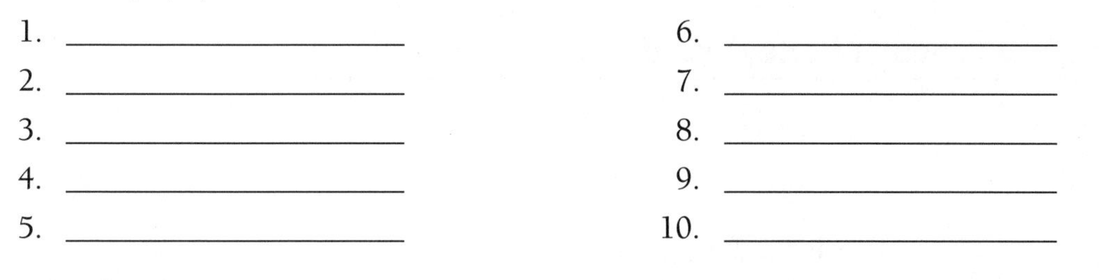

1. ____________________
2. ____________________
3. ____________________
4. ____________________
5. ____________________
6. ____________________
7. ____________________
8. ____________________
9. ____________________
10. ____________________

EXERCISE 7-21.

INSTRUCTIONS

Write the appropriate term in each blank.

olecranon	carpal bones	clavicle	ulna	radius
metacarpal bones	phalanges	scapula	humerus	

1. The anatomical name for the collarbone ____________
2. The five bones in the palm of the hand ____________
3. The medial forearm bone (in the anatomical position) ____________
4. The upper part of the ulna, which forms the point of the elbow ____________
5. The 14 small bones that form the framework of the fingers on each hand ____________
6. The bone located on the thumb side of the forearm ____________
7. The bone containing the supraspinous and infraspinous fossae ____________

EXERCISE 7-22.

INSTRUCTIONS

Write the appropriate term in each blank.

greater trochanter	patella	tibia	calcaneus	pubis
fibula	ilium	ischium	acetabulum	

1. The deep socket in the hip bone that holds the head of the femur ____________
2. The most inferior bone in the pelvis ____________
3. The lateral bone of the leg ____________
4. A bone that is wider and more flared in females ____________
5. The scientific name for the kneecap ____________
6. The largest of the tarsal bones; the heel bone ____________
7. The large, rounded projection at the upper and lateral portion of the femur ____________

12. Compare the structure of the female pelvis and the male pelvis.

EXERCISE 7-23.

INSTRUCTIONS

Write "male" or "female" in the spaces below to make each statement true.

1. The pelvic outlet is narrower in the _____ than in the _______.
2. The angle of the pubic arch is broader in the ________ than in the _________.
3. The sacrum and coccyx are shorter and less curved in the _______ than in the _______.
4. The ilia are narrower in the ______ than in the _______.

13. Describe five types of bone disorders.

EXERCISE 7-24.

INSTRUCTIONS

Write the appropriate term in each blank.

osteomyelitis	scoliosis	osteoporosis	chondrosarcoma
kyphosis	lordosis	osteopenia	osteosarcoma

1. An excessive concave curvature of the thoracic spine ____________
2. A lateral curvature of the vertebral column ____________
3. A mild reduction in bone density levels ____________
4. A bone infection caused by pus-producing bacteria ____________
5. A common disorder in older women resulting from abnormal bone metabolism ____________
6. A malignant tumor originating in cartilage ____________
7. An excessive lumbar curve ____________

14. Name and describe eight types of fractures.

EXERCISE 7-25: Types of Fractures (Text Fig. 7-28)

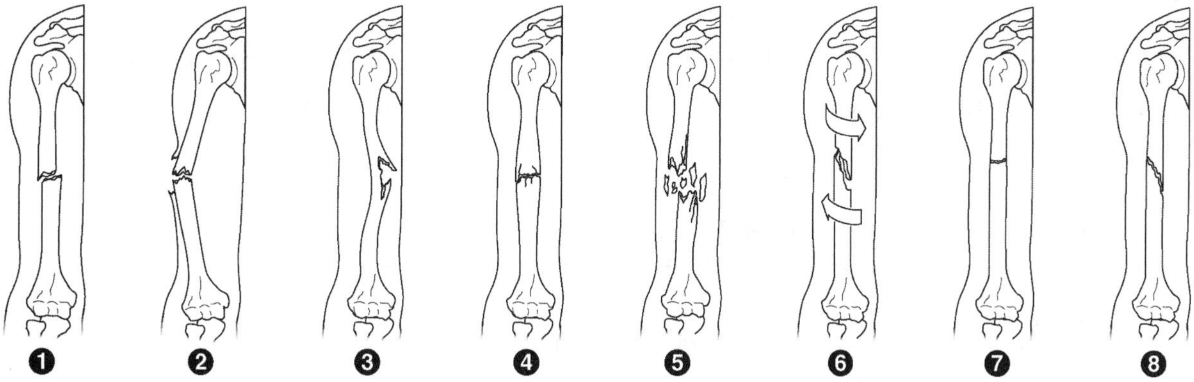

1 2 3 4 5 6 7 8

Label each of the types of fractures pictured above.

1. ____________________
2. ____________________
3. ____________________
4. ____________________
5. ____________________
6. ____________________
7. ____________________
8. ____________________

15. Describe how the skeleton changes with age.

EXERCISE 7-26.

INSTRUCTIONS

Write the appropriate term in each blank, from the following list: protein, intervertebral discs, collagen, calcium, intercostal cartilages.

In older adults, bones are weaker because of a loss in ______ (1) salts and a general decline in the manufacture of ____________ (2). Height may be reduced because the ________________ (3) become thinner. The chest may become smaller because the ______________________ (4) become calcified and less flexible. The reduction in levels of the protein _____________ (5) in tendons and ligaments makes movement more difficult.

16. Describe the three types of joints.

EXERCISE 7-27.

INSTRUCTIONS

Write the most appropriate term in each blank. Each term will be used once.

cartilaginous joint	articulation	diarthrosis	synarthrosis
fibrous joint	synovial joint	amphiarthrosis	

1. The region of union of two or more bones; a joint ________________
2. A slightly moveable joint, defined by its function ________________
3. A freely moveable joint, defined by its function ________________
4. An immovable joint, defined by its function ________________
5. A joint held together by fibrous connective tissue ________________
6. A joint held together by cartilage ________________
7. A joint in which there is a fluid-filled space between the bones ________________

17. Describe the structure of a synovial joint and give six examples of synovial joints.

EXERCISE 7-28: The Knee Joint (Text Fig. 7-30)

INSTRUCTIONS

1. Write the name of each labeled part on the numbered lines in different colors. Use a dark color for part 7, which can be outlined.
2. Color the different structures on the diagram with the corresponding color. Try to color every structure in the figure with the appropriate color. For instance, structure number 2 is found in two locations.

1. ____________________
2. ____________________
3. ____________________
4. ____________________
5. ____________________
6. ____________________
7. ____________________
8. ____________________
9. ____________________
10. ____________________
11. ____________________
12. ____________________
13. ____________________

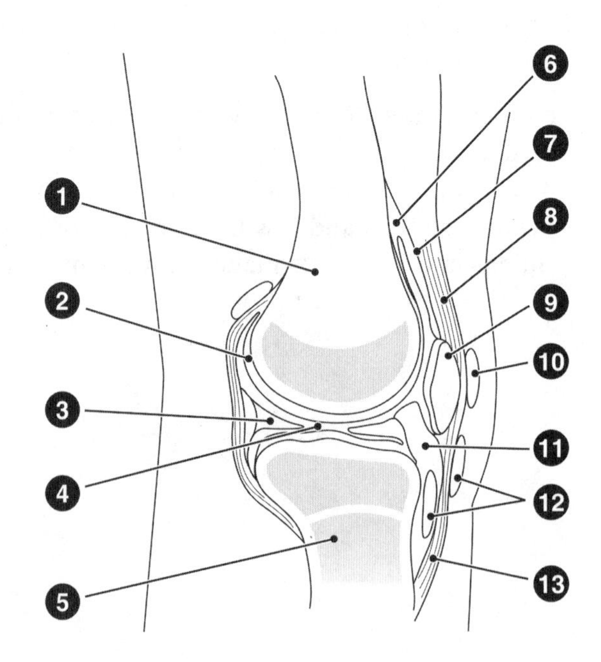

EXERCISE 7-29: Types of Synovial Joints (Text Table 7-3)

INSTRUCTIONS

Label each of the different types of synovial joints.

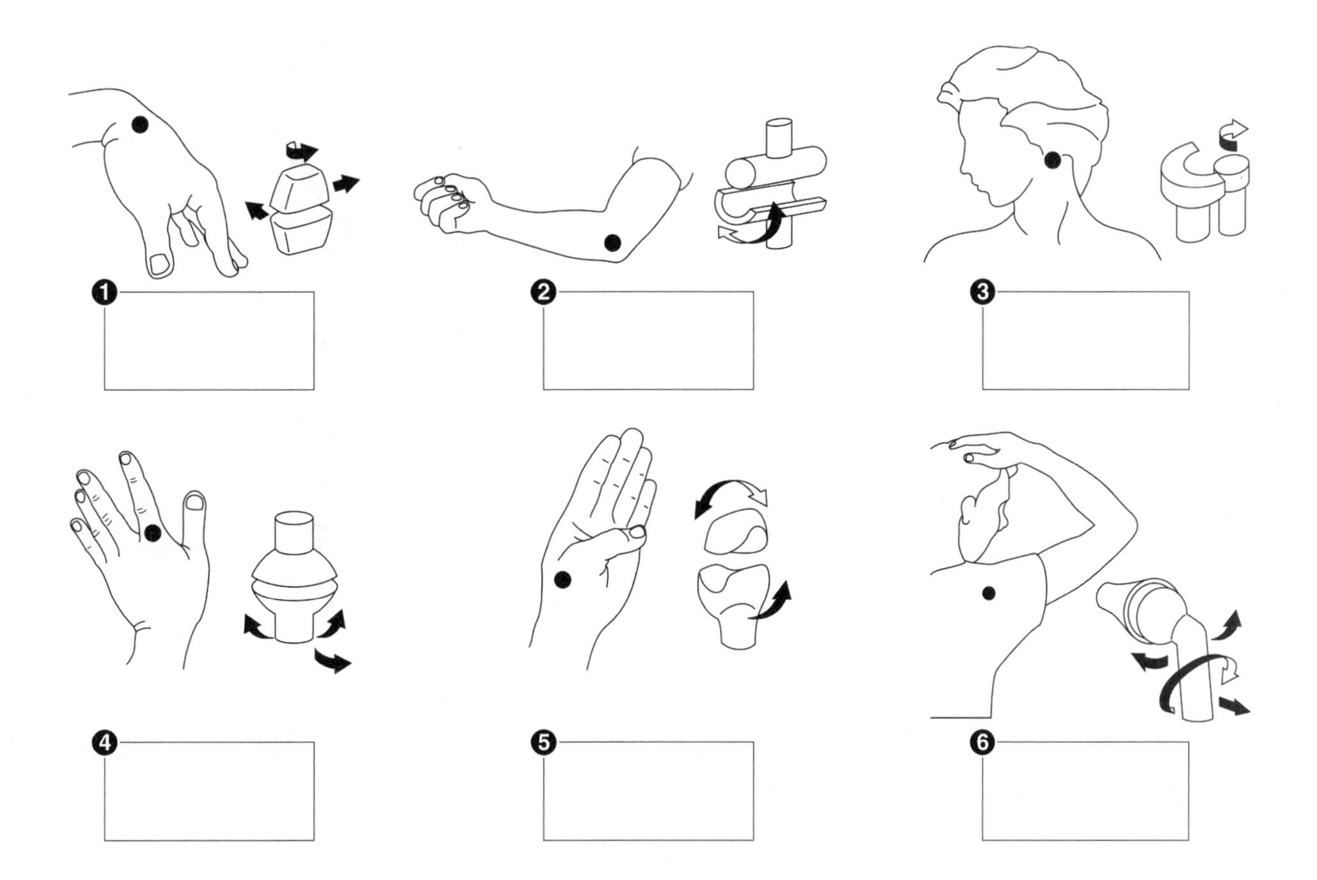

18. Demonstrate six types of movement that occur at synovial joints.

EXERCISE 7-30: Movements at Synovial Joints (Text Fig. 7-31)

INSTRUCTIONS

Label each of the illustrated motions with the correct term for that movement.

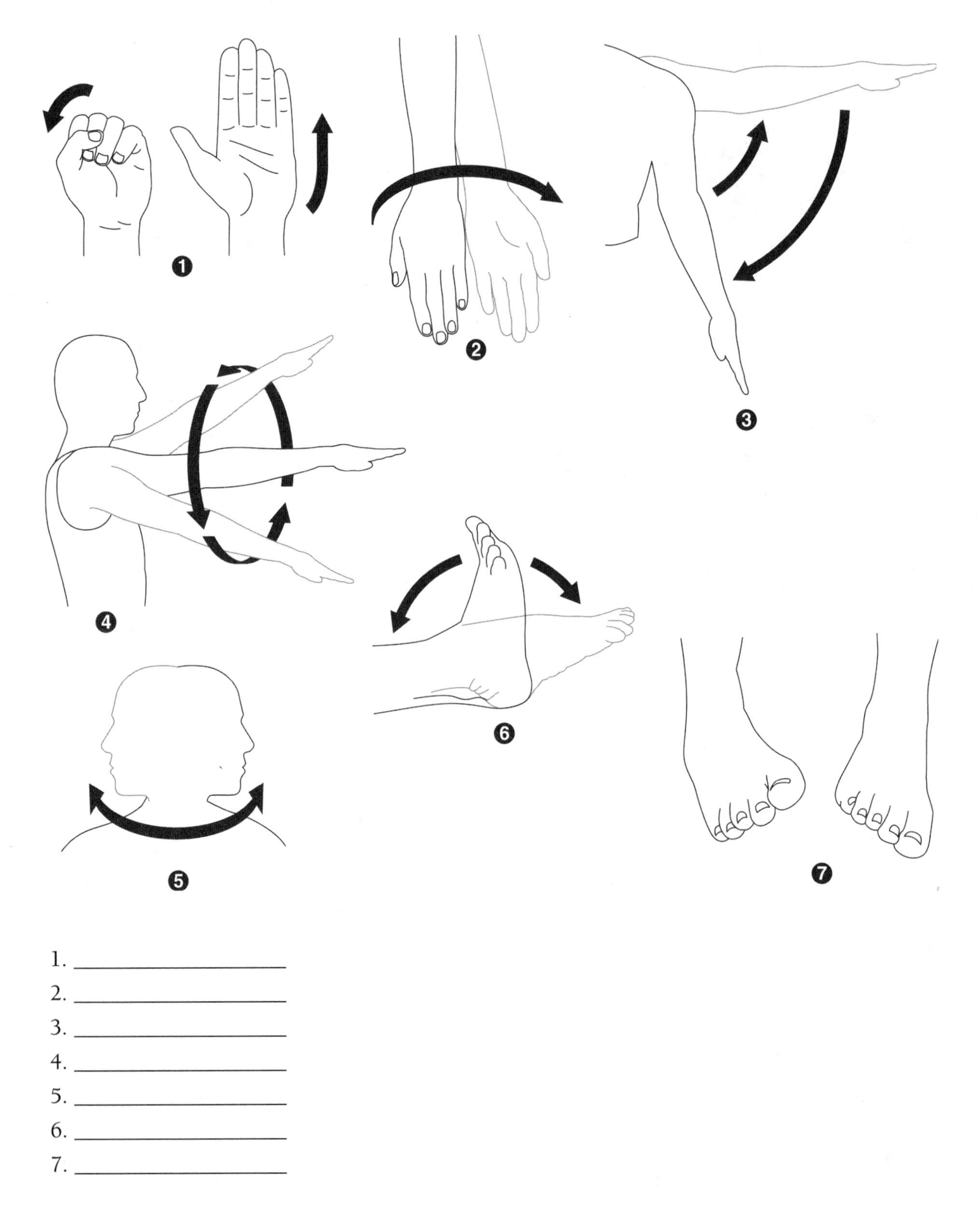

1. ______________________
2. ______________________
3. ______________________
4. ______________________
5. ______________________
6. ______________________
7. ______________________

EXERCISE 7-31.

INSTRUCTIONS

Write the appropriate term in each blank.

flexion	rotation	abduction	extension	adduction
supination	circumduction	dorsiflexion	plantar flexion	

1. A movement that increases the angle between two bones ______________
2. Movement away from the midline of the body ______________
3. Motion around a central axis ______________
4. A bending motion that decreases the angle between two parts ______________
5. Movement toward the midline of the body ______________
6. The act of turning the palm up or forward ______________
7. The act of pointing the toes downward ______________

19. Describe four types of arthritis.

EXERCISE 7-32.

INSTRUCTIONS

Write the appropriate term in each blank.

rheumatoid arthritis   osteoarthritis   septic arthritis   gout

1. Another name for degenerative joint disease ______________
2. Arthritis caused by overproduction of uric acid ______________
3. A crippling inflammatory disease of joints ______________
4. Joint inflammation caused by bacteria ______________

20. List some causes of backache.

EXERCISE 7-33.

INSTRUCTIONS

Which of the following are common causes of backache? Circle all that apply.

a. kidney problems
b. variations in the position of the uterus
c. intervertebral disc disorders
d. osteoarthritis
e. lifting a large weight by holding it close to the body
f. using the legs, not the back, to lift a heavy load

21. Describe methods used to correct diseased joints.

EXERCISE 7-34.

INSTRUCTIONS

Write the appropriate term in each blank.

arthrocentesis   arthroplasty   arthroscope

1. The removal of excess fluid from the joint cavity ________
2. An instrument to identify and repair joint problems ________
3. Joint replacement ________

22. Show how word parts are used to build words related to the skeleton.

EXERCISE 7-35.

INSTRUCTIONS

Complete the following table by writing the correct word part or meaning in the space provided. Write a word that contains each word part in the Example column.

Word Part	Meaning	Example
1. ________	lack of	________
2. -clast	________	________
3. ________	rib	________
4. amphi-	________	________
5. arthr/o	________	________
6. ________	away from	________
7. ________	around	________
8. ________	toward, added to	________
9. dia-	________	________
10. pariet/o	________	________

Making the Connections

The following concept map deals with bone structure. Complete the concept map by filling in the appropriate term or phrase that describes the indicated structure or process.

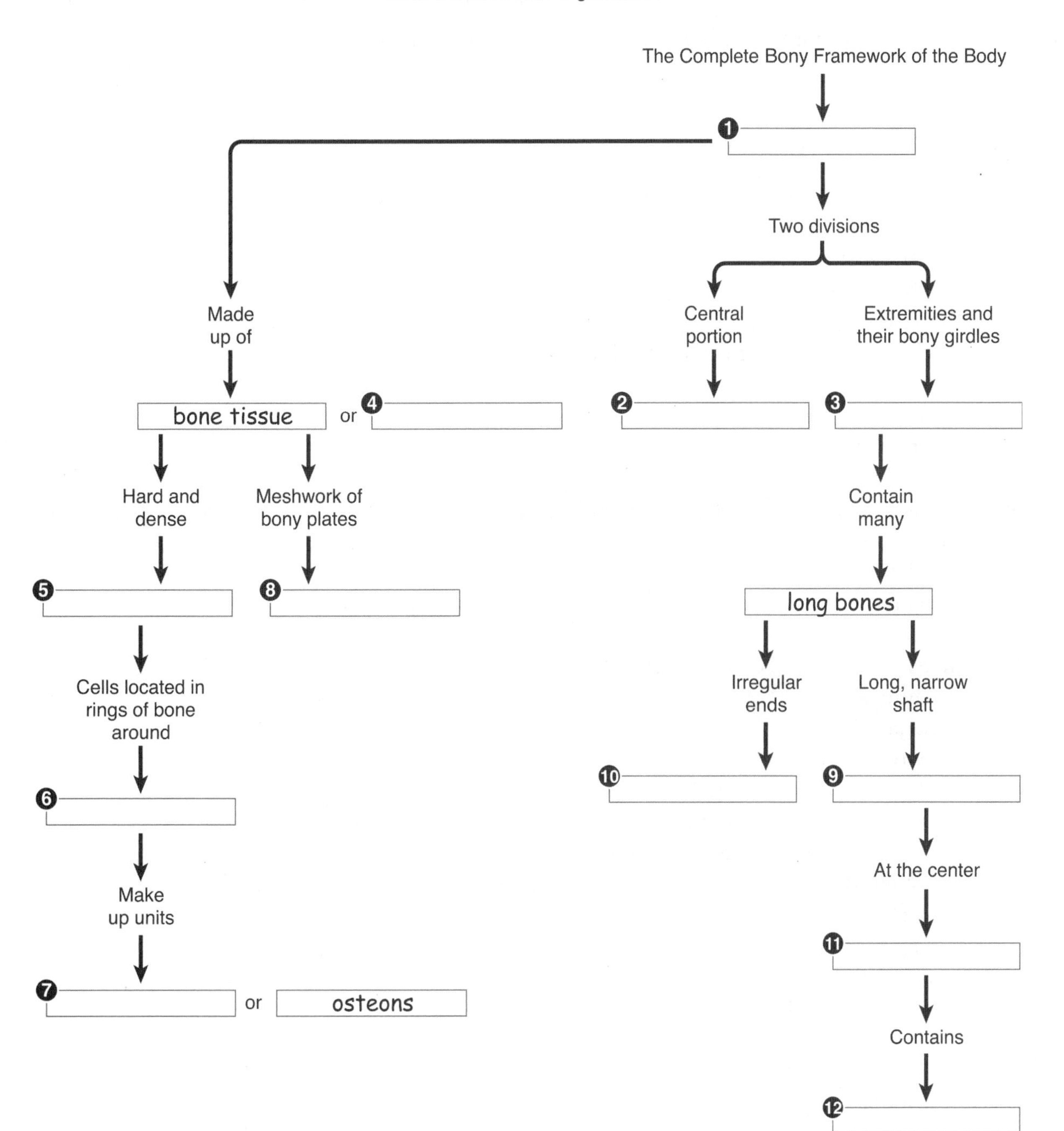

Optional Exercise: Make your own concept map, based on the different bone markings. Choose your own terms to incorporate into your map or use the following list: bone markings, projections, depressions, head, process, condyle, crest, spine, foramen, sinus, fossa, meatus, sella turcica, mastoid sinus, foramen magnum, acromion, intervertebral foramina, supraspinous fossa, and scapula spine. Try to find an example of each bone marking.

Testing Your Knowledge

Building Understanding

I. Multiple Choice

Select the best answer and write the letter of your choice in the blank.

1. The bone cells that synthesize new bone matrix are called 1. _______
   a. osteoblasts
   b. osteocytes
   c. osteoclasts
   d. osteons
2. A suture is an example of an immovable joint also called a(n) 2. _______
   a. synovial joint
   b. diarthrosis
   c. synarthrosis
   d. amphiarthrosis
3. Which of the following is a projection? 3. _______
   a. process
   b. fossa
   c. foramen
   d. sinus
4. The back of the hard palate is formed by the 4. _______
   a. vomer bone
   b. palatine bones
   c. hyoid bone
   d. mandible
5. The os coxae is a fused bone consisting of the ilium, ischium, and 5. _______
   a. femur
   b. acetabulum
   c. sacrum
   d. pubis
6. The patella is the largest of a type of bone that develops within a tendon or a joint capsule. It is described as 6. _______
   a. sesamoid
   b. axial
   c. tarsal
   d. symphysis
7. The shoulder girdle consists of the the clavicle and the 7. _______
   a. sternum
   b. tibia
   c. scapula
   d. os coxae
8. Ribs that individually attach to the sternum are called the 8. _______
   a. false ribs
   b. floating ribs
   c. xiphoid ribs
   d. true ribs

9. The foramen magnum is 9. _______
   a. a large hole in a hip bone near the symphysis pubis
   b. the curved rim along the top of the hip bone
   c. a hole between vertebrae that allows for passage of a spinal nerve
   d. a large opening at the base of the skull through which the spinal cord passes
10. Which of the following is an example of a cartilaginous joint? 10. _______
    a. condyloid
    b. pivot
    c. saddle
    d. pubic symphysis

II. Completion Exercise

Write the word or phrase that correctly completes each sentence.

1. The first cervical vertebra is called the _____________.
2. The bat-shaped bone that extends behind the eyes and also forms part of the base of the skull is called the _____________.
3. The bone located between the eyes that extends into the nasal cavity, eye sockets, and cranial floor is called the _____________.
4. The hard bone matrix is composed mainly of salts of the element _____________.
5. The part of the skull that encloses the brain is the _____________.
6. A joint between bones of the skull is a(n) _____________.
7. Pivot, hinge, and gliding joints are examples of freely movable joints, also called _____________.
8. Swimming the overhead crawl requires a broad circular movement at the shoulder that is a combination of simpler movements. This combined motion is called _____________.
9. When you bend your foot upward to walk on your heels, the position of the foot is technically called _____________.
10. In the embryo, most of the developing bones are made of _____________.
11. The type of bone tissue that makes up the shaft of a long bone is called _____________.
12. The skull, vertebrae, ribs, and sternum make up the division of the skeleton called the _____________.

Understanding Concepts

I. True/False

For each question, write *T* for true or *F* for false in the blank to the left of each number. If a statement is false, correct it by replacing the underlined term and write the correct statement in the blank below the question.

_______ 1. The shaft of a long bone contains yellow marrow.

____

_______ 2. The ethmoid bone is in the axial skeleton.

____

_______ 3. Moving a bone toward the midline is abduction.

____

_______ 4. The second cervical vertebra is called the axis.

____

_______ 5. Increasing the angle at a joint is extension.

____

_______ 6. There are six pairs of false ribs.

____

_______ 7. A mature bone cell is an osteocyte.

____

_______ 8. Immovable joints are called synovial joints.

____

_______ 9. The ends of a long bone are composed mainly of spongy bone.

____

_______ 10. Excessive lumbar curve of the spine is called scoliosis.

____

_______ 11. The medial malleolus is found at the distal end of the fibula.

____

_______ 12. A malignant bone tumor arising in cartilage is called a chondrosarcoma.

____

II. Practical Applications

Ms. M, aged 67, suffered a serious fall at a recent bowling tournament. As a physician assistant trainee, you are responsible for her preliminary evaluation.

1. Her right forearm is bent at a peculiar angle. You suspect a fracture to the radius or to the ____________.
2. The broken bone does not project through the skin. Fractures without an open wound are called ____________.
3. An x-ray reveals a single break at an angle across the bone. This type of fracture is called a(n) ____________.
4. The arm x-ray also reveals a number of fractures in the wrist bones, which are also called the ____________.
5. The wrist bones are fractured in many places. This type of fracture is called a(n) ____________.
6. Ms. M also reports pain in the hip region. The hip joint consists of the femur and a deep socket called the ____________.
7. An x-ray reveals a crack in the "sitting bone" that support the weight of the trunk when sitting. This bone is called the ____________.
8. The large number of fractures Ms. M suffered suggests that she may have a bone disorder. Changes in her bone mass can be detected using a test called a(n) ____________.
9. The test confirms that she has a significant reduction in bone mass and bone protein, leading to the diagnosis of ____________.
10. The physician prescribes a new medication designed to increase the activity of cells that synthesize new bone tissue. These cells are called ____________.

III. Short Essays

1. What is the function of the fontanels?

____

____

____

2. Describe the four curves of the adult spine and explain the purpose of these curves.

____

____

____

3. What is the difference between true ribs, false ribs, and floating ribs?

____

____

____

Conceptual Thinking

1. The following questions relate to the knee joint.
   A. Classify the knee joint in terms of the **degree** of movement permitted.

   ____

   B. Classify the knee joint based on the **types** of movement permitted.

   ____

   C. Classify the knee joint in terms of the material between the adjoining bones.

   ____

   D. List the bones that articulate within the capsule of the knee joint.

   ____

   E. List the types of movement that can occur at the knee joint.

   ____

Expanding Your Horizons

The human skeleton has evolved from that of four-legged animals. Unfortunately, the adaptation is far from perfect; thus, our upright posture causes problems like backache and knee injuries. If you could design the human skeleton from scratch, what would you change? A *Scientific American* article suggests some improvements.

Resources

1. Olshansky JS, Carnes BA, Butler RN. If humans were built to last. Sci Am 2001; 284:50–55.

The Muscular System

Overview

There are three basic types of muscle tissue: skeletal, smooth, and cardiac. This chapter focuses on **skeletal muscle**, which is usually attached to bones. Skeletal muscle is also called **voluntary muscle**, because it is normally under conscious control. The muscular system is composed of more than 650 individual muscles.

Skeletal muscles are activated by electrical impulses from the nervous system. A nerve fiber makes contact with a muscle cell at the **neuromuscular junction**. The neurotransmitter **acetylcholine** transmits the signal from the neuron to the muscle cell by producing an electrical change called the **action potential** in the muscle cell membrane. The action potential causes the release of **calcium** from the endoplasmic reticulum into the muscle cell cytoplasm. Calcium enables two types of intracellular filaments inside the muscle cell, made of **actin** and **myosin**, to contact each other. The myosin filaments pull the actin filaments closer together, resulting in muscle contraction. **ATP** is the direct source of energy for the contraction. To manufacture ATP, the cell must have adequate supplies of **glucose** and **oxygen** delivered by the blood. A reserve supply of glucose is stored in muscle cells in the form of **glycogen**. Additional oxygen is stored by a muscle cell pigment called **myoglobin**.

When muscles do not receive enough oxygen, as during strenuous activity, they can produce a small amount of ATP and continue to function for a short period. As a result, however, the cells produce **lactic acid**, which may contribute to muscle fatigue. A person must then rest and continue to inhale oxygen, which is used to convert the lactic acid into other substances. The amount of oxygen needed for this purpose is referred to as the **oxygen debt**.

Muscles usually work in groups to execute a body movement. The muscle that produces a given movement is called the **prime mover**; the muscle that produces the opposite action is the **antagonist**.

Muscles act with the bones of the skeleton as lever systems, in which the joint is the pivot point or fulcrum. Exercise and proper body mechanics help maintain muscle health and effectiveness. Continued activity delays the undesirable effects of aging.

This chapter contains some challenging concepts, particularly in respect to muscle contractions, and many muscles to memorize. Try to learn the muscle names and actions by using your own body. You should be familiar with the different movements and the anatomy of joints from Chapter 7 before you tackle this chapter.

Addressing the Learning Outcomes

1. Compare the three types of muscle tissue.

EXERCISE 8-1.

INSTRUCTIONS

Write the appropriate term in each blank.

cardiac muscle	skeletal muscle	smooth muscle	fascicle	ligament
endomysium	perimysium	epimysium	tendon	

1. A cordlike structure that attaches a muscle to bone ______________
2. A bundle of muscle fibers ______________
3. A connective tissue layer surrounding muscle fiber bundles ______________
4. Muscle under voluntary control ______________
5. The only muscle type that does not have visible striations ______________
6. An involuntary muscle containing intercalated disks ______________
7. The innermost layer of the deep fascia that surrounds the entire muscle ______________

EXERCISE 8-2: Structure of a Skeletal Muscle (Text Fig. 8-1)

INSTRUCTIONS

Label each of the indicated parts.

1. ______________
2. ______________
3. ______________
4. ______________
5. ______________
6. ______________
7. ______________
8. ______________
9. ______________

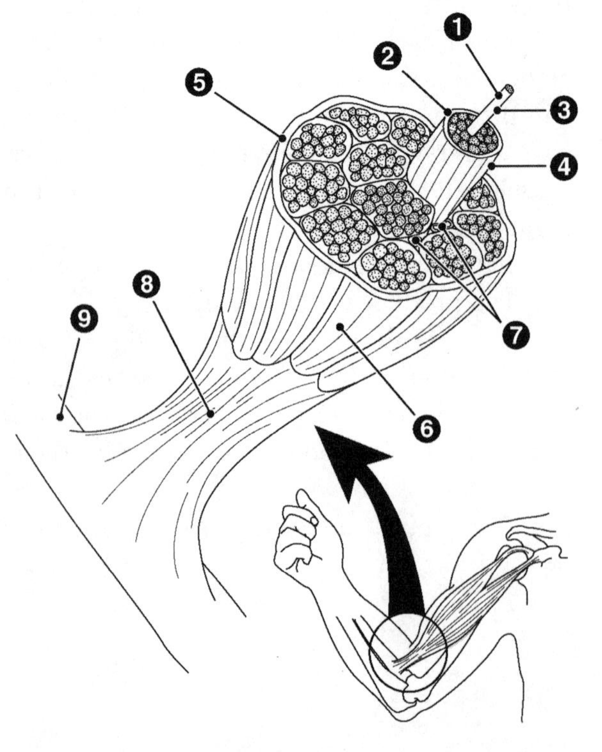

2. Describe three functions of skeletal muscle.

EXERCISE 8-3.

INSTRUCTIONS

List three functions of skeletal muscle in the spaces below.

1. ________________________
2. ________________________
3. ________________________

3. Briefly describe how skeletal muscles contract.

EXERCISE 8-4: Neuromuscular Junction (Text Fig. 8-3)

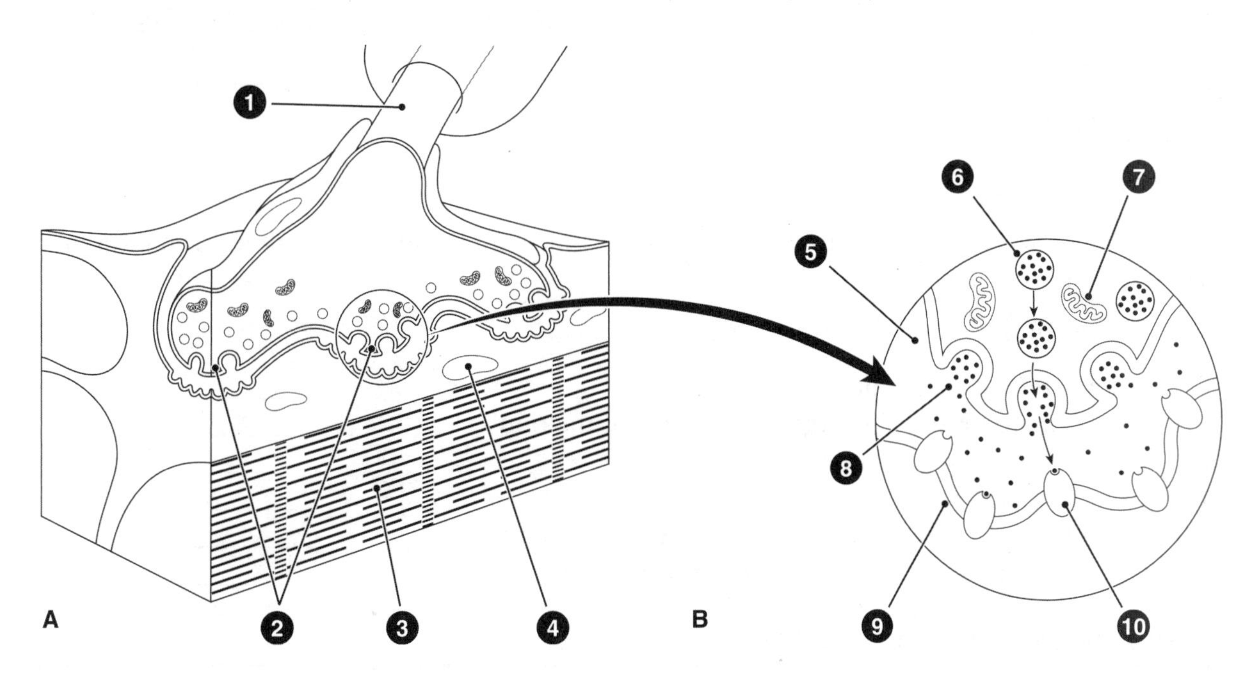

INSTRUCTIONS

Label each of the indicated parts.

1. ________________________
2. ________________________
3. ________________________
4. ________________________
5. ________________________
6. ________________________
7. ________________________
8. ________________________
9. ________________________
10. ________________________

EXERCISE 8-5.

INSTRUCTIONS

Write the appropriate term in each blank.

synaptic cleft   motor end plate   motor unit   actin
myosin   troponin   sarcomere

1. The protein that makes up muscle's lighter, thin filaments ________________
2. The protein that interacts with actin to form crossbridges ________________
3. The membrane of the muscle cell that binds ACh ________________
4. The space between the neuron and the muscle cell ________________
5. A single neuron and all of the muscle fibers it stimulates ________________
6. A protein that binds calcium during muscle contraction ________________

4. List the substances needed in muscle contraction and describe the function of each.

EXERCISE 8-6.

INSTRUCTIONS

In the blanks, write the name of the substance that is accomplishing each action. Each term may be used more than once.

ATP   calcium   acetylcholine

1. Substance released into the synpatic cleft ________________
2. The immediate source of energy for muscle contraction ________________
3. Binds to troponin when muscle contracts ________________
4. Used to detach the myosin head ________________
5. Pumped back into the smooth ER when muscle relaxes ________________
6. Causes an action potential when it binds the motor end plate ________________

5. Define the term *oxygen debt.*

EXERCISE 8-7.

INSTRUCTIONS

Circle all correct answers. (There are two.)
Oxygen debt:

A. Occurs when muscles are operating aerobically.
B. Occurs when muscles are operating anaerobically.
C. Results from the accumulation of glycogen in working muscles.
D. Is the amount of oxygen required to convert accumulated lactic acid into other substances.

6. Describe three compounds stored in muscle that are needed to generate energy in highly active muscle cells.

EXERCISE 8-8.

INSTRUCTIONS

Write the appropriate term in each blank.

glycogen creatine phosphate myoglobin

1. A compound similar to ATP that can be used to generate ATP ____________
2. A polysaccharide that can be used to generate glucose ____________
3. A compound that stores additional oxygen ____________

7. Cite the effects of exercise on muscles.

EXERCISE 8-9.

INSTRUCTIONS

In the blank following each statement, write (T) if it is true and (F) if it is false.

1. Resistance exercise causes muscle hypertrophy. ____________
2. Blood vessels constrict in actively contracting muscles. ____________
3. Weight lifting is the most efficent way to improve endurance. ____________
4. Regular exercise increase the number of capillaries in muscles. ____________
5. Regular exercise decreases the number of mitochondria in muscles. ____________

8. Compare isotonic and isometric contractions (see Exercise 8-10).

9. Explain how muscles work in pairs to produce movement.

EXERCISE 8-10.

INSTRUCTIONS

Write the appropriate term in each blank.

origin	prime mover	antagonist	synergist
isotonic	isometric	insertion	

1. A muscle acting as a helper to accomplish a particular movement ____________
2. The muscle attachment joined to a moving part of the body ____________
3. The muscle attachment joined to a more fixed part of the body ____________
4. The muscle that produces a given movement ____________
5. A muscle that relaxes during a given movement ____________
6. A contraction in which the muscle shortens but muscle tension remains the same ____________
7. A contraction in which muscle tension increases but muscle length is unchanged ____________

EXERCISE 8-11: Muscle Attachment to Bones (Text Fig. 8-7)

INSTRUCTIONS

Label each of the indicated parts.

1. ______________________
2. ______________________
3. ______________________
4. ______________________
5. ______________________
6. ______________________
7. ______________________
8. ______________________

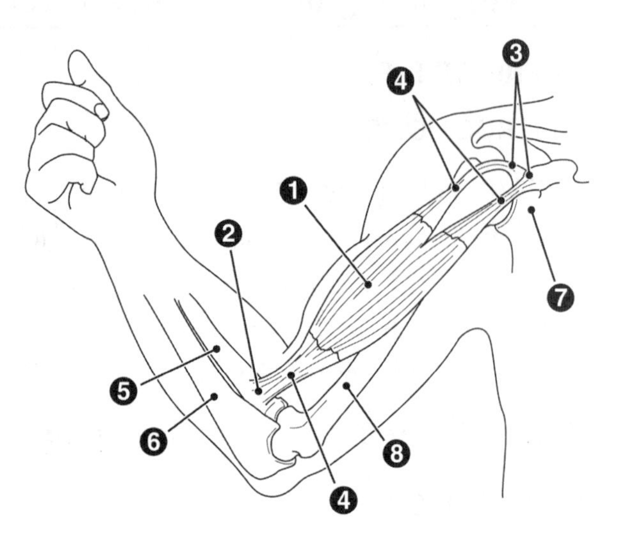

10. Compare the workings of muscles and bones to lever systems.

EXERCISE 8-12.

INSTRUCTIONS

For each of the following muscle actions, state which class of lever (1st, 2nd, or 3rd) is the most applicable.

1. Nodding the head _____
2. Performing a biceps curl ____
3. Standing on tiptoes ____

11. Explain how muscles are named.

EXERCISE 8-13.

INSTRUCTIONS

For each muscle name, write the characteristic(s) used for the name. Choose between the following 6 options: location, size, shape, direction of fibers, number of heads, action. The number of blanks indicates how many characteristics apply to each muscle. Note that femoris means thigh, brachii means arm, carpii means wrist, teres means long and round.

1. trapezius ______________________
2. quadriceps femoris ______________________ ______________________
3. rectus abdominus ______________________ ______________________
4. flexor carpii ______________________ ______________________
5. teres minor ______________________ ______________________

12. Name some of the major muscles in each muscle group and describe the main function of each.

EXERCISE 8-14: Superficial Muscles: Anterior View (Text Fig. 8-9)

INSTRUCTIONS

1. Write the name of each labeled muscle on the numbered lines in different colors.
2. Color the different muscles on the diagram with the correspon-ding color.

1. ____________________
2. ____________________
3. ____________________
4. ____________________
5. ____________________
6. ____________________
7. ____________________
8. ____________________
9. ____________________
10. ____________________
11. ____________________
12. ____________________
13. ____________________
14. ____________________
15. ____________________
16. ____________________
17. ____________________
18. ____________________
19. ____________________
20. ____________________
21. ____________________
22. ____________________
23. ____________________
24. ____________________
25. ____________________

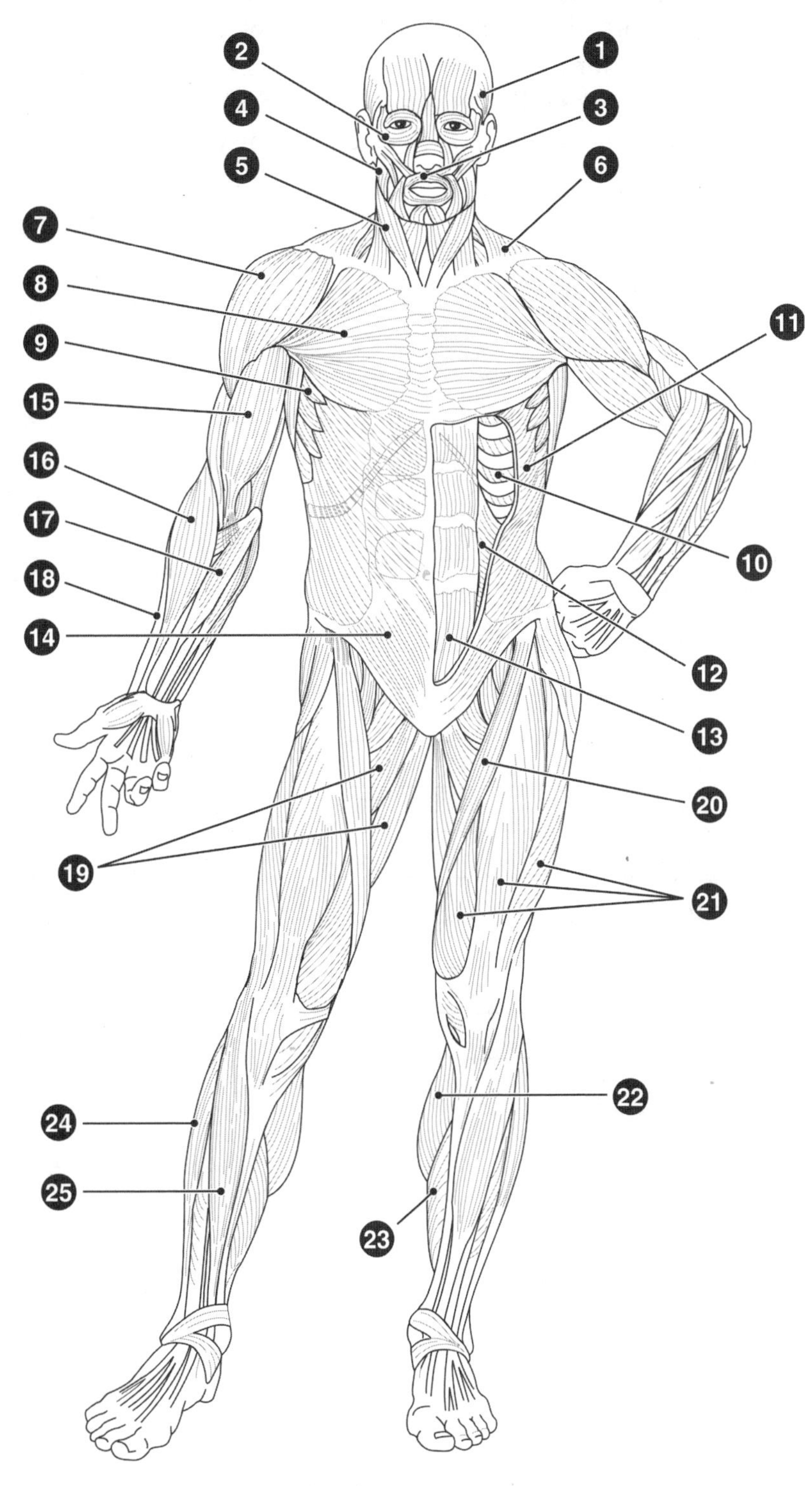

Anterior view

EXERCISE 8-15: Superficial Muscles: Posterior View (Text Fig. 8-10)

INSTRUCTIONS

1. Write the name of each labeled muscle or tendon on the numbered lines in different colors. If possible, use the same color you used for the muscle in Exercise 8-14.
2. Color the different muscles and tendons on the diagram with the corresponding color.

1. ______________________
2. ______________________
3. ______________________
4. ______________________
5. ______________________
6. ______________________
7. ______________________
8. ______________________
9. ______________________
10. ______________________
11. ______________________
12. ______________________
13. ______________________
14. ______________________
15. ______________________
16. ______________________
17. ______________________
18. ______________________
19. ______________________
20. ______________________

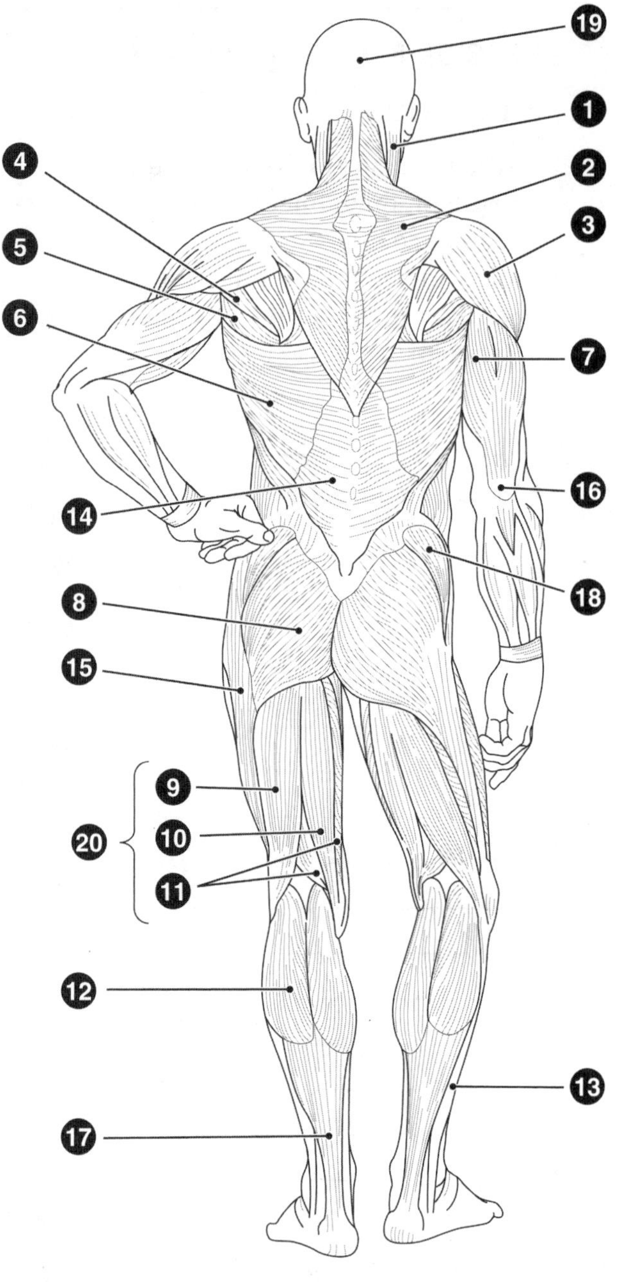

Posterior view

EXERCISE 8-16: Muscles of the Head (Text Fig. 8-11)

INSTRUCTIONS

1. Write the name of each labeled muscle or tendon on the numbered lines in different colors. If possible, use the same color you used for the muscle in Exercises 8-14 and 8-15.
2. Color the different muscles and tendons on the diagram with the corresponding color.

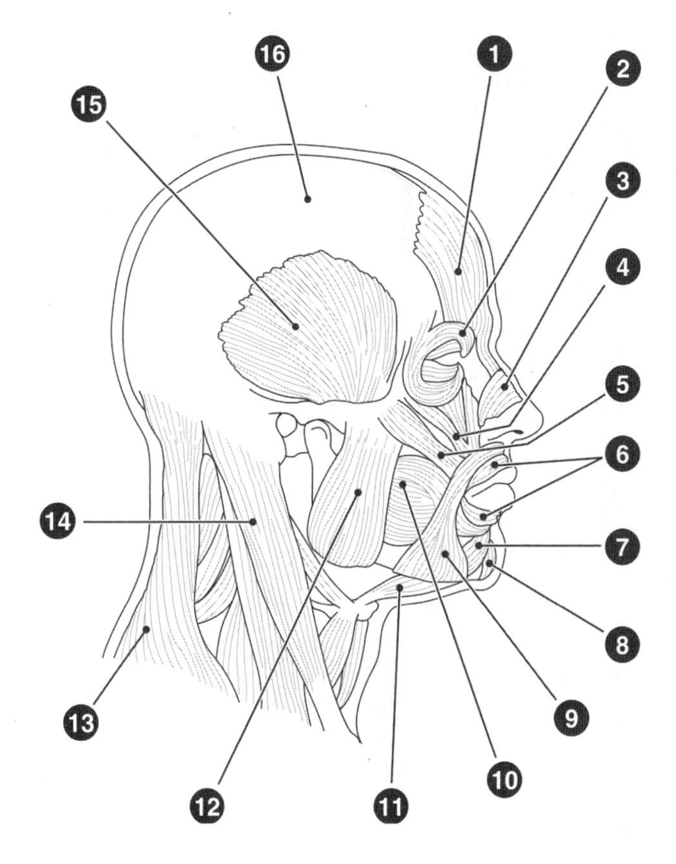

1. ______________________
2. ______________________
3. ______________________
4. ______________________
5. ______________________
6. ______________________
7. ______________________
8. ______________________
9. ______________________
10. ______________________
11. ______________________
12. ______________________
13. ______________________
14. ______________________
15. ______________________
16. ______________________

EXERCISE 8-17: Muscles of the Thigh: Anterior View (Text Fig. 8-16A)

INSTRUCTIONS

1. Write the name of each labeled muscle, tendon, or bone on the numbered lines in different colors. If possible, use the same color you used for the muscle in Exercise 8-14.
2. Color the different structures on the diagram with the corresponding color.

1. ______________________
2. ______________________
3. ______________________
4. ______________________
5. ______________________
6. ______________________
7. ______________________
8. ______________________
9. ______________________
10. ______________________
11. ______________________

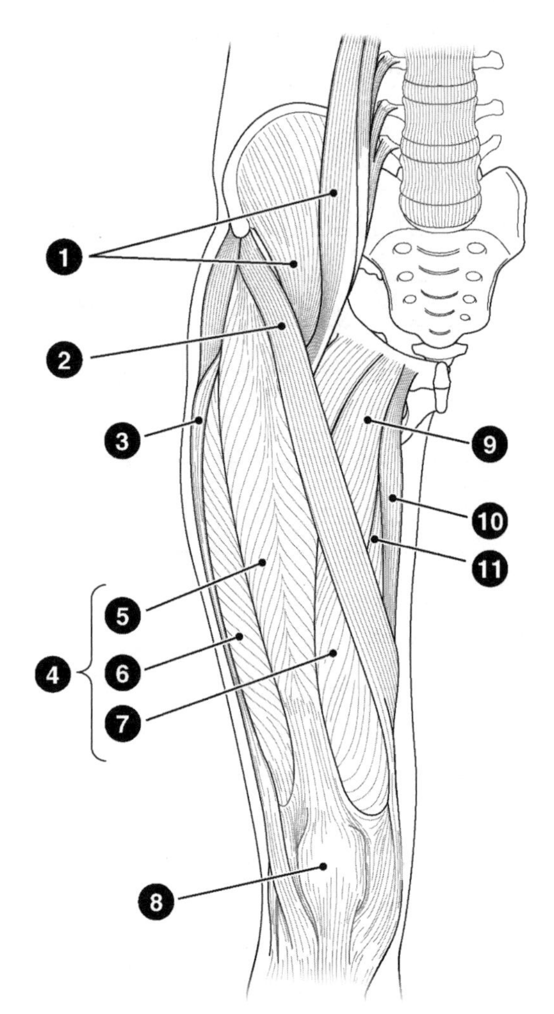

EXERCISE 8-18.

INSTRUCTIONS

Write the appropriate term in each blank.

sternocleidomastoid	buccinator	masseter	trapezius	orbicularis oris
deltoid	rotator cuff	latissimus dorsi	orbicularis oculi	brachialis

1. The muscle capping the shoulder and upper arm ________________
2. A deep muscle group that supports the shoulder joint ________________
3. A muscle that closes the eye ________________
4. The muscle that makes up the fleshy part of the cheek ________________
5. A muscle that closes the jaw ________________
6. A muscle on the side of the neck that flexes the head ________________
7. The main flexor of the forearm ________________
8. A triangular muscle on the back of the neck and the upper back that extends the head ________________

EXERCISE 8-19.

INSTRUCTIONS

Write the appropriate term in each blank.

triceps brachii	serratus anterior	brachioradialis	biceps brachii	trapezius
intercostals	levator ani	erector spinae	latissimus dorsi	

1. A large muscle of the middle and lower back that inserts in the humerus and extends the arm at the shoulder behind the back ________________
2. The muscle in the pelvic floor that aids in defecation ________________
3. A muscle on the front of the arm that flexes the elbow and supinates the hand ________________
4. The large muscle on the back of the arm that extends the elbow ________________
5. A chest muscle inferior to the axilla that moves the scapula forward ________________
6. A deep muscle that extends the vertebral column ________________
7. Muscles between the ribs that can enlarge the thoracic cavity ________________

EXERCISE 8-20.

INSTRUCTIONS

Write the appropriate term in each blank.

rectus abdominis	transversus abdominis	gluteus maximus	gluteus medius	rectus femoris
iliopsoas	adductor longus	gracilis	biceps femoris	

1. Part of the quadriceps femoris muscle ________________
2. The muscle that forms much of the fleshy part of the buttock ________________
3. A deep muscle of the buttock that abducts the thigh ________________
4. A vertical muscle covering the anterior surface of the abdomen ________________
5. A muscle that aids in pressing the thighs together ________________
6. A muscle extending from the pubic bone to the tibia that adducts the thigh at the hip ________________
7. A powerful flexor of the thigh that arises from the ilium ________________

EXERCISE 8-21.

INSTRUCTIONS

Write the appropriate term in each blank.

sartorius	gastrocnemius	soleus	peroneus longus
tibialis anterior	quadriceps femoris	semimembranosus	flexor digitorum group

1. The thin muscle that travels down and across the medial surface of the thigh ______________
2. The chief muscle of the calf of the leg ______________
3. The muscle that inverts and dorsiflexes the foot ______________
4. Muscles that flex the toes ______________
5. The muscle that everts the foot ______________
6. A deep muscle that plantar flexes the foot at the ankle ______________

13. Describe how muscles change with age.

EXERCISE 8-22.

INSTRUCTIONS

List two changes that occur in aging muscles.

1. ____
2. ____

14. List the major muscular disorders.

EXERCISE 8-23.

INSTRUCTIONS

Write the appropriate term in each blank.

fibromyositis	myalgia	bursitis	carpal tunnel syndrome
muscular dystrophy	strain	sprain	fibrositis

1. Inflammation of connective tissue and muscle tissue ______________
2. A muscle injury resulting from overuse or overstretching ______________
3. A group of disorders, seen more frequently in male children, that cause progressive weakness and paralysis ______________
4. An injury resulting from tearing of the ligaments around a joint ______________
5. Inflammation of a fluid-filled sac near a joint ______________
6. A term that means muscle pain ______________

15. Show how word parts are used to build words related to the muscular system.

EXERCISE 8-24.

INSTRUCTIONS

Complete the following table by writing the correct word part or meaning in the space provided. Write a word that contains each word part in the Example column.

Word Part	Meaning	Example
1. ________	muscle	________
2. brachi/o	________	________
3. ________	strength	________
4. erg/o	________	________
5. -algia	________	________
6. ________	four	________
7. ________	absent, lack of	________
8. ________	flesh	________
9. vas/o	________	________
10. iso-	________	________

Making the Connections

The following concept map deals with substances and structures required for muscle contraction. Each pair of terms is linked together by a connecting phrase into a sentence. The sentence should be read in the direction of the arrow. Complete the concept map by filling in the appropriate term or phrase. There is one right answer for each term. However, there are many correct answers for the connecting phrases.

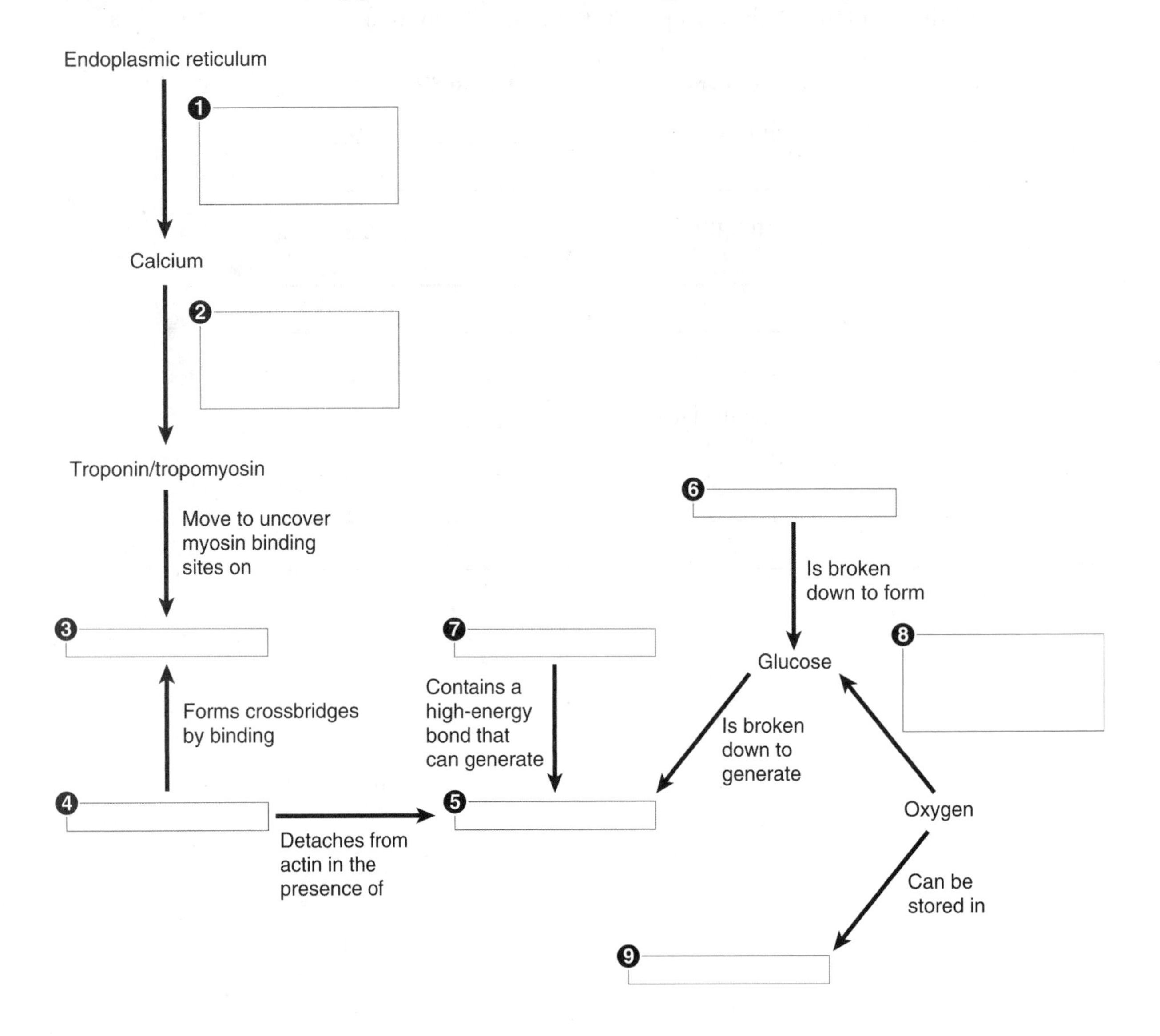

Optional Exercise: Make your own concept map, based on the events of muscle contraction. Choose your own terms to incorporate into your map or use the following list: neuron, acetylcholine, neuromuscular junction, snaptic cleft, motor end plate, myosin, actin, endoplasmic reticulum, calcium, sarcomere, troponin/tropomyosin, and ATP.

Testing Your Knowledge

Building Understanding

I. Multiple Choice

Select the best answer and write the letter of your choice in the blank.

1. Which of the following statements is NOT true of skeletal muscle? 1. _______
   a. The cells are long and threadlike.
   b. It is involuntary.
   c. It is described as striated.
   d. The cells are multinucleated.
2. When muscles and bones act together in the body as a lever system, the pivot point or fulcrum of the system is the 2. _______
   a. joint
   b. tendon
   c. ligament
   d. myoglobin
3. The quadriceps muscles act to 3. _______
   a. flex the thigh
   b. extend the leg
   c. adduct the leg
   d. abduct the thigh
4. Soreness and stiffness in a muscle resulting from overuse is probably a 4. _______
   a. sprain
   b. strain
   c. break
   d. tendinitis
5. Which of the following is NOT a muscle of the hamstring group? 5. _______
   a. biceps femoris
   b. rectus femoris
   c. semimembranosus
   d. semitendinosus
6. The connective tissue layer around individual muscle fibers is called the 6. _______
   a. epimysium
   b. perimysium
   c. superficial fascia
   d. endomysium
7. During muscle contraction, ATP binds to 7. _______
   a. tropomyosin
   b. myosin
   c. actin
   d. troponin

II. Completion Exercise

Write the word or phrase that correctly completes each sentence.

1. Normally, muscles are in a partially contracted state, even when not in use. This state of mild constant tension is called ____________________.
2. The hemoglobin-like compound that stores oxygen in muscle is ____________________.
3. The muscle attachment that is usually relatively fixed is called its ____________________.
4. A contraction that generates tension but does not shorten the muscle is called ____________________.
5. The band of connective tissue that attaches the gastrocnemius muscle to the heel is the ____________________.
6. The muscles of the pelvic floor together form the ____________________.
7. A muscle that must relax during a given movement is called the ____________________.
8. A term that means muscle pain is ____________________.
9. The muscular partition between the thoracic and abdominal cavities is the ____________________.
10. A superficial muscle of the neck and upper back acts at the shoulder. This muscle is the ____________________.
11. The large muscle of the upper chest that flexes the arm across the body is the ____________________.
12. A spasm of the visceral muscles is known as ____________________.
13. The muscle responsible for dorsiflexion and inversion of the foot is the ____________________.

Understanding Concepts

I. True/False

For each question, write *T* for true or *F* for false in the blank to the left of each number. If a statement is false, correct it by replacing the <u>underlined</u> term and write the correct statement in the blank below the question.

________ 1. The triceps brachii <u>flexes</u> the arm at the elbow.

____

________ 2. Muscles enter into oxygen debt when they are functioning <u>anaerobically</u>.

____

________ 3. In an <u>isometric</u> contraction, muscle tension increases but the muscle does not shorten.

____

________ 4. A term that describes inflammation of connective tissues is <u>myositis</u>.

____

_______ 5. The neurotransmitter used at the neuromuscular junction is norepinephrine.

____

_______ 6. A contracting subunit of skeletal muscle is called a(n) crossbridge.

____

_______ 7. The storage form of glucose is called creatine phosphate.

____

_______ 8. The element that binds to the troponin and tropomyosin complex is calcium.

____

II. Practical Applications

Study each discussion. Then write the appropriate word or phrase in the space provided.

➤ Group A

Ms. J is sitting at her desk studying for her anatomy final exam.

1. She has excellent posture, with her back straight. The deep back muscle responsible for her erect posture is the ________________.
2. Despite her excellent posture, Ms. J has developed a muscle spasm that has fixed her head in a flexed, rotated position. This condition is called wryneck, or ________________.
3. Ms. J has a cheerful disposition, and she likes to whistle while she works. The cheek muscle involved in whistling is the ________________.
4. Ms. J is furiously writing notes, and her hand is flexed around the pen. The muscle groups that flex the hand are called the ________________.
5. Ms. J leans on her left elbow while she reads her notes, resulting in inflammation of the bursa over the point of the elbow. This disorder is called student's elbow, or ________________.
6. After several hours of intense studying, Ms. J takes a break. Stretching, she straightens her leg at the knee joint. The muscle that accomplishes this action is the ________________.

➤ Group B

Mr. L, aged 72, is a retired concert pianist.

1. He reports numbness and weakness in his right hand. This is symptomatic of a repetitive use disorder called ________________.
2. Mr. L also experiences pain when he sits. The healthcare worker suspects a problem with the fluid-filled sacs near his sit bones. This inflammatory disorder is called ________________.
3. The healthcare worker recommends that he undertakes a regular exercise program, incorporating stretching, aerobic exercise, and resistance training. This type of varied program is called interval training, or ________________.
4. Mr. L asks if the resistance training will give him bigger muscles. He is told that his muscle cells may increase in size, a change called ________________.

III. Short Essays

1. Mr. Q has embarked on an exercise program based solely on jogging. Name three distinct changes that will occur in his muscles.

____

____

____

2. Ms. L has suffered an embarrassing (but relatively painless) fall. Mr. L is staring at her with his mouth open. Ms. L asks him to activate two muscles to close his jaw. Name these two muscles.

____

____

____

Conceptual Thinking

1. Ms. S is competing in an endurance event at the Olympics. She is consuming a special herbal product that reputedly degrades lactic acid in the absence of oxygen. Explain the potential benefit of this product, discussing the production and effects of lactic acid.

____

____

____

2. While attending the ballet, you notice a dancer raising her heel to stand on her tiptoes.
   A. What is the name of this action (i.e., adduction)? ________________
   B. What is the prime mover for this action? ______________________
   C. Name a synergistic muscle involved in this action. ________________
   D. Name an antagonist muscle involved. _______________
   E. Which class of lever is represented by this action? _________________
   F. Name the fulcrum and resistance, and describe the relative positions of the fulcrum, resistance, and effort.

____

____

____

Expanding Your Horizons

Are world-class athletes born or made? It is no coincidence that athletic performance tends to run in families. Genetic influences on muscular function are discussed in an article in *Scientific American*. This information could be used to screen for elite athletes—perhaps children of the future will know which sports they can excel in based on their genetic profile. This possibility is discussed in a special *Scientific American* issue entitled "Building the Elite Athlete."

Resources

1. Andersen JL, Schjerling P, Saltin B. Muscle, genes and athletic performance. *Sci Am* 2000; 283:48–55.
2. Olshansky JS, Carnes BA, Butler RN. If humans were built to last. *Sci Am* 2001; 284:50–55.
3. Taubes G. Toward Molecular Talent Scouting. Scientific American Presents: Building the Elite Athlete 2000; 11:26–31.

UNIT IV Coordination and Control

CHAPTER 9

The Nervous System: The Spinal Cord and Spinal Nerves

Overview

The nervous system is the body's coordinating system—receiving, sorting, and controlling responses to both internal and external changes (stimuli). The nervous system as a whole is divided structurally into the **central nervous system** (**CNS**), made up of the brain and the spinal cord, and the **peripheral nervous system** (**PNS**), made up of the cranial and spinal nerves. The PNS connects all parts of the body with the CNS. The brain and cranial nerves are the subject of Chapter 10. Functionally, the nervous system is divided into the somatic (voluntary) system and the autonomic (involuntary) system.

The nervous system functions by means of the **nerve impulse**, an electrical current or **action potential** that spreads along the membrane of the **neuron** (nerve cell). Each neuron is composed of a cell body and fibers, which are threadlike extensions from the cell body. A **dendrite** is a fiber that carries impulses toward the cell body, and an **axon** is a fiber that carries impulses away from the cell body. Some axons are covered with a sheath of fatty material called **myelin**, which insulates the fiber and speeds conduction along the fiber. In the PNS, neuron fibers are collected in bundles to form **nerves.** Bundles of fibers in the CNS are called **tracts**. Nerve cells make contact at a junction called a **synapse.** Here, a nerve impulse travels across a very narrow cleft between the cells by means of a chemical referred to as a **neurotransmitter.** Neurotransmitters are released from axons of presynaptic cells to be picked up by receptors in the membranes of responding cells, the postsynaptic cells.

A neuron may be classified as either a sensory (afferent) type, which carries impulses toward the central nervous system, or a motor (efferent) type, which carries impulses away from the central nervous system. **Interneurons** are connecting neurons within the central nervous system.

The basic functional pathway of the nervous system is the **reflex arc**, in which an impulse travels from a receptor, along a sensory neuron to a synapse or synapses in the central nervous system, and then along a motor neuron to an effector organ that carries out a response. The spinal cord carries impulses to and from the brain. It is also a center for simple reflex activities in which responses are coordinated within the cord.

The **autonomic nervous system** controls unconscious activities. This system regulates the actions of glands, smooth muscle, and the heart muscle. The autonomic nervous system has two divisions, the **sympathetic nervous system** and the **parasympathetic nervous system**, which generally have opposite effects on a given organ.

Addressing the Learning Outcomes

1. Describe the organization of the nervous system according to structure and function.

EXERCISE 9-1: Anatomic Divisions of the Nervous System (Text Fig. 9-1)

INSTRUCTIONS

Label the parts and divisions of the nervous system shown below.

1. ______________________
2. ______________________
3. ______________________
4. ______________________
5. ______________________
6. ______________________

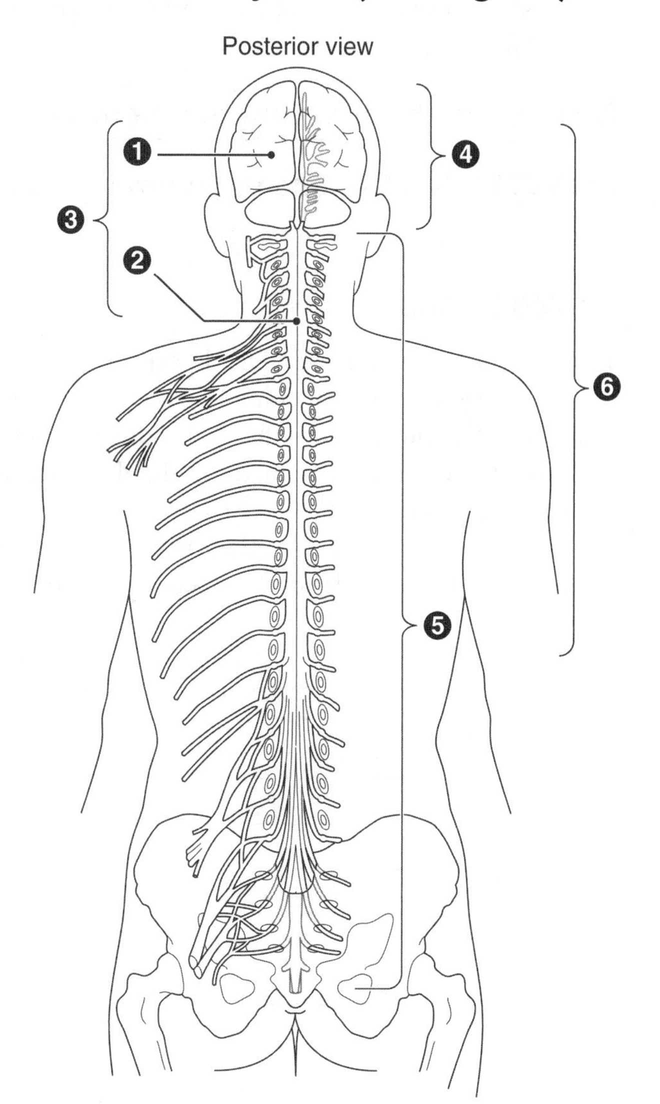

EXERCISE 9-2.

INSTRUCTIONS

Write the appropriate term in each blank.

central nervous system	peripheral nervous system	somatic nervous system
sympathetic nervous system	parasympathetic nervous system	autonomic nervous system

1. The functional division of the nervous system that is also called the visceral nervous system ____________
2. The functional division of the nervous system that controls skeletal muscles ____________
3. The system that promotes the fight-or-flight response ____________
4. The system that stimulates the activity of the digestive tract ____________
5. The structural division of the nervous system that includes the brain ____________
6. The structural division of the nervous system that includes the cranial nerves ____________

2. Describe the structure of a neuron.

EXERCISE 9-3: The Motor Neuron (Text Fig. 9-2)

INSTRUCTIONS

1. Write the name of each labeled part on the numbered lines in different colors. Structures 4 to 6 will not be colored, so write their names in black.
2. Color the different structures on the diagram with the corresponding color.
3. Add large arrows showing the direction the nerve impulse will travel, from the dendrites to the muscle.

1. ____________
2. ____________
3. ____________
4. ____________
5. ____________
6. ____________
7. ____________
8. ____________

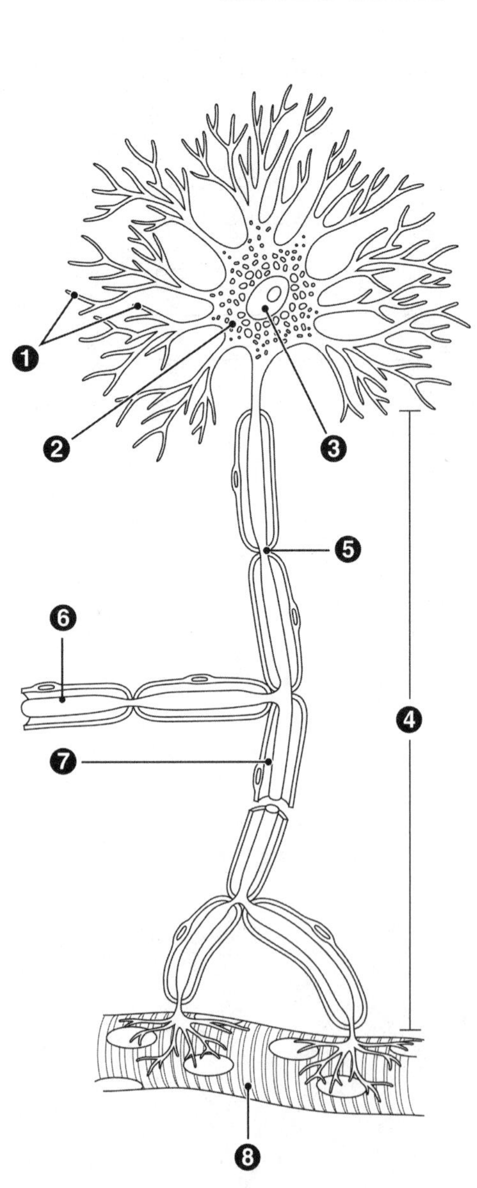

EXERCISE 9-4.

INSTRUCTIONS

Write the appropriate term in each blank.

dendrite	neurilemma	axon
white matter	gray matter	node

1. A nerve cell fiber that carries impulses away from the cell body ________________
2. The part of a neuron that receives a stimulus ________________
3. The sheath around some neuron fibers that aids in regeneration ________________
4. A gap in the neuron sheath ________________
5. The portion of the spinal cord made up of myelinated axons ________________

3. Describe how neuron fibers are built into a nerve.

EXERCISE 9-5.

INSTRUCTIONS

Write the appropriate term in each blank.

afferent neurons	efferent neurons	neuron	tract
endoneurium	perineurium	epineurium	nerve

1. The scientific name for a nerve cell ________________
2. A bundle of neuron fibers located outside the central nervous system ________________
3. A bundle of neuron fibers located within the central nervous system ________________
4. Neurons that conduct impulses towards the brain ________________
5. Neurons that conduct impulses away from the brain ________________
6. The coating of individual nerve fibers ________________
7. The coating of an entire nerve ________________

4. Explain the purpose of neuroglia.

EXERCISE 9-6.

INSTRUCTIONS

List five functions of neuroglia in the spaces below.

1. ________________________________
2. ________________________________
3. ________________________________
4. ________________________________
5. ________________________________

5. Diagram and describe the steps in an action potential.

EXERCISE 9-7.

INSTRUCTIONS

In the space below, draw an action potential tracing as illustrated in Fig. 9-7. On your diagram, indicate which event is occurring (depolarization, repolarization, or resting) and which ion is moving (sodium [Na] or potassium [K], if any.

EXERCISE 9-8.

INSTRUCTIONS

Write the appropriate term in each blank.

depolarization   action potential   repolarization Na^+ K^+
resting

1. The step in which the membrane potential returns to rest ______________
2. The ion that crosses the neuron membrane to cause depolarization ______________
3. The result of positive ions entering the neuron ______________
4. A sudden change in membrane potential that is transmitted along axons ______________
5. The ion that leaves the neuron to cause repolarization ______________

6. Briefly describe the transmission of a nerve impulse.

EXERCISE 9-9.

INSTRUCTIONS

Place the following events in the order in which they occur.

A. The action potential opens sodium channels in adjacent portions of the membrane.
B. Sodium entry into the cell causes another action potential in the adjacent portion of the membrane.
C. A stimulus initiates an action potential.
D. Open sodium channels let sodium enter the cell.

1. _____
2. _____
3. _____
4. _____

7. Explain the role of myelin in nerve conduction.

EXERCISE 9-10: Formation of a Myelin Sheath (Text Fig. 9-4)

INSTRUCTIONS

1. Write the name of each labeled part on the numbered lines in different colors. Structures 4 and 7-9 will not be colored, so write their names in black.
2. Color the different structures on the diagram with the corresponding color. Make sure you color the structure in all parts of the diagram. For instance, structure 3 is visible in 3 locations.

1. ____________________
2. ____________________
3. ____________________
4. ____________________
5. ____________________
6. ____________________
7. ____________________
8. ____________________
9. ____________________

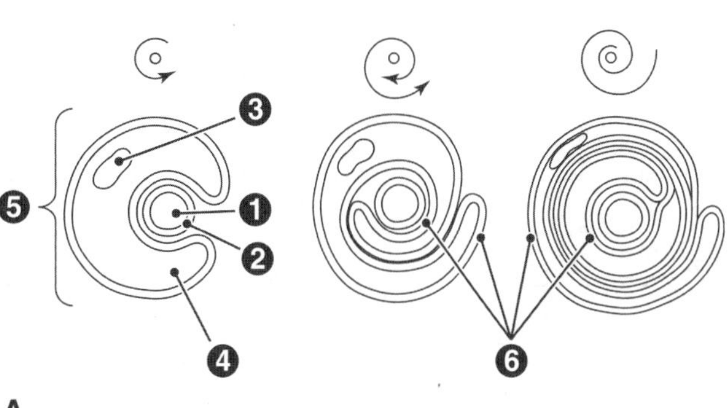

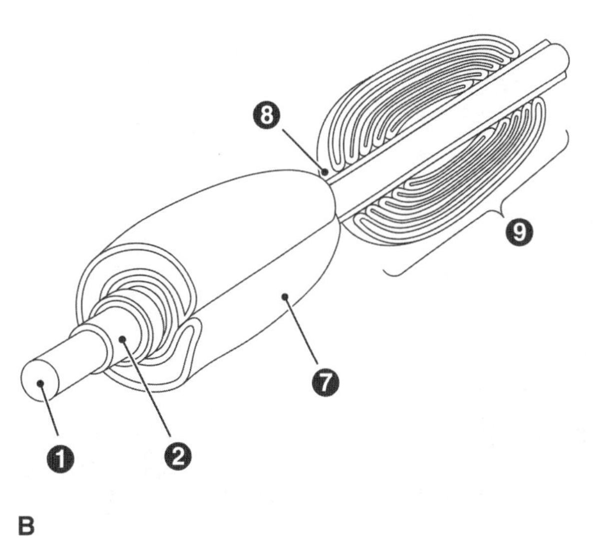

EXERCISE 9-11.

INSTRUCTIONS

Label each of the following statements as true (T) or false (F).

1. Action potentials occur in axon regions surrounded by myelin. ____________
2. Action potential transmission is faster in myelinated neurons. ____________
3. Action potentials occur at nodes. ____________

8. Briefly describe transmission at a synapse.

EXERCISE 9-12: A Synapse (Text Fig. 9-9)

INSTRUCTIONS

Label the parts of the synapse shown below.

1. ____________________
2. ____________________
3. ____________________
4. ____________________
5. ____________________
6. ____________________
7. ____________________
8. ____________________
9. ____________________
10. ____________________

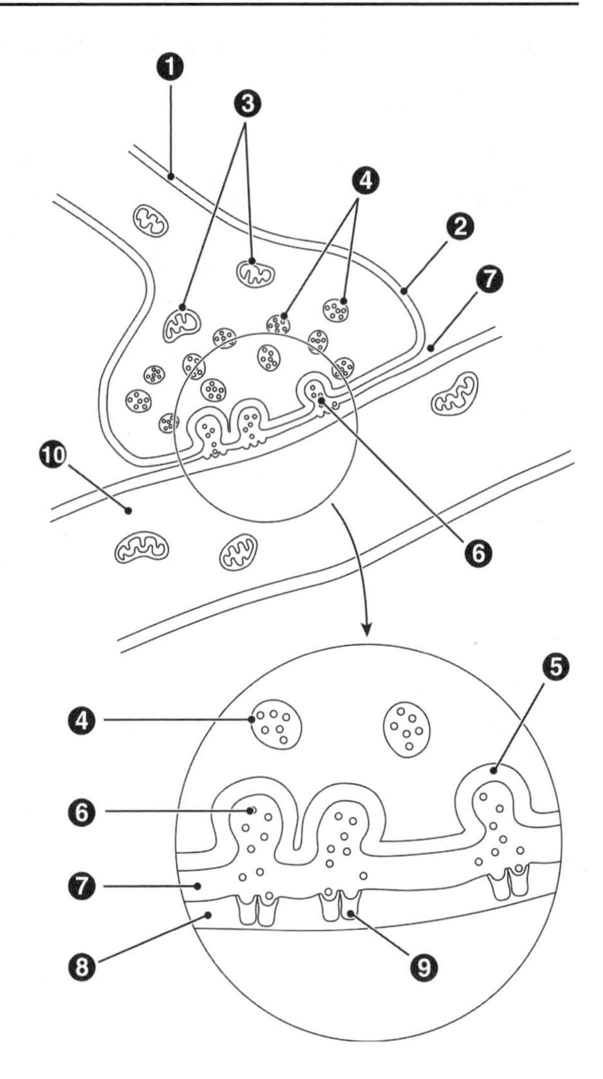

EXERCISE 9-13.

INSTRUCTIONS

Place the following synaptic events in order.

A. Neurotransmitter molecules bind receptors in the postsynaptic membrane.
B. Neurotransmitter molecules are released into the synaptic cleft.
C. The nerve impulse arrives at the end of the presynaptic neuron.
D. Vesicles containing neurotransmitter fuse with the cell membrane.
E. The activity of the postsynaptic cell is altered.

1. ____
2. ____
3. ____
4. ____
5. ____

9. Define *neurotransmitter* and give several examples of neurotransmitters.

EXERCISE 9-14.

INSTRUCTIONS

List three examples of neurotransmitters in the spaces below.

1. ______________________
2. ______________________
3. ______________________

10. Describe the distribution of gray and white matter in the spinal cord.

EXERCISE 9-15: Spinal Cord (Text Fig. 9-12)

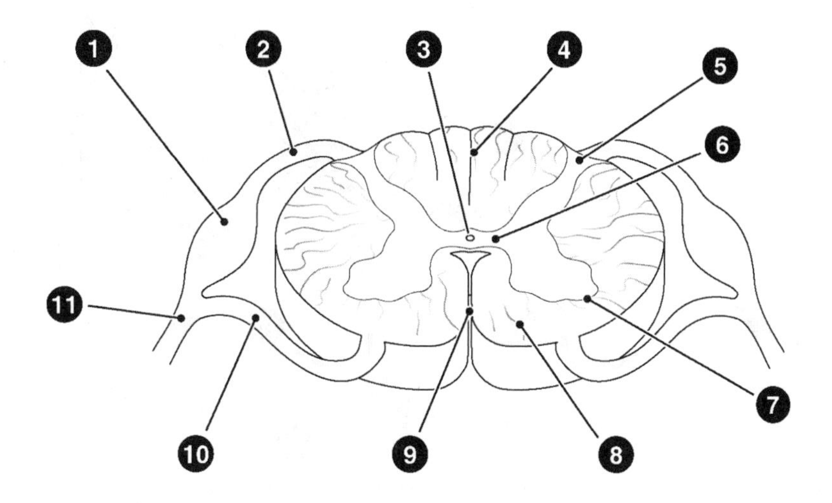

INSTRUCTIONS

1. Write the name of each labeled part on the numbered lines. Use the following color scheme:
   - 1 and 2: red
   - 3, 4, 9: different dark colors
   - 5: pink
   - 6: any light color
   - 7: light bluer
   - 10: medium blue
   - 11: purple
2. Color or outline the different structures on the diagram with the corresponding color.

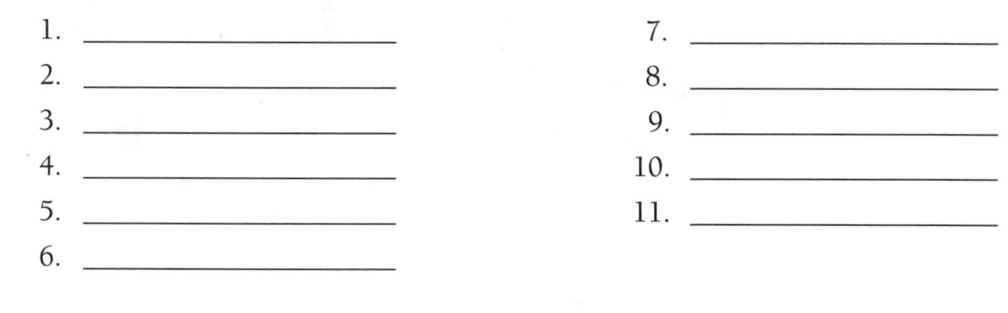

1. ______________________
2. ______________________
3. ______________________
4. ______________________
5. ______________________
6. ______________________
7. ______________________
8. ______________________
9. ______________________
10. ______________________
11. ______________________

11. List the components of a reflex arc.

EXERCISE 9-16: Reflex Arc (Text Fig. 9-13)

INSTRUCTIONS

1. Write the names of the five components of a reflex arc on the numbered lines 1 to 5 in different colors, and color the components with the corresponding color. Follow the color scheme provided below.
2. Write the names of the parts of the spinal cord on numbered lines 6 to 12 in different colors, and color the structures with the appropriate color. Follow the color scheme provided below.

   Color Scheme
   - 1, 2, 6, 7, 8: red
   - 3: green
   - 4, 10: medium blue
   - 5: purple
   - 9: do not color (write name in black)
   - 11: pink
   - 12: light blue

1. ____________________
2. ____________________
3. ____________________
4. ____________________
5. ____________________
6. ____________________
7. ____________________
8. ____________________
9. ____________________
10. ____________________
11. ____________________
12. ____________________

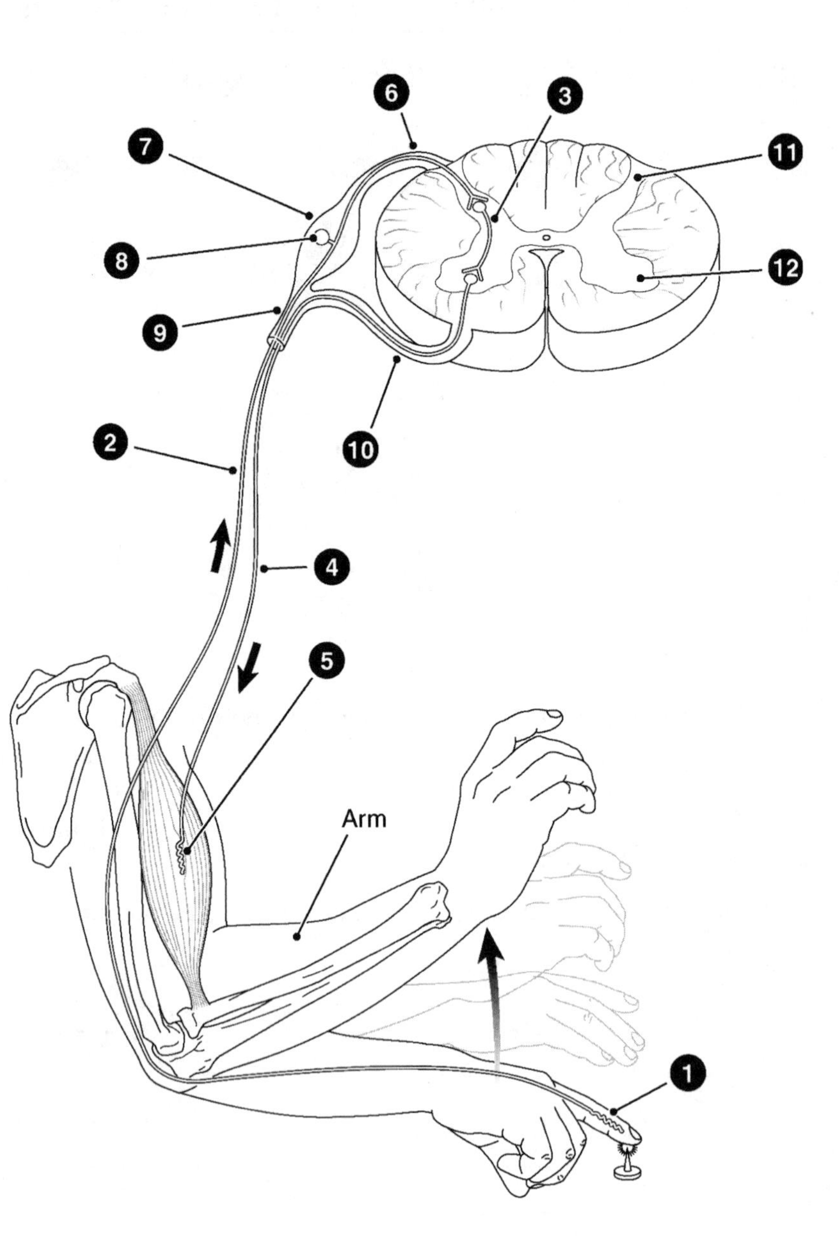

12. Define a simple reflex and give several examples of reflexes.

EXERCISE 9-17.

INSTRUCTIONS

Fill in the blanks in the discussion below using the following terms: patellar reflex, simple reflex, somatic reflexes, and autonomic reflexes.

A ______________________ (1) describes any rapid automatic response involving very few neurons. Reflexes involving skeletal muscles are called ______________________ (2); reflexes involving smooth muscle or glands are ______________________ (3). An example of the former type is the ____________________ (4), a stretch reflex that involves striking a tendon at the top of the leg.

13. Describe and name the spinal nerves and three of their main plexuses.

EXERCISE 9-18: Spinal Cord (Text Fig. 9-11)

INSTRUCTIONS

1. Label the parts of the central nervous system (structures 1 to 5).
   - Spinal nerves are named based on the site they emerge from the spinal column. Write the names of the nerve groups on the appropriate line (lines 6 to 10).
   - Some of the anterior branches of the spinal nerves interlace to form plexuses. Identify the plexuses (structures 11 to 13).
   - Label the specific nerves (structures 14 to 20).
2. Color or outline the different structures on the diagram with the corresponding color.

1. ____________________
2. ____________________
3. ____________________
4. ____________________
5. ____________________
6. ____________________
7. ____________________
8. ____________________
9. ____________________
10. ____________________
11. ____________________
12. ____________________
13. ____________________
14. ____________________
15. ____________________
16. ____________________
17. ____________________
18. ____________________
19. ____________________
20. ____________________

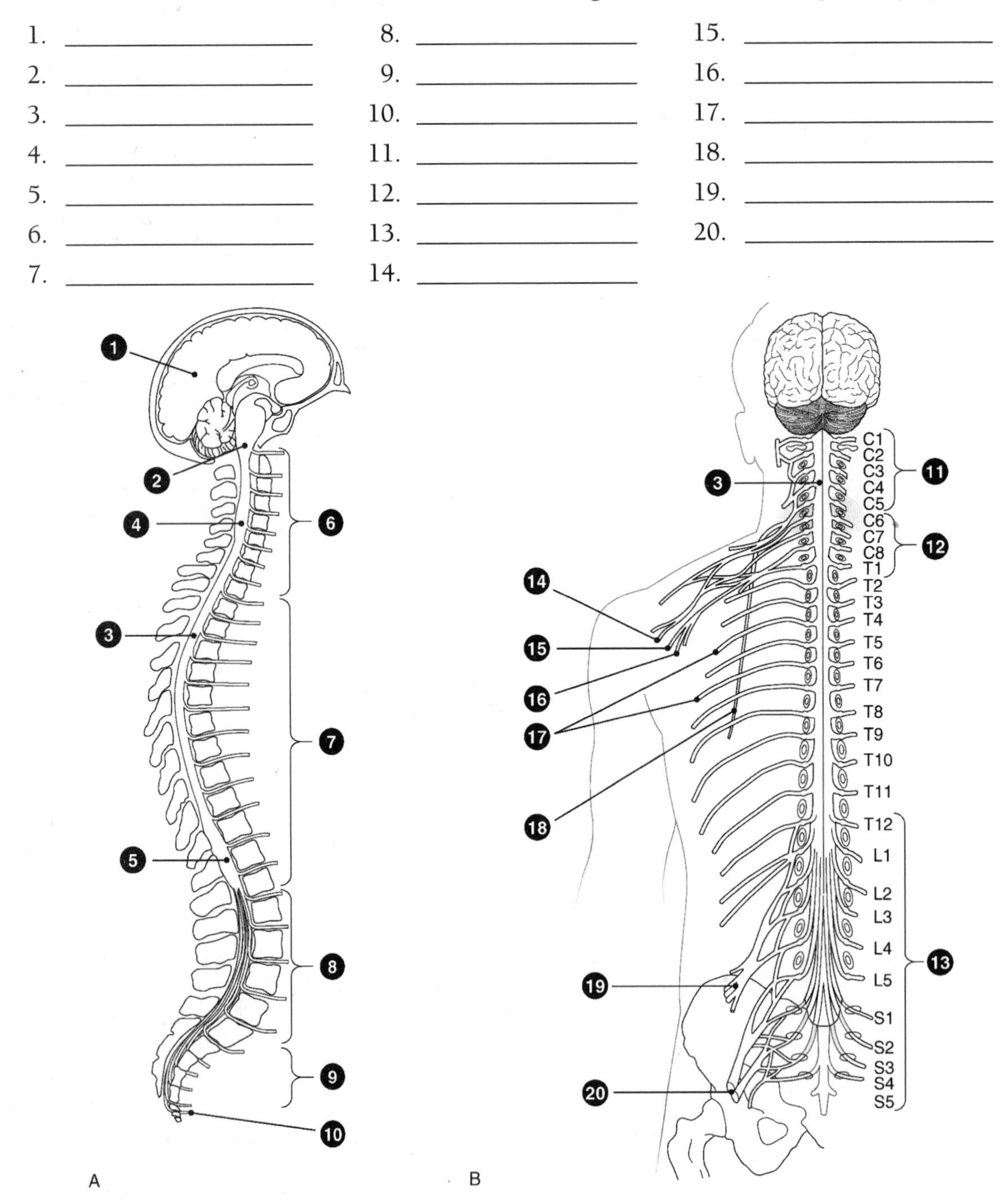

14. Compare the location and functions of the sympathetic and parasympathetic nervous systems.

EXERCISE 9-19.

INSTRUCTIONS

Do each of the following statements apply to the sympathetic nervous system (S) or the parasympathetic nervous system (P)?

1. Also described as the adrenergic system ______
2. Also described as the cholinergic system ______
3. Motor neurons originate in the thoracolumbar region of the spinal cord ______
4. Motor neurons originate in the craniosacral region of the spinal cord ______
5. Activation causes the pupils to dilate ______
6. Activation causes blood vessels in digestive organs to dilate ______
7. Terminal ganglia located in or near the effector ______
8. Ganglia located near the spinal cord or in collateral ganglia ______
9. Activation decreases kidney activity ______
10. Activation stimulates the sweat glands ______

15. Explain the role of cellular receptors in the action of neurotransmitters.

EXERCISE 9-20.

INSTRUCTIONS

Write the appropriate term in each blank.

muscarinic receptor   nicotinic receptor   adrenergic receptor

1. binds norepinephrine ______
2. binds acetylcholine and induces muscle contraction ______
3. acetylcholine receptor found on effector organs of the parasympathetic system ______

16. Describe several disorders of the spinal cord and of the spinal nerves

EXERCISE 9-21.

INSTRUCTIONS

Write the appropriate term in each blank.

paraplegia	shingles	peripheral neuritis	poliomyelitis
multiple sclerosis	hemiplegia	monoplegia	amyotrophic lateral sclerosis

1. Degeneration of nerves supplying the extremities ____________
2. A disease associated with the destruction of myelin sheaths ____________
3. Paralysis of one arm ____________
4. A viral disease resulting in paralysis ____________
5. A viral disease caused by herpes zoster ____________
6. Loss of sensation and motion in the lower part of the body ____________
7. A disease associated with selective destruction of motor neurons ____________

17. Show how word parts are used to build words related to the nervous system.

EXERCISE 9-22.

INSTRUCTIONS

Complete the following table by writing the correct word part or meaning in the space provided. Write a word that contains each word part in the Example column.

Word Part	Meaning	Example
1. ______	sheath	______
2. para-	______	______
3. ______	four	______
4. soma-	______	______
5. hemi-	______	______
6. ______	paralysis	______
7. ______	nerve, nervous tissue	______
8. ______	remove	______
9. aut/o	______	______
10. post-	______	______

Making the Connections

The following concept map deals with the organization of the nervous system. Each pair of terms is linked together by a connecting phrase into a sentence. The sentence should be read in the direction of the arrow. Complete the concept map by filling in the appropriate term or phrase. There is one right answer for each term. However, there are many correct answers for the connecting phrases (2, 8, 9, and 12).

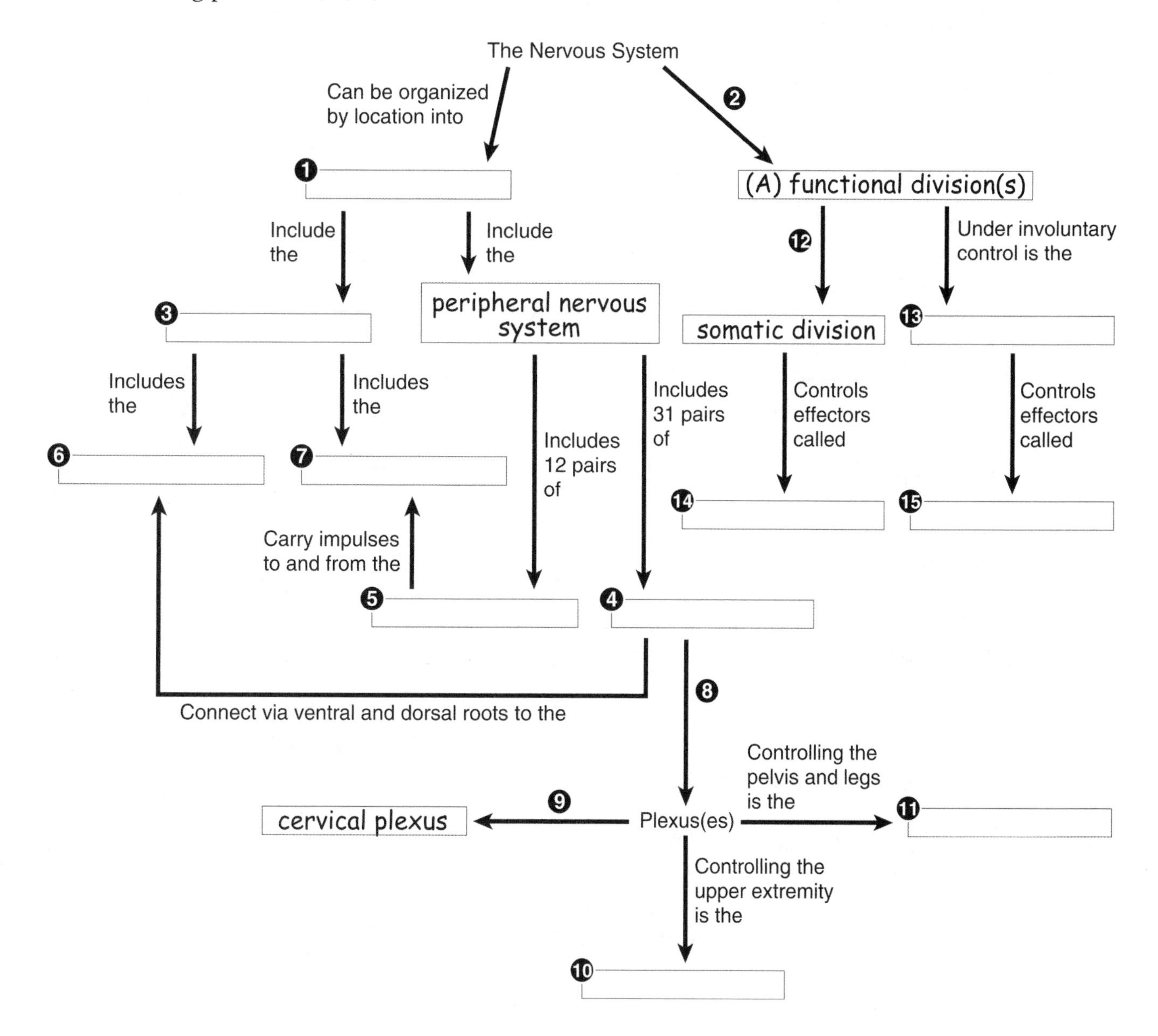

Optional Exercise: Make your own concept map, based on the structures of the spinal cord and the components of a reflex loop. Choose your own terms to incorporate into your map, or use the following list: dorsal root ganglion, gray matter, white matter, ventral root ganglion, afferent fibers, efferent fibers, sensory fibers, motor fibers, receptor, muscle, gland, and effector.

Testing Your Knowledge

Building Understanding

I. Multiple Choice

Select the best answer and write the letter of your choice in the blank.

1. The skin region supplied by a single spinal nerve is called a(n) 1. _______
   a. dermatome
   b. ganglion
   c. plexus
   d. synapse
2. The voluntary nervous system controls 2. _______
   a. visceral muscle
   b. skeletal muscle
   c. glands
   d. cardiac muscle
3. Which of the following are effectors of the nervous system? 3. _______
   a. sensory neurons and ganglia
   b. muscles and glands
   c. synapses and dendrites
   d. receptors and neurotransmitters
4. Cell bodies of sensory neurons are collected in ganglia in the 4. _______
   a. dorsal root of the spinal nerve
   b. sympathetic chain
   c. ventral horn of the spinal cord
   d. effector organ
5. Which of the following substances is a neurotransmitter? 5. _______
   a. myelin
   b. actin
   c. epinephrine
   d. sebum
6. A lumbar tap is done to remove 6. _______
   a. blood
   b. cerebrospinal fluid
   c. intracellular fluid
   d. interstitial fluid
7. Neurons that relay impulses within the spinal cord are called 7. _______
   a. afferent neurons
   b. motor neurons
   c. interneurons
   d. mixed nerves

II. Completion Exercise

Write the word or phrase that correctly completes each sentence.

1. Fibers that carry impulses towards the neuron cell body are called ________________.
2. The portion of the spinal cord made up of cell bodies and unmyelinated axons is called the ________________.
3. The fatty material that covers some axons is called ________________.
4. Dilation of the bronchial tubes is increased by the part of the autonomic nervous system called the ________________.
5. The junction between two neurons is called a(n) ________________.
6. The network of spinal nerves that supplies the pelvis and legs is the ________________.
7. The brain and spinal cord together are referred to as the ________________.
8. The neurotransmitter used at cholinergic synapses is ________________.
9. The small channel in the center of the spinal cord that contains cerebrospinal fluid is the ________________.
10. A nerve cell is also called a(n) ________________.
11. The bridge of gray matter connecting the right and left horns of the spinal cord is the ________________.

Understanding Concepts

I. True/False

For each question, write *T* for true or *F* for false in the blank to the left of each number. If a statement is false, correct it by replacing the underlined term and write the correct statement in the blank below the question.

_______ 1. Motor impulses leave the <u>dorsal</u> horn of the spinal cord.

____

_______ 2. <u>Poliomyelitis</u> results from damage to the myelin sheath around axons in the spinal cord.

____

_______ 3. A <u>tract</u> is a bundle of neuron fibers within the central nervous system.

____

_______ 4. The <u>parasympathetic</u> system has terminal ganglia.

____

_______ 5. The <u>brachial plexus</u> controls the shoulder and arm.

____

_______ 6. The parasympathetic system is <u>adrenergic</u>.

____

_______ 7. The spinal nerves are part of the <u>central</u> nervous system.

____

_______ 8. Neurotransmitters bind to specific proteins on the postsynaptic cell called <u>transporters</u>.

____

_______ 9. At a synapse, a neurotransmitter is released from the <u>postsynaptic</u> cell.

____

_______ 10. A reflex arc that passes through the spinal cord but not the brain is called a <u>spinal</u> reflex.

____

II. Practical Applications

Study each discussion. Then write the appropriate word or phrase in the space provided.

1. Mr. W, a patient with diabetes mellitus for 10 years, complained of pain and numbness of his feet. In observing Mr. W walk, the physician noted there was weakness in the muscles responsible for dorsiflexion of the foot. These symptoms are caused by a degenerative disorder of nerves to the extremities. This disorder is known as _______________
2. The nerves that supply the foot are found in a plexus called the _______________
3. The physician pricked Mr. W's foot with a needle. Mr. W did not feel the needle prick, suggesting that there was a problem with the nerves that carry impulses to the brain. These nerves are called _______________
4. The physician tapped below Mr. W's knee to elicit a knee-jerk response. The tendon she struck was the _______________
5. The effector in this reflex arc is the _______________

III. Short Essays

1. List the events that occur in an action potential.

____

____

____

2. What are neuroglia, and what are some functions of neuroglia?

____

____

____

3. Compare and contrast poliomyelitis and amyotrophic lateral sclerosis. Name one similarity and one difference.

___

___

___

Conceptual Thinking

1. Ms. J is teaching English in Japan. She dines on a local delicacy called pufferfish, and shortly thereafter her lips become numb. She later discovers that pufferfish contain a toxin that blocks sodium channels. Explain why her lips are numb.

___

___

___

2. Dopamine is a neurotransmitter involved in feelings of pleasure. Cocaine blocks the reuptake of dopamine. Use this information to discuss how cocaine affects mood.

___

___

___

Expanding Your Horizons

Paralysis resulting from spinal cord injuries is generally thought to be irreversible, because neurons in the central nervous system do not easily regenerate. However, there is new hope for patients with spinal cord injuries. New therapies are discussed in a recent *Scientific American* article.

Resources

1. McDonald JW. Repairing the damaged spinal cord. *Sci Am* 1999; 281:64–73.

CHAPTER 10

The Nervous System: The Brain and Cranial Nerves

Overview

The brain consists of the two cerebral hemispheres, diencephalon, brainstem, and cerebellum. Each cerebral hemisphere is covered by a layer of gray matter, the **cerebral cortex**, which is further divided into four lobes (the frontal, parietal, temporal, and occipital lobes). Specific functions have been localized to the different lobes. For instance, the interpretation of visual images is performed by an area of the occipital lobe. The diencephalon consists of the **thalamus**, an important relay station for sensory impulses, and the **hypothalamus**, which plays an important role in homeostasis. The brainstem links the spinal cord to the brain and regulates many involuntary functions necessary for life, whereas the cerebellum is involved in coordination and balance. The **limbic system** is not found in a specific brain division. It consists of several structures located between the cerebrum and diencephalon that are involved in emotion, learning, and memory.

The brain and spinal cord are covered by three layers of fibrous membranes called the **meninges**. The **cerebrospinal fluid** (CSF) also protects the brain and spinal cord by providing support and cushioning. The CSF is produced by the choroid plexuses (capillary networks) in four ventricles (spaces) within the brain.

Connected with the brain are 12 pairs of **cranial nerves**, most of which supply structures in the head. Most of these, like all the spinal nerves, are mixed nerves containing both sensory and motor fibers. A few of the cranial nerves contain only sensory fibers, whereas others are motor in function.

Addressing the Learning Outcomes

1. Give the location and functions of the four main divisions of the brain.

EXERCISE 10-1: Brain, Sagittal Section (Text Fig. 10-1)

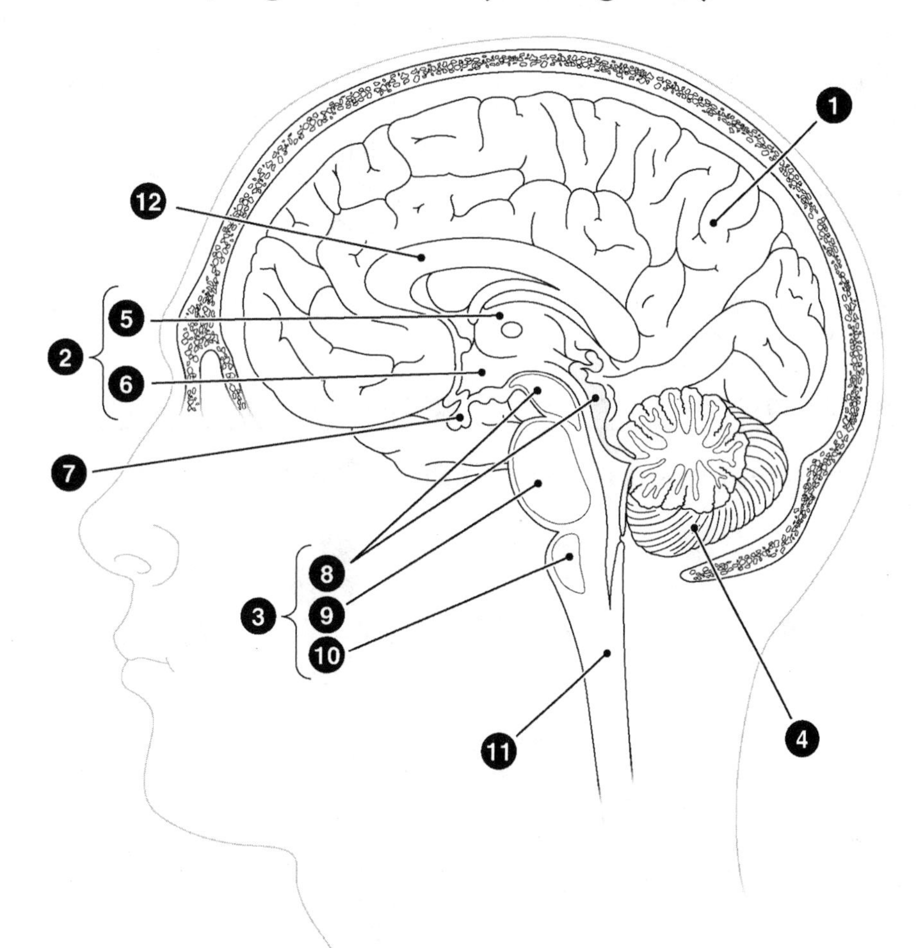

INSTRUCTIONS

1. Write the names of the four labeled brain divisions in lines 1 to 4, using four different colors. Use red for no. 2 and blue for no. 3. DO NOT COLOR THE DIAGRAM YET.
2. Write the name of each labeled structure on the appropriate numbered line in different colors. Use different shades of red for structures 5 and 6 and different shades of blue for structures 8 to 10.
3. Color each structure on the diagram with the corresponding color.

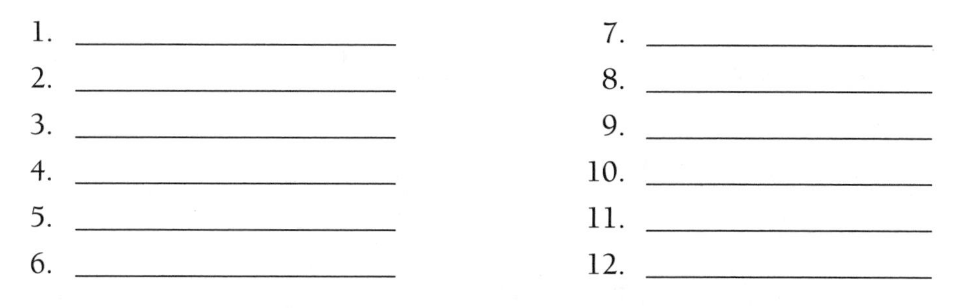

1. ____________________ 7. ____________________
2. ____________________ 8. ____________________
3. ____________________ 9. ____________________
4. ____________________ 10. ____________________
5. ____________________ 11. ____________________
6. ____________________ 12. ____________________

EXERCISE 10-2.

INSTRUCTIONS

Write the appropriate term in each blank.

lobe	hemisphere	cerebrum	cerebellum
brainstem	meninges	diencephalon	

1. Each half of the cerebrum ____________
2. The "little" brain that coordinates voluntary muscle movements ____________
3. An individual subdivision of the cerebrum that regulates specific functions ____________
4. The portion of the brain that contains the thalamus and hypothalamus ____________
5. Connects the spinal cord with the brain ____________
6. The largest part of the brain ____________

2. Name and describe the three meninges.

EXERCISE 10-3: Meninges and Related Parts (Text Fig. 10-3)

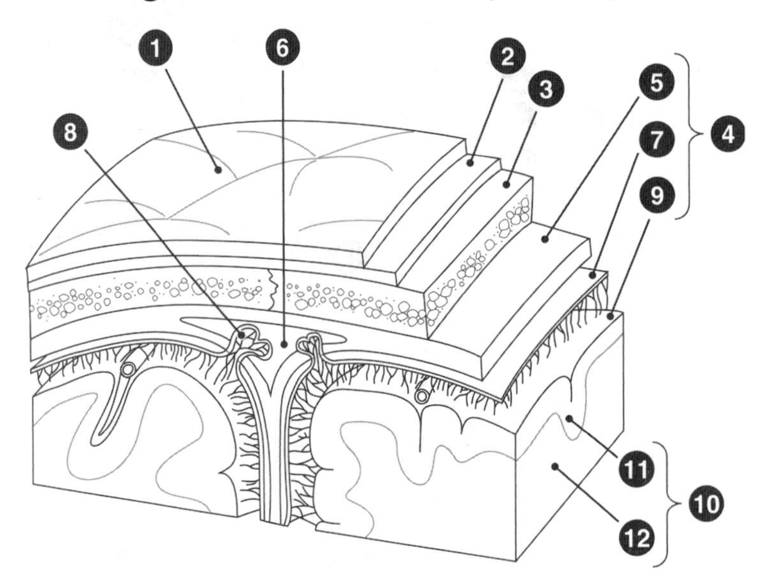

INSTRUCTIONS

1. Write the name of each labeled part on the numbered lines in different colors. Use the same color for structures 7 and 8. Write the name of structures 9, 10, and 12 in black.
2. Color the structures on the diagram with the corresponding color. Do not color structures 4, 10, and 12.

1. ____________
2. ____________
3. ____________
4. ____________
5. ____________
6. ____________
7. ____________
8. ____________
9. ____________
10. ____________
11. ____________
12. ____________

3. Cite the function of cerebrospinal fluid and describe where and how this fluid is formed.

EXERCISE 10-4: Flow of Cerebrospinal Fluid (Text Fig. 10-4)

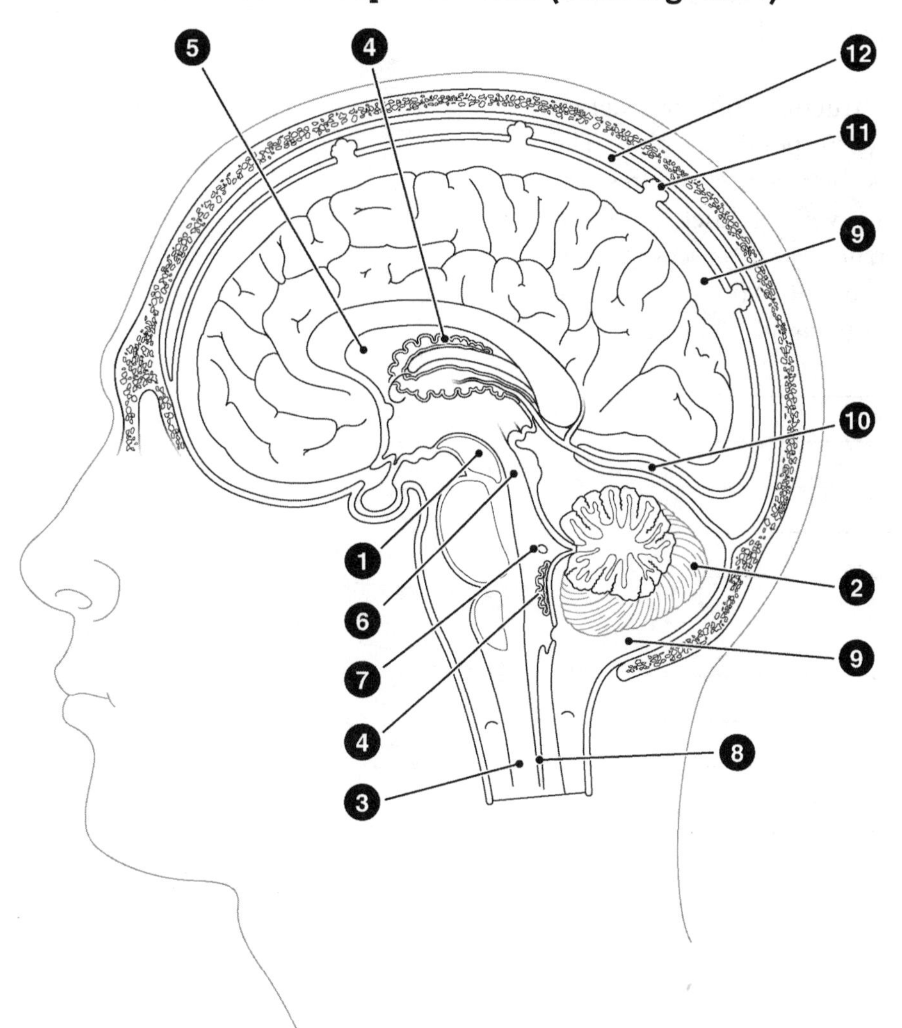

INSTRUCTIONS

1. Write the name of each labeled part on the numbered lines in different colors. Use light colors for structures 5 to 12.
2. Color the structures on the diagram with the corresponding color. The boundaries between structures 5 to 12 (inclusive) are not always well defined. For instance, structure 6 is continuous with structure 7. You can overlap your colors to signify this fact.
3. Draw arrows to indicate the direction of CSF flow.

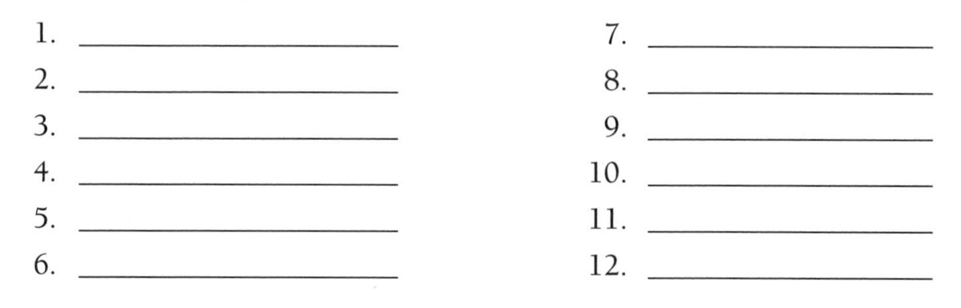

EXERCISE 10-5: Ventricles of the Brain (Text Fig. 10-5)

INSTRUCTIONS

1. Write the name of each labeled part on the numbered lines in different colors.
2. Color the structures on the diagram with the corresponding color. The boundaries between structures are not always well defined. For instance, structure 2 is continuous with structure 4. You can overlap your colors to signify this fact.

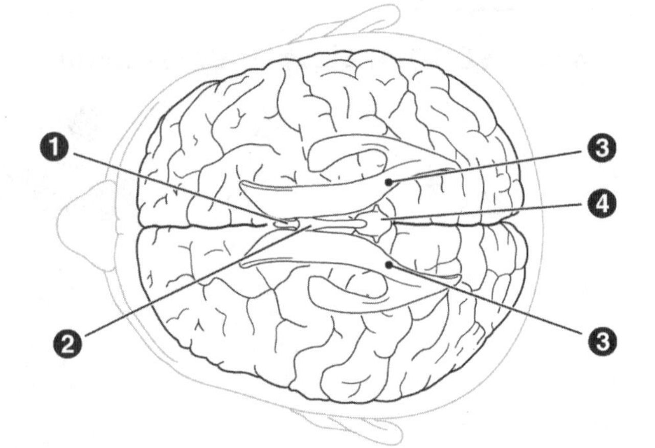

1. ____________________
2. ____________________
3. ____________________
4. ____________________
5. ____________________
6. ____________________
7. ____________________
8. ____________________
9. ____________________

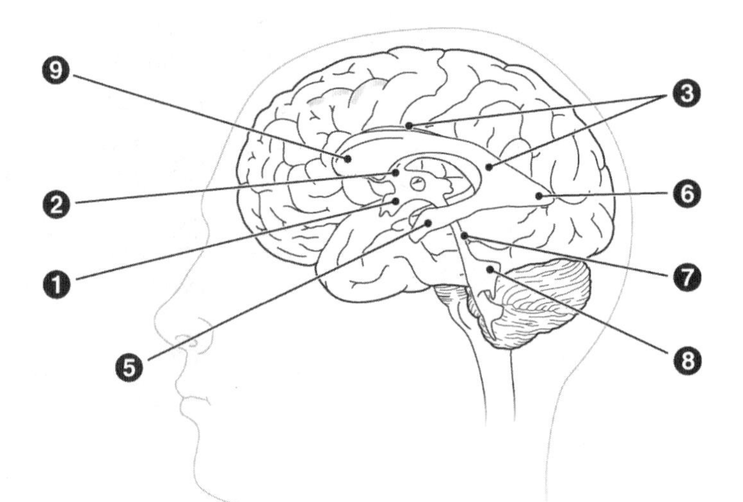

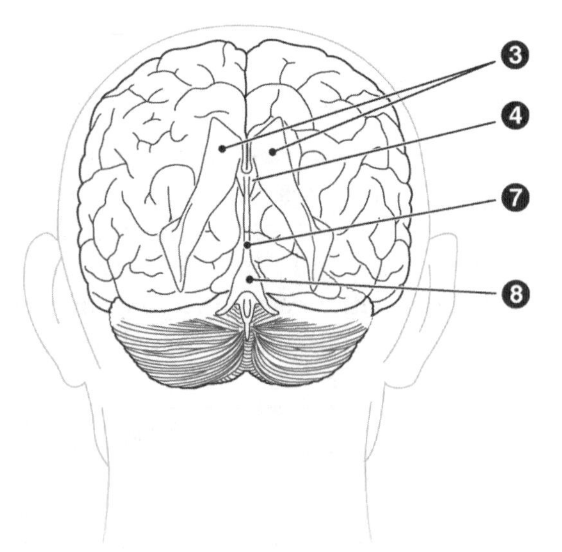

EXERCISE 10-6.

INSTRUCTIONS

Write the appropriate term in each blank.

dura mater	pia mater	arachnoid	choroid plexus	cerebral aqueduct
subarachnoid space	arachnoid villi	ventricle	dural sinus	

1. The weblike middle meningeal layer ________
2. Venous channel between the two outermost meninges ________
3. The innermost layer of the meninges, the delicate membrane in which there are many blood vessels ________
4. The area in which cerebrospinal fluid collects before its return to the blood ________
5. The vascular network in a ventricle that forms cerebrospinal fluid ________
6. The projections in the dural sinuses through which CSF is returned to the blood ________
7. The outermost layer of the meninges, which is the thickest and toughest ________

4. Name and locate the lobes of the cerebral hemispheres (also see Exercise 10-8).

EXERCISE 10-7.

INSTRUCTIONS

Write the appropriate term in each blank.

gyrus	central sulcus	lateral sulcus	basal ganglia
dopamine	corpus callosum	internal capsule	cortex

1. A shallow groove that separates the temporal lobe from the frontal and parietal lobes ________
2. Masses of gray matter deep within the cerebrum that help regulate body movement and the muscles of facial expression ________
3. A band of white matter that carries impulses between the cerebrum and the brainstem ________
4. An elevated portion of the cerebral cortex ________
5. The thin layer of gray matter on the surface of the cerebrum ________
6. A band of myelinated fibers that bridges the two cerebral hemispheres ________
7. The neurotransmitter used by the basal nuclei neurons ________

5. Cite one function of the cerebral cortex in each lobe of the cerebrum.

EXERCISE 10-8: Functional Areas of the Cerebral Cortex (Text Fig. 10-8)

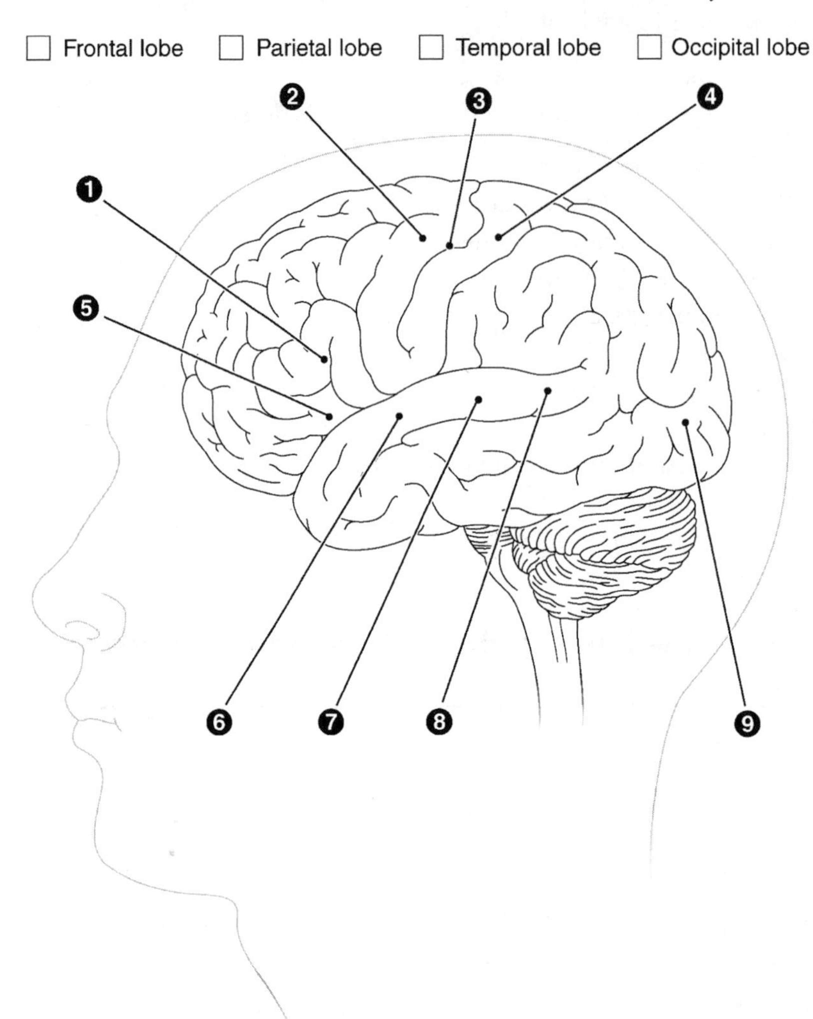

INSTRUCTIONS

1. Color the boxes next to the cerebral lobe names as follows: frontal lobe, pink; parietal lobe, light purple; temporal lobe, light blue; occipital lobe, light green.
2. Lightly color the four cerebral lobes on the diagram with corresponding colors.
3. Write the names of structures 1 to 9 on the appropriate lines in different colors. For all structures except 3, use a darker color than the one used for the corresponding cerebral lobe. For instance, a structure found in the frontal lobe could be colored red. Use the same color for structures 6 to 8. Use a dark color for structure 3.
4. Color or outline the structures on the diagram with the corresponding color.

1. ____________________
2. ____________________
3. ____________________
4. ____________________
5. ____________________
6. ____________________
7. ____________________
8. ____________________
9. ____________________

EXERCISE 10-9.

INSTRUCTIONS

Write the appropriate term in each blank.

temporal lobe parietal lobe occipital lobe frontal lobe

1. The portion of the cerebral cortex where visual impulses from the retina are interpreted ____________
2. The portion of the cerebral cortex where auditory impulses are interpreted ____________
3. Location of a sensory area for interpretation of pain, touch, and temperature ____________
4. The lobe controlling voluntary muscles ____________

6. Name two divisions of the diencephalon and cite the functions of each (see Exercise 10-10).

7. Locate the three subdivisions of the brainstem and give the functions of each.

EXERCISE 10-10.

INSTRUCTIONS

Write the appropriate term in each blank.

thalamus	medulla oblongata	pons	midbrain
vasomotor center	cardiac center	limbic system	hypothalamus

1. The portion of the brainstem composed of myelinated nerve fibers that connects to the cerebellum ____________
2. The superior portion of the brainstem ____________
3. The part of the brain between the pons and the spinal cord ____________
4. The region of the diencephalon that acts as a relay center for sensory stimuli ____________
5. The region consisting of portions of the cerebrum and diencephalon that is involved in emotional states and behavior ____________
6. Nuclei that regulate the contraction of smooth muscle in blood vessel walls ____________
7. The portion of the brain controlling the autonomic nervous system ____________

8. Describe the cerebellum and cite its functions.

EXERCISE 10-11.

INSTRUCTIONS

List three functions of the cerebellum in the spaces below.

1. ______________________________
2. ______________________________
3. ______________________________

9. Name some techniques used to study the brain.

EXERCISE 10-12.

INSTRUCTIONS

Write the appropriate term in each blank.

MRI CT PET EEG

1. Technique that produces a picture of brain activity levels in the different parts of the brain ____________
2. Technique that measures electric currents in the brain ____________
3. X-ray technique that provides photos of bone, cavities, and lesions ____________
4. Technique used to visualize soft tissue, such as scar tissue, hemorrhages, and tumors that does not use x-rays ____________

10. Cite the names and functions of the 12 cranial nerves.

EXERCISE 10-13: Cranial Nerves (Text Fig. 10-18)

INSTRUCTIONS

1. Write the number and name of each labeled cranial nerve on the numbered lines in different colors. Use the same color for structures 1 and 2.
2. Color the nerves on the diagram with the corresponding color.

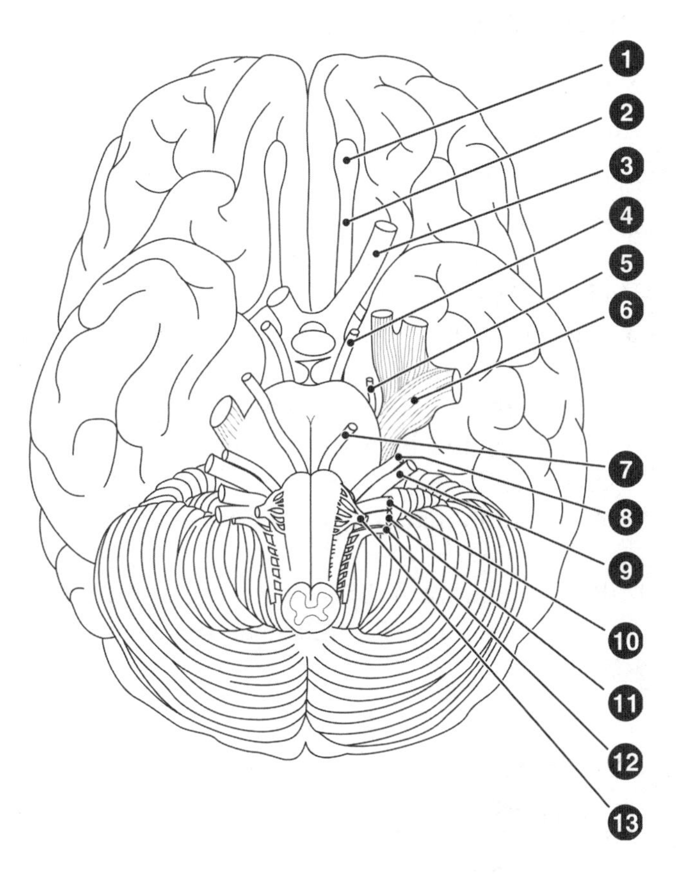

1. ____________________
2. ____________________
3. ____________________
4. ____________________
5. ____________________
6. ____________________
7. ____________________
8. ____________________
9. ____________________
10. ____________________
11. ____________________
12. ____________________
13. ____________________

EXERCISE 10-14.

INSTRUCTIONS

Write the appropriate term in each blank.

glossopharyngeal nerve	optic nerve	vagus nerve	trochlear nerve	abducens nerve
vestibulocochlear nerve	facial nerve	trigeminal nerve	accessory nerve	

1. A motor nerve controlling the trapezius, sternocleidomastoid, and larynx muscles ____________
2. The nerve that controls contraction of a single eye muscle ____________
3. The nerve that carries visual impulses from the eye to the brain ____________
4. The most important sensory nerve of the face and head ____________
5. The nerve that supplies most of the organs in the thoracic and abdominal cavities ____________
6. The nerve that supplies the muscles of facial expression ____________
7. The nerve that carries sensory impulses for hearing and equilibrium ____________

11. List some disorders that involve the brain, its associated structures, or the cranial nerves.

EXERCISE 10-15.

INSTRUCTIONS

Write the appropriate term in each blank.

Alzheimer disease	glioma	hydrocephalus	epidural hematoma	epilepsy
Parkinson disease	aphasia	encephalitis	cerebrovascular accident	subdural hematoma

1. A brain tumor derived from neuroglia ________________
2. A chronic brain disorder that usually can be diagnosed by electroencephalography ________________
3. Damage to brain tissue caused by a blood clot, ruptured vessel, or embolism ________________
4. Loss of the power of expression by speech or writing ________________
5. A degenerative brain disorder associated with the development of amyloid ________________
6. A condition that may result from obstruction of the normal flow of CSF ________________
7. Bleeding between the dura mater and the skull ________________
8. The general term for inflammation of the brain ________________

12. Show how word parts are used to build words related to the nervous system.

EXERCISE 10-16.

INSTRUCTIONS

Complete the following table by writing the correct word part or meaning in the space provided. Write a word that contains each word part in the Example column.

Word Part	Meaning	Example
1. ________	cut	________________
2. chori/o	________	________________
3. ________	tongue	________________
4. encephal/o	________	________________
5. cerebr/o	________	________________
6. ________	head	________________
7. ________	speech, ability to talk	________________
8. ________	lateral, side	________________
9. gyr/o	________	________________
10. -rhage	________	________________

Making the Connections

The following concept map deals with the structure of the brain. Each pair of terms is linked together by a connecting phrase into a sentence. The sentence should be read in the direction of the arrow. Complete the concept map by filling in the appropriate term or phrase. There is one right answer for each term. However, there are many correct answers for the connecting phrases (4, 8, 11, 14, 17). There are many other connections that could be made on this map. For instance, how could you connect structures 2 and 16? Can you add more terms to the map?

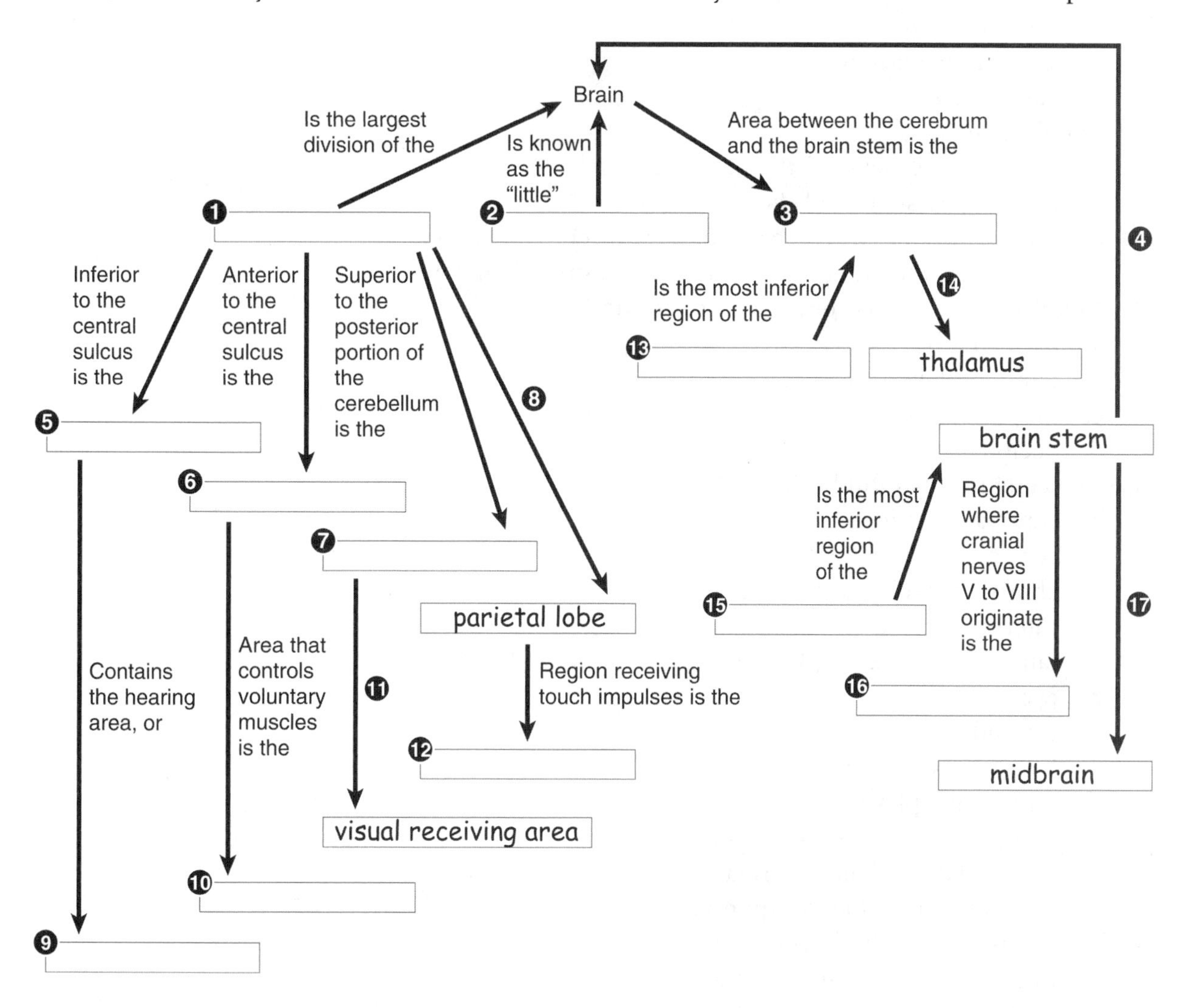

Optional Exercise: Make your own concept map, based on structures involved in the synthesis and movement of cerebrospinal fluid. Choose your own terms to incorporate into your map, or use the following list: ventricle, choroid plexus, lateral ventricles, third ventricle, fourth ventricle, foramina, horns, cerebral aqueduct, spinal cord, hydrocephalus, cerebrospinal fluid, dural sinuses, subarachnoid space, and arachnoid villi.

Testing Your Knowledge

Building Understanding

I. Multiple Choice

Select the best answer and write the letter of your choice in the blank.

1. The shallow groove lying between the frontal and parietal lobe is the — 1. ______
   a. lateral sulcus
   b. central sulcus
   c. longitudinal fissure
   d. basal nuclei
2. The dura mater is — 2. ______
   a. the innermost layer of the meninges
   b. the outermost layer of the meninges
   c. the network of vessels that produces cerebrospinal fluid
   d. the part of the brain that connects with the spinal cord
3. Impulses for the sense of taste travel to the — 3. ______
   a. parietal lobe
   b. temporal lobe
   c. hippocampus
   d. occipital lobe
4. The cerebrospinal fluid is formed in the — 4. ______
   a. cerebral aqueduct
   b. central sulcus
   c. choroid plexus
   d. internal capsule
5. The abducens nerve supplies the — 5. ______
   a. eye
   b. ear and pharynx
   c. face and salivary gland
   d. tongue and pharynx
6. Multi-infarct dementia is the result of — 6. ______
   a. accumulation of an abnormal protein
   b. ischemia (lack of blood supply)
   c. obstruction of the flow of CSF
   d. infection of the brain coverings
7. The reticular formation is — 7. ______
   a. a region of the limbic system that controls wakefulness and sleep
   b. a deep groove that divides the cerebral hemispheres
   c. the part of the temporal lobe concerned with the sense of smell
   d. the fifth lobe of the cerebrum

II. Completion Exercise

Write the word or phrase that correctly completes each sentence.

1. A severe pain disorder affecting the fifth cranial nerve is known as tic douloureux or ____________________.
2. The four chambers within the brain where cerebrospinal fluid is produced are the ____________________.
3. Sounds are interpreted in the area of the temporal lobe called the ____________________.
4. The region of the diencephalon that helps maintain homeostasis (e.g., water balance, appetite, and body temperature) and controls the autonomic nervous system is the ____________________.
5. The storage of information to be recalled at a later time is called ____________________.
6. Records of the electrical activity of the brain can be made with an instrument called a(n) ____________________.
7. Damage to the facial nerve resulting in paralysis is called ____________________.
8. Extensions of the lateral ventricles into the cerebral lobes are called ____________________.
9. Except for the first two pairs, all the cranial nerves arise from the ____________________.
10. The three layers of membranes that surround the brain and spinal cord are called the ____________________.
11. The clear liquid that helps to support and protect the brain and spinal cord is ____________________.
12. The number of pairs of cranial nerves is ____________________.
13. The area of the brainstem concerned with eye-tracking reflexes involved in reading is the ____________________.
14. The middle portion of the cerebellum is called the ____________________.

Understanding Concepts

I. True/False

For each question, write *T* for true or *F* for false in the blank to the left of each number. If a statement is false, correct it by replacing the underlined term and write the correct statement in the blank below the question.

_______ 1. The pia mater is the middle layer of the meninges.

____

_______ 2. The interventricular foramina form channels between the lateral ventricles and the fourth ventricle.

____

_______ 3. The primary motor cortex is found in the parietal lobe.

____

_______ 4. The internal capsule consists of myelinated fibers linking the cerebral hemispheres with the brainstem.

____

_______ 5. The <u>parietal lobe</u> lies posterior to the central sulcus and superior to the occipital lobe.

____

_______ 6. The raised areas on the surface of the cerebrum are called <u>gyri</u>.

____

_______ 7. <u>Alzheimer disease</u> results from a lack of dopamine in the substantia nigra.

____

_______ 8. The <u>hypoglossal</u> nerve transmits sensory information from the tongue.

____

II. Practical Applications

Study each discussion. Then write the appropriate word or phrase in the space provided.

1. Mr. B, age 87, drove his car into a ditch. When the paramedics arrived, he was unable to speak, a disorder called ________________________.
2. Mr. B also experienced paralysis on his right side, indicating a problem within the largest division of the brain, the ________________________.
3. The right side of Mr. B's face droops, and he cannot control his facial expressions. The facial muscles are controlled by the cranial nerve numbered ________________________.
4. A computed tomography scan at the hospital revealed the rupture of a cerebral blood vessel, resulting in bleeding, or ________________________.
5. The rupture occurred in the left side of the brain. The anatomical name for the left side of the brain is the left ________________________.
6. The physician diagnosed Mr. B with a type of brain injury called ________________________.
7. Mr. B's speech disorder indicates that the bleed affected his motor speech area, also called ________________________.

III. Short Essays

1. Describe the structures that protect the brain and spinal cord.

____

____

____

2. List some functions of the structures in the diencephalon.

____

____

____

3. List the name, number, and sensory information conveyed for each of the purely sensory cranial nerves.

____

____

____

Conceptual Thinking

1. Mr. S, age 42, has difficulty initiating movements. Other symptoms include hand tremors, frequent falls, and limb and joint rigidity. What disorder does Mr. S have? How would you explain this disorder to his family? What are some possible treatments?

____

____

____

2. Describe the journey of cerebrospinal fluid (CSF), beginning with its synthesis and ending with its entry into the circulatory system.

____

____

____

Expanding Your Horizons

Do you use herbal supplements like gingko biloba to boost your learning power? As discussed in this *Scientific American* article, gingko biloba does boost memory—to the same extent as a candy bar! Mental exercise may be the best way to improve your academic performance.

Resources

1. Gold PE, Cahill L, Wenk GL. *The lowdown on Gingko biloba.* Sci Am 2003;288:86–91.
2. Holloway M. *The mutable brain.* Sci Am 2003;289:78–85.

CHAPTER 11

The Sensory System

Overview

The sensory system enables us to detect changes taking place both internally and externally. These changes are detected by specialized structures called receptors. Any change that acts on a receptor to produce a response in the nervous system is termed a stimulus. The special senses, so called because the receptors are limited to a few specialized sense organs in the head, include the senses of vision, hearing, equilibrium, taste, and smell. The receptors of the eye are the rods and cones located in the retina. The receptors for both hearing (the organ of Corti) and equilibrium (the vestibule and semicircular canals) are located within the inner ear. Receptors for the chemical senses of taste and smell are located on the tongue and in the upper part of the nose, respectively. The general senses are scattered throughout the body; they respond to touch, pressure, temperature, pain, and position. Receptors for the sense of position, known as proprioceptors, are found in muscles, tendons, and joints. The nerve impulses generated in a receptor cell by a stimulus must be carried to the central nervous system by way of a sensory (afferent) neuron. Here, the information is processed and a suitable response is made. Disorders of the eye and ear are common. They are associated with aging, infection, environmental factors, inherited malfunctions, and injury.

This chapter is quite challenging, because it contains both difficult concepts and large amounts of detail. You can use concept maps to assemble all of the details into easy-to-remember frameworks.

Addressing the Learning Outcomes

1. Describe the function of the sensory system.

EXERCISE 11-1.

INSTRUCTIONS

Fill in the blanks in the following paragraph using these terms:

central nervous system, homeostasis, sensory neuron, sensory receptor

The sensory system protects people by detecting changes in the internal and external environment that threaten to disrupt ____________________ (1), which is the maintenance of a constant internal environment. The change is detected by a ____________________ (2), which sends an impulse through a ________________ (3) to the ______________ (4).

2. Differentiate between the special and general senses and give examples of each.

EXERCISE 11-2.

INSTRUCTIONS

Classify each of the following senses as general senses (G) or special senses (S).

1. sense of position _______
2. smell ________
3. vision ________
4. touch ______
5. temperature _____
6. equilibrium _______

3. Describe the structure of the eye.

EXERCISE 11-3: The Eye (Text Fig. 11-3)

INSTRUCTIONS

1. Write the name of each labeled part on the numbered lines in different colors. Use the same color for structures 3 and 4 and structures 6 to 9 (inclusive). Write the name of structures 1 and 2 in black, because they will not be colored.
2. Color the different structures on the diagram with the corresponding color. Some structures are present in more than one location on the diagram. Try to color all of a particular structure in the appropriate color. For instance, only one of the suspensory ligaments is labeled, but color both suspensory ligaments.

1. ____________________
2. ____________________
3. ____________________
4. ____________________
5. ____________________
6. ____________________
7. ____________________
8. ____________________
9. ____________________
10. ____________________
11. ____________________
12. ____________________
13. ____________________
14. ____________________
15. ____________________
16. ____________________

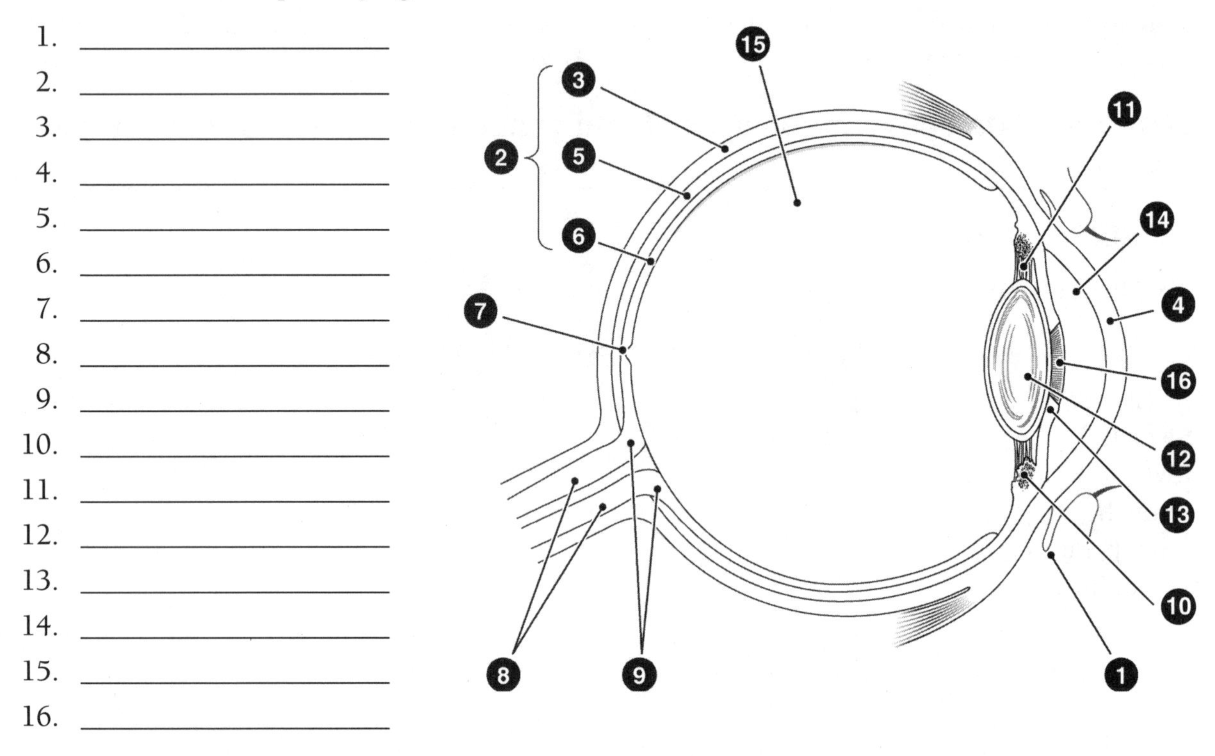

4. List and describe the structures that protect the eye.

EXERCISE 11-4: The Lacrimal Apparatus (Text Fig. 11-2)

INSTRUCTIONS

Label the indicated parts.

1. ____________________
2. ____________________
3. ____________________
4. ____________________
5. ____________________
6. ____________________
7. ____________________

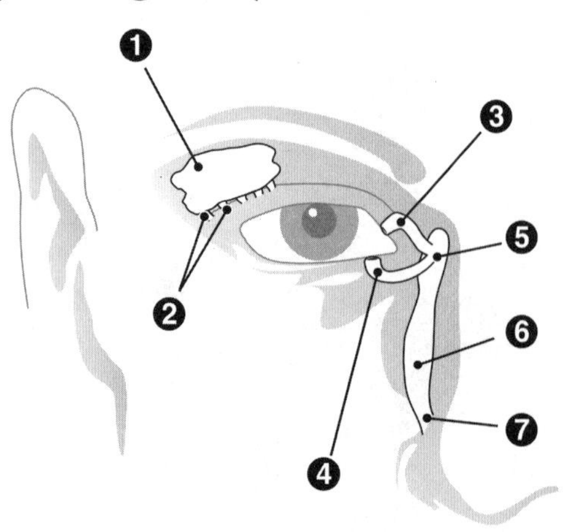

5. Define *refraction* and list the refractive parts of the eye.

EXERCISE 11-5.

INSTRUCTIONS

List 4 eye structures that bend (refract) light in the spaces below.

1. ______________________
2. ______________________
3. ______________________
4. ______________________

6. Differentiate between the rods and the cones of the eye.

EXERCISE 11-6.

INSTRUCTIONS

Write the appropriate term in each blank below.

cone	cornea	rhodopsin	sclera
optic disk	retina	rod	fovea centralis

1. A vision receptor that is sensitive to color ______________
2. The part of the eye that light rays pass through first as they enter the eye ______________
3. Another name for the blind spot, the region where the optic nerve connects with the eye ______________
4. The innermost coat of the eyeball, the nervous tissue layer that includes the receptors for the sense of vision ______________
5. A vision receptor that functions well in dim light ______________
6. A pigment needed for vision ______________
7. The depressed area in the retina that is the point of clearest vision ______________

7. Compare the functions of the extrinsic and intrinsic muscles of the eye.

EXERCISE 11-7: Extrinsic Muscles of the Eye (Text Fig. 11-6)

INSTRUCTIONS

1. Write the name of each labeled muscle on the numbered lines in different colors.
2. Color the different muscles on the diagram with the corresponding color.

1. ______________________
2. ______________________
3. ______________________
4. ______________________
5. ______________________

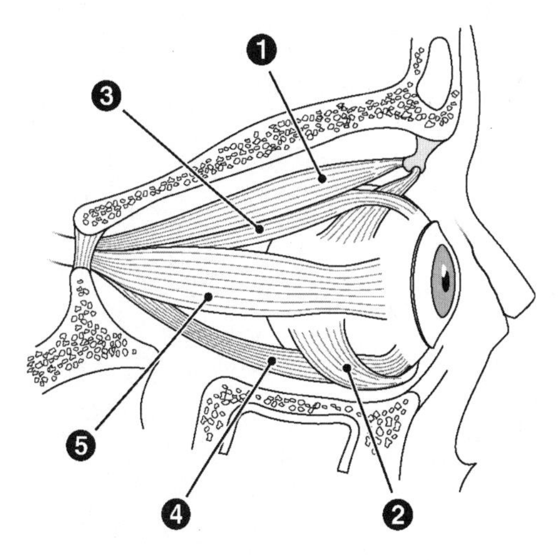

EXERCISE 11-8.

INSTRUCTIONS

Write the appropriate term in each blank.

aqueous humor	vitreous body	lens	ciliary muscle
choroid	conjunctiva	pupil	iris

1. The structure that alters the shape of the lens for accommodation ________________
2. The watery fluid that fills much of the eyeball in front of the crystalline lens ________________
3. The vascular, pigmented middle tunic of the eyeball ________________
4. Structure with two sets of muscle fibers that regulate the amount of light entering the eye ________________
5. The jellylike material located behind the crystalline lens that maintains the spherical shape of the eyeball ________________
6. The central opening of the iris ________________
7. The membrane that lines the eyelids ________________

8. Describe the nerve supply to the eye.

EXERCISE 11-9: Nerves of the Eye (Text Fig. 11-10)

INSTRUCTIONS

Label the indicated nerves.

1. ________________
2. ________________
3. ________________
4. ________________
5. ________________
6. ________________

(also see Exercise 11-15)

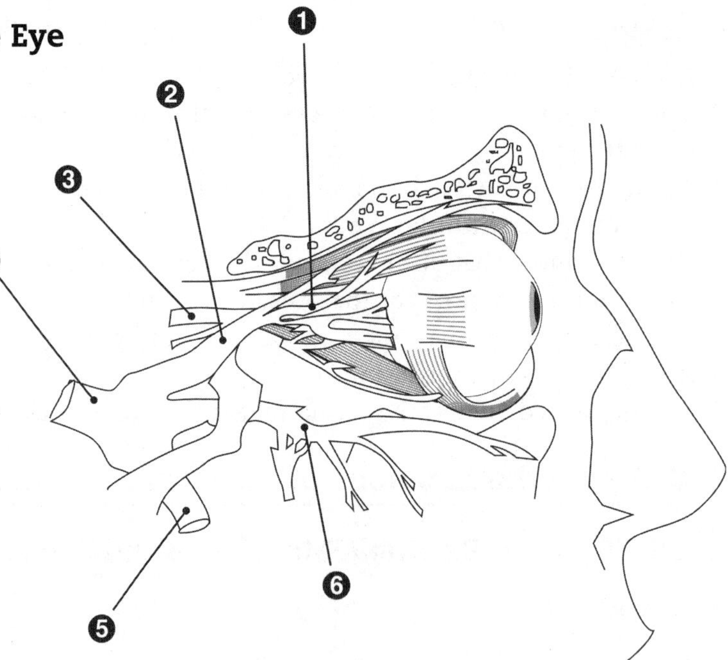

9. Describe the three divisions of the ear.

EXERCISE 11-10: The Ear (Text Fig. 11-12)

INSTRUCTIONS

1. Write the names of the three ear divisions on the appropriate lines (1 to 3).
2. Write the names of the labeled parts on the numbered lines in different colors.
3. Color each part with the corresponding color.

1. ______________________
2. ______________________
3. ______________________
4. ______________________
5. ______________________
6. ______________________
7. ______________________
8. ______________________
9. ______________________
10. ______________________
11. ______________________
12. ______________________
13. ______________________
14. ______________________

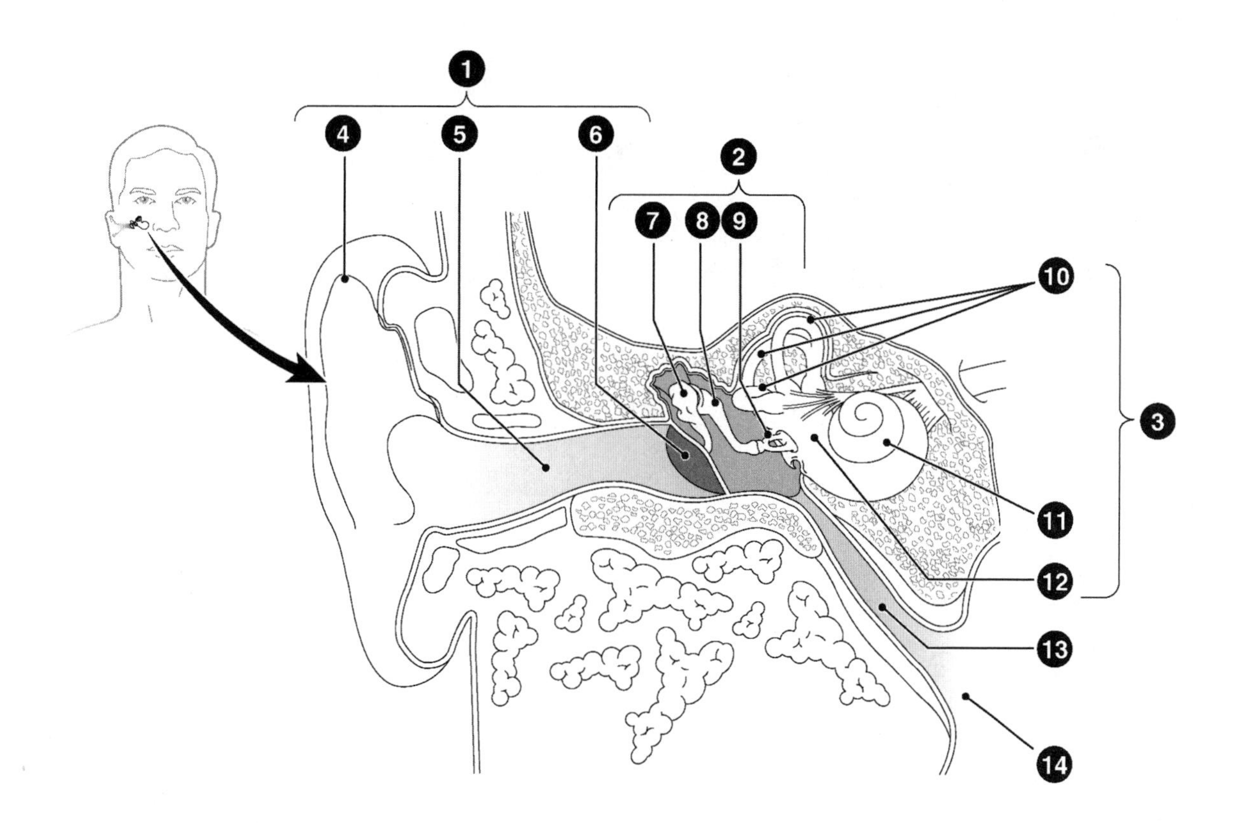

EXERCISE 11-11: The Inner Ear (Text Fig. 11-14)

INSTRUCTIONS

Label the indicated parts.

1. ____________________
2. ____________________
3. ____________________
4. ____________________
5. ____________________
6. ____________________
7. ____________________
8. ____________________

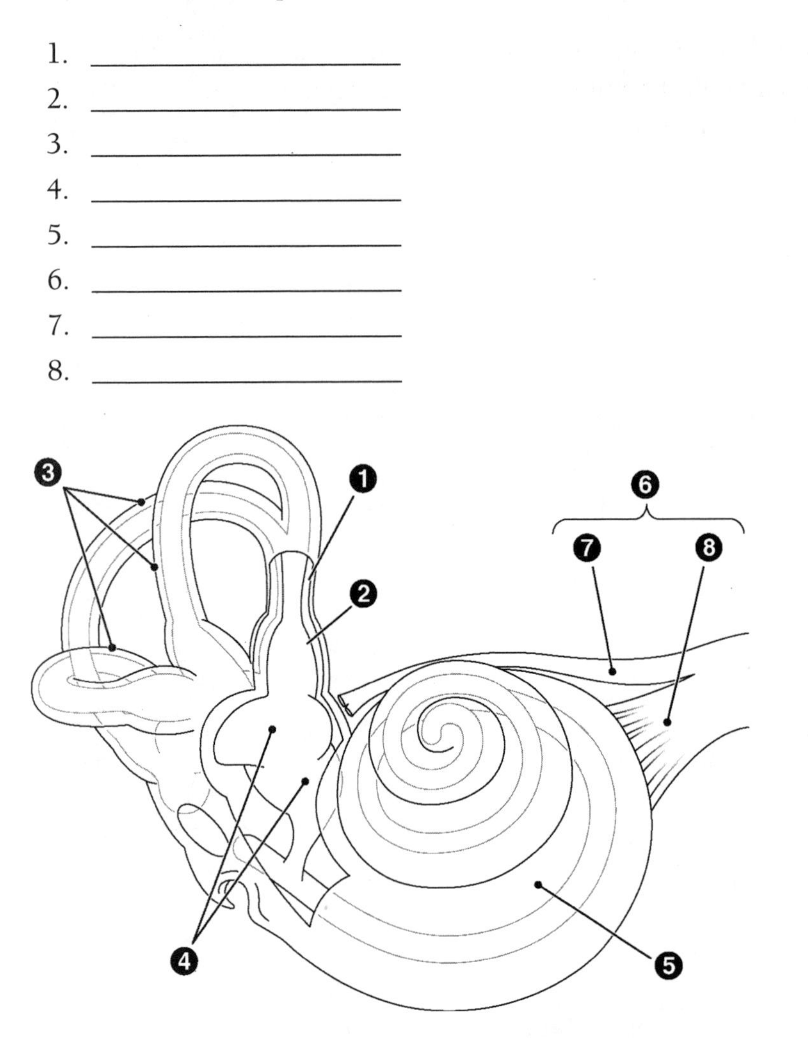

10. Describe the receptor for hearing and explain how it functions.

EXERCISE 11-12: Cochlea and Organ of Corti (Text Fig. 11-15)

INSTRUCTIONS

1. Write the name of each labeled part on the numbered lines. Use colors for structures 3 to 7, 11, and 12. Use black for the other structures.
2. Color structures 3 to 7, 11, and 12 with the corresponding color.

1. ____________________
2. ____________________
3. ____________________
4. ____________________
5. ____________________
6. ____________________
7. ____________________
8. ____________________
9. ____________________
10. ____________________
11. ____________________
12. ____________________
13. ____________________

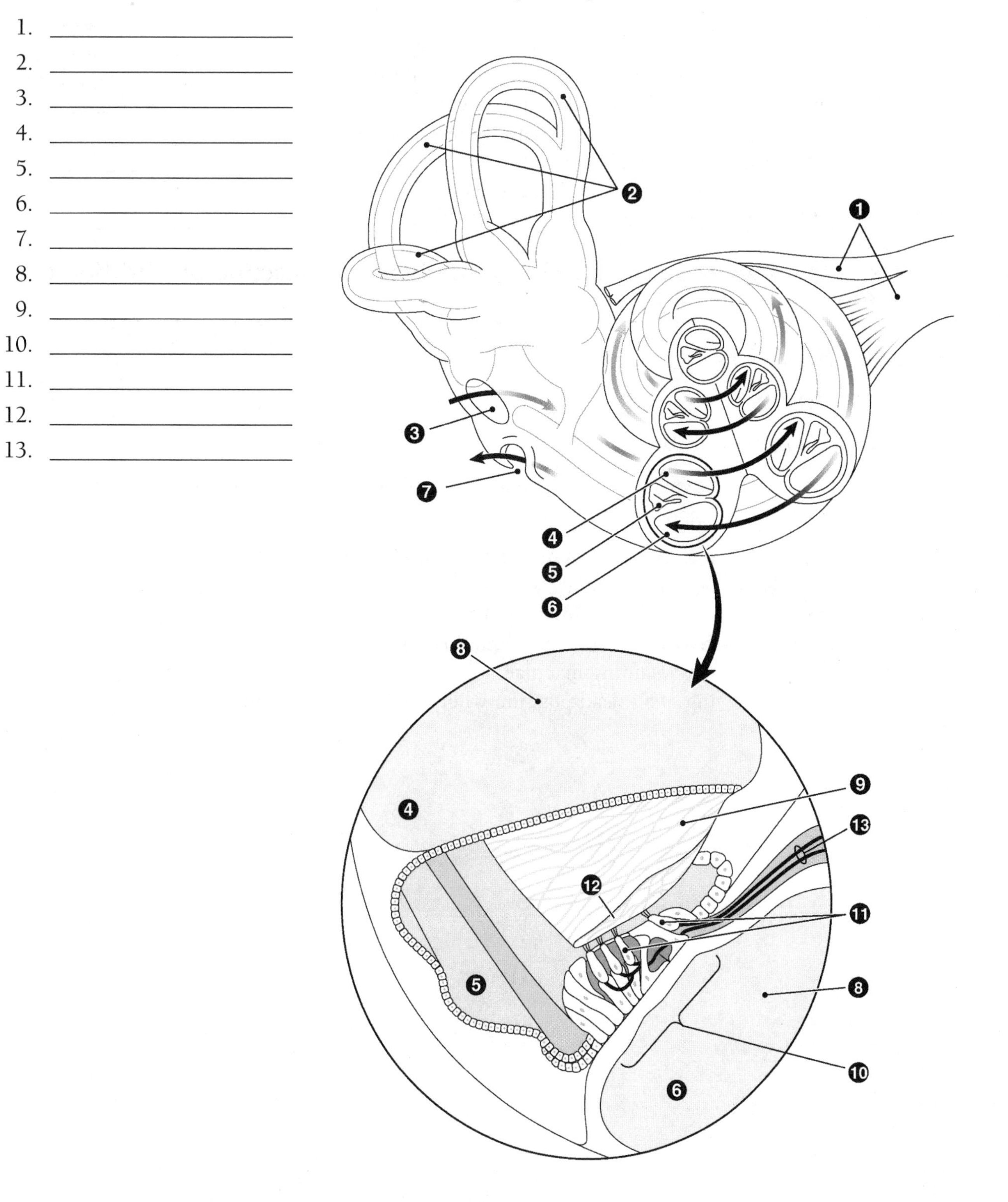

EXERCISE 11-13.

INSTRUCTIONS

Write the appropriate term in each blank.

oval window	organ of Corti	malleus	eustachian tube	bony labyrinth
perilymph	incus	pinna	cochlear duct endolymph	

1. The fluid contained within the membranous labyrinth of the inner ear ____________
2. The bone that interacts with the tympanic membrane ____________
3. Another name for the projecting part, or auricle, of the ear ____________
4. The channel connecting the middle ear cavity with the pharynx ____________
5. The fluid of the inner ear contained within the bony labyrinth and surrounding the membranous labyrinth ____________
6. Ciliated receptor cells that detect sound waves ____________
7. The skeleton of the inner ear ____________

11. Compare static and dynamic equilibrium and describe the location and function of these receptors.

EXERCISE 11-14.

INSTRUCTIONS

Write the appropriate term in each blank.

vestibule	dynamic equilibrium	semicircular canals	cristae
cochlear duct	static equilibrium	otoliths	

1. The sense of knowing the position of the head in relation to gravity ____________
2. Small crystals that activate maculae ____________
3. The sense organ involved in dynamic equilibrium ____________
4. The receptor cells involved in dynamic equilibrium ____________
5. Two small chambers containing maculae ____________
6. The sense of knowing one's head position when the body is spinning ____________

12. Explain the function of proprioceptors.

EXERCISE 11-15.

INSTRUCTIONS

Write the appropriate term in each blank.

kinesthesia	proprioception	tactile corpuscle	cochlear nerve
vestibular nerve	oculomotor nerve	ophthalmic nerve	equilibrium
optic nerve	free nerve endings		

1. The branch of the vestibulocochlear nerve that carries hearing impulses ________________
2. The nerve that carries visual impulses from the retina to the brain ________________
3. The branch of the fifth cranial nerve that carries impulses of pain, touch, and temperature from the eye to the brain ________________
4. The largest of the three cranial nerves that carry motor fibers to the eyeball muscles ________________
5. The sense of knowing the position of one's body and the relative positions of different muscles ________________
6. The sense of body movement ________________
7. Receptors that detect changes in temperature ________________

13. List several methods for treatment of pain.

EXERCISE 11-16.

INSTRUCTIONS

Write the appropriate term in each blank.

NSAID narcotic anesthetic endorphin analgesic

1. Term describing any drug that relieves pain ________________
2. A substance produced by the brain that relieves pain ________________
3. Drug that acts on the CNS to alter pain perception, such as morphine ________________
4. Drug that acts locally to reduce inflammation ________________

14. Describe sensory adaptation and explain its value.

EXERCISE 11-17.

INSTRUCTIONS

Define "sensory adaptation" in the space below.

____

____

15. List some disorders of the sensory system.

EXERCISE 11-18.

INSTRUCTIONS

Write the appropriate term in each blank.

macular degeneration	strabismus	glaucoma	myopia	hyperopia
ophthalmia neonatorum	cataract	trachoma	astigmatism	

1. A serious eye infection of the newborn that can be prevented with a suitable antiseptic ________________
2. The scientific name for nearsightedness, in which the focal point is in front of the retina and distant objects appear blurred ________________
3. The visual defect caused by irregularity in the curvature of the lens or cornea ________________
4. Condition in which the eyes do not work together because the muscles do not coordinate ________________
5. Condition caused by continued high pressure of the aqueous humor, which may result in destruction of the optic nerve fibers ________________
6. The scientific name for farsightedness, in which light rays are not bent sharply enough to focus on the retina when viewing close objects ________________
7. A chronic eye infection for which antibiotics and proper hygiene have reduced the incidence of reinfection and blindness ________________

EXERCISE 11-19.

INSTRUCTIONS

Write the appropriate term in each blank.

otitis media	otitis externa	conductive hearing loss
otosclerosis	presbycusis	sensorineural hearing loss

1. The scientific name for swimmer's ear ________________
2. A hereditary bone disorder that prevents normal vibration of the stapes ________________
3. Slow, progressive hearing loss associated with aging ________________
4. Hearing loss resulting from damage to the cochlea or to nerves associated with hearing ________________
5. Infection and inflammation of the middle ear cavity ________________

16. Show how word parts are used to build words related to the sensory system.

EXERCISE 11-20.

INSTRUCTIONS

Complete the following table by writing the correct word part or meaning in the space provided. Write a word that contains each word part in the Example column.

Word Part	Meaning	Example
1. presby-	________	________________
2. ________	stone	________________
3. -opia	________	________________
4. -stomy	________	________________
5. ________	drum	________________
6. ________	yellow	________________
7. propri/o-	________	________________
8. ________	pain	________________
9. -esthesia	________	________________
10. ________	hearing	________________

Making the Connections

The concept map on the next page deals with the structure and function of the eye. Each pair of terms is linked together by a connecting phrase into a sentence. The sentence should be read in the direction of the arrow. Complete the concept map by filling in the appropriate term or phrase. There is one right answer for each term. However, there are many correct answers for the connecting phrases (2, 9).

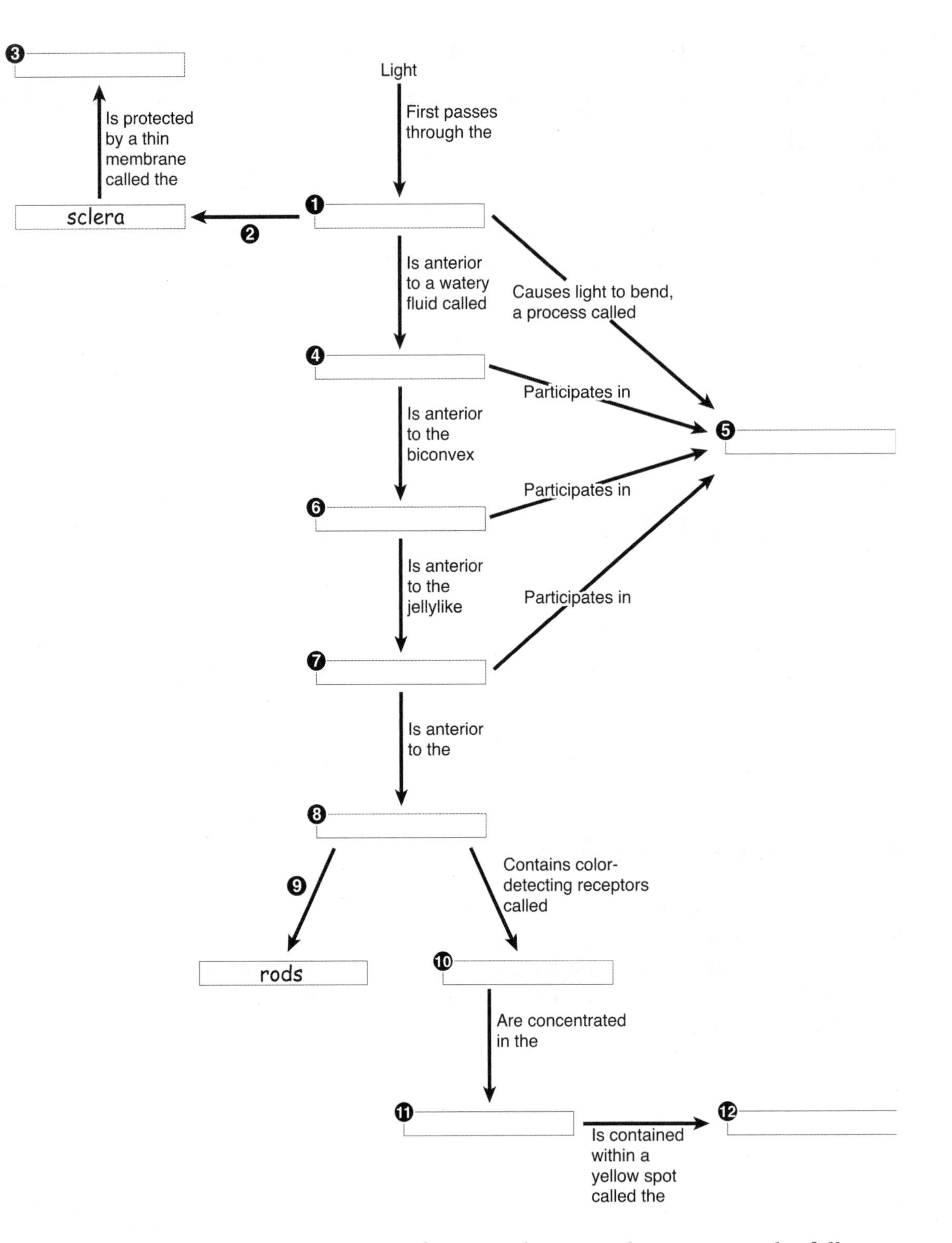

Optional Exercise: Construct a concept map of terms relating to the ear using the following terms and any others you would like to include: tympanic membrane, stapes, malleus, incus, pinna, bony labyrinth, organ of Corti, oval window, round window, cochlear duct, tectorial membrane, and cochlear nerve. You may also want to construct concept maps relating to the other special senses (equilibrium, taste, smell) and the general senses (touch, pressure, temperature, proprioception).

Testing Your Knowledge

Building Understanding

I. Multiple Choice

1. A physician who specializes in disorders of the eye is a(n) 1. _______
 a. ophthalmologist
 b. internist
 c. allergist
 d. orthopedic surgeon
2. A term related to the sense of taste is 2. _______
 a. tactile
 b. gustatory
 c. proprioceptive
 d. thermal
3. Alterations in the lens' shape to allow for near or far vision is called 3. _______
 a. accommodation
 b. convergence
 c. divergence
 d. dark adaptation
4. The term *lacrimation* refers to the secretion of 4. _______
 a. mucus
 b. wax
 c. tears
 d. aqueous humor
5. Painkillers that are released from certain regions of the brain are 5. _______
 a. narcotics
 b. endorphins
 c. anaesthetics
 d. nonsteroidal anti-inflammatory drugs
6. A person who lacks cones in the retina will suffer from 6. _______
 a. blindness
 b. color blindness
 d. glaucoma
 d. trachoma
7. The organ of Corti is the receptor for 7. _______
 a. taste
 b. smell
 c. hearing
 d. equilibrium
8. A cataract is 8. _______
 a. an irregularity in the cornea's shape
 b. an infection of the conjunctiva
 c. an abnormally short eyeball
 d. loss of lens transparency

9. Inflammation of the membrane lining the eyelid is called 9. _______
   a. otitis
   b. conjunctivitis
   c. retinitis
   d. glaucoma

II. Completion Exercise

1. The transparent portion of the sclera is the ______________
2. The glands that secrete ear wax are called ______________
3. The nerve endings that aid in judging position and changes in location of body parts are the ______________
4. The sense of position is partially governed by equilibrium receptors in the internal ear, including two small chambers in the vestibule and the three ______________
5. The tactile corpuscles are the receptors for the sense of ______________
6. Any drug that relieves pain is called a(n) ______________
7. When you enter a darkened room, it takes a while for the rods to begin to function. This interval is known as the period of ______________
8. The receptor tunic (layer) of the eye is the ______________
9. The bending of light rays as they pass through the media of the eye is ______________

Understanding Concepts

I. True/False

For each question, write *T* for true or *F* for false in the blank to the left of each number. If a statement is false, correct it by replacing the underlined term and write the correct statement in the blank below the question.

_______ 1. Extrinsic eye muscles control the diameter of the pupil.

____

_______ 2. There are seven extrinsic muscles connected to each eye.

____

_______ 3. The iris is an intrinsic muscle of the eye.

____

_______ 4. The sense of temperature is a general sense.

____

_______ 5. The rods of the eye function in bright light and detect color.

____

_______ 6. When the eyes are exposed to a bright light, the pupils <u>constrict</u>.

____

_______ 7. The scientific name for nearsightedness is <u>hyperopia</u>.

____

_______ 8. The ciliary muscle <u>contracts</u> to allow thickening of the lens.

____

_______ 9. The sense of smell is also called <u>olfaction</u>.

____

II. Practical Applications

Study each discussion. Then write the appropriate word or phrase in the space provided.

➤ Group A

Baby L was brought in by his mother because he awakened crying and holding the right side of his head. He had been suffering from a cold, but now he seemed to be in pain. Complete the following descriptions relating to his evaluation and treatment.

1. Examination revealed a bulging red eardrum. The eardrum is also called the _______________.
2. The cause of Baby L's painful bulging eardrum was an infection of the middle ear, a condition called _______________.
3. Antibiotic treatment of Baby L's middle ear infection was begun, because this early treatment usually prevents complications. In this case, however, it was necessary to cut the eardrum to prevent its rupture. Another name for this surgical procedure is _______________.
4. The mother was warned that Baby L may be particularly susceptible to middle ear infections. To prevent further damage to his eardrum, a special tube was inserted. This tube is called a(n) _______________.
5. Baby L will have to be careful in the future, because repeated middle ear infections can lead to a type of hearing loss called _______________.
6. Baby L was returned to the emergency room the next day because he was falling down repeatedly. The physician suspected a problem with his sense of balance, or _______________.
7. Baby L's mother asked how an ear infection could affect balance. The physician explained that two structures were located within the inner ear that are involved with balance, named the semicircular canals and the _______________.
8. In particular, the physician feared that the middle ear infection had spread to the fluid within the membranous labyrinth. This fluid is called _______________.

➤ Group B

Sixty-year-old Mr. S had ridden his scooter over some broken glass. A fragment of glass bounced up and flew into one eye. Complete the following descriptions relating to his evaluation and treatment.

1. Examination by the eye specialist showed that there was a cut in the transparent window of the eye, the ________________.
2. On further examination of Mr. S, the colored part of the eye was seen to protrude from the wound. This part of the eye is the ________________.
3. Mr. S's treatment included antiseptics, anesthetics, and suturing of the wound. Medication was instilled in the saclike structure at the anterior of the eyeball. This sac is lined with a thin epithelial membrane, the ________________.
4. The eye specialist evaluated Mr. S's vision in his uninjured eye. Like virtually all elderly adults, Mr. S was shown to have difficulties with near vision. This condition is called ________________.
5. The eye specialist also observed that the pressure in his aqueous humor was abnormally high. This finding signifies that Mr. S suffers from ________________.
6. Mr. S returned to the emergency room 1 week later with a severe infection in the injured eye. Despite proper wound care and several changes of antibiotics, the damaging infection persisted. The eye specialist reluctantly decided to remove the eyeball, a procedure called ________________.

➤ Group C

You are conducting hearing tests at a senior citizens' home. During the course of the afternoon, you encounter the following patients. Complete the following descriptions relating to the evaluation and treatment of hearing loss.

1. Mrs. B complained of some hearing loss and a sense of fullness in her outer ear. Examination revealed that her ear canal was plugged with hardened ear wax, which is scientifically called ________________.
2. Mr. J, age 72, complained of gradually worsening hearing loss, although he had no symptoms of pain or other ear problems. Examination revealed that his hearing loss was due to nerve damage. The cranial nerve that carries hearing impulses to the brain is called the ________________.
3. In particular, the endings of this nerve were damaged. These nerve endings are located in the spiral-shaped part of the inner ear, a part of the ear that is known as the ________________.
4. Mr. J's hearing loss, because it reflects nerve damage, is known as ________________.
5. Mrs. C complained of hearing loss that resembled the type from which her aunt and her mother suffered. She requested surgical treatment, which is often successful in such cases. This disorder, in which bony changes prevent the stapes from vibrating normally, is called ________________.

III. Short Essays

1. Describe several different structural forms of sensory receptors and give examples of each.

____

____

____

2. Describe some changes that occur in the sensory receptors with age.

____

____

____

3. List three methods to relieve pain that do not involve administration of drugs.

____

____

____

Conceptual Thinking

1. You have probably been sitting in a chair for quite a while, yet you have not been constantly aware of your legs contacting the chair. Why not?

____

____

____

2. Write your name at the bottom of this sheet of paper. Explain the contributions of different sensory receptors that were required to successfully complete that simple task. For instance, proprioceptors are required to indicate the fingers' location at every moment.

____

____

____

Expanding Your Horizons

Imagine if you could taste a triangle, or hear blue. This is reality for individuals with a disorder called synesthesia. Read about some exceptional artists that suffer from this disorder, and how synesthesia has helped us understand how the brain processes sensory information in the article below.

Here is an exercise you can do to find your own blind spot. Draw a cross (on the left) and a circle (on the right) on a piece of paper that are separated by a handwidth. Focus on the cross and notice (but do not focus on) the circle. Move the paper closer and further away until the circle disappears. Weird activities to investigate your blind spot can be found at the website http://serendip.brynmawr.edu/bb/blindspot1.html.

Resources

1. Ramachandran VS, Hubbard EM. Hearing colors, tasting shapes. *Sci Am* 2003; 288:52–59.

CHAPTER 12

The Endocrine System: Glands and Hormones

Overview

The endocrine system and the nervous system are the main coordinating and controlling systems of the body. Chapter 9 discusses how the nervous system uses chemical and electrical stimuli to control very rapid, short-term responses. This chapter discusses the endocrine system, which uses specific chemicals called hormones to induce short- or long-term changes in the growth, development, and function of specific cells. Although some hormones act locally, most travel in the blood to distant sites and exert their effects on any cell (the target cell) that contains a specific receptor for the hormone. Although hormones are produced by many tissues, certain glands, called endocrine glands, specialize in hormone production. These endocrine glands include the pituitary (hypophysis), thyroid, parathyroids, adrenals, pancreas, gonads, thymus, and pineal. Together, these glands comprise the endocrine system. The activity of endocrine glands is regulated by negative feedback, other hormones, nervous stimulation, and/or biological rhythms. One of the most important endocrine glands is the pituitary gland, which comprises the anterior and posterior pituitary. The posterior pituitary gland is made of nervous tissue—it contains the axons and terminals of neurons that have their cell bodies in a part of the brain called the hypothalamus. Hormones are synthesized in the hypothalamus and released from the posterior pituitary. The anterior pituitary secretes a number of hormones that act on other endocrine glands. The cells of the anterior pituitary are controlled in part by releasing hormones made in the hypothalamus. These releasing hormones pass through the blood vessels of a portal circulation to reach the anterior pituitary. Hormones are also made outside the traditional endocrine glands. Other structures that secrete hormones include the stomach, small intestine, kidney, heart, skin, and placenta. Hormones are extremely potent, and small variations in hormone concentrations can have significant effects on the body.

This chapter contains a lot of details for you to learn. Try to summarize the material using concept maps and summary tables. You should also understand positive and negative feedback (Chapter 1) before you tackle the concepts in this chapter.

Addressing the Learning Outcomes

1. Compare the effects of the nervous system and the endocrine system in controlling the body.

EXERCISE 12-1.

INSTRUCTIONS

Identify the following characteristics as belonging to the nervous system (N) or the endocrine system (E).

1. controls rapid responses ___
2. used to regulate growth ___
3. uses chemical stimuli only ___
4. uses both chemical and electrical stimuli ____

2. Describe the functions of hormones.

EXERCISE 12-2.

INSTRUCTIONS

Define the following terms:

1. receptor ________________________________
2. target tissue ________________________________
3. hormone ________________________________

3. Discuss the chemical composition of hormones.

EXERCISE 12-3.

INSTRUCTIONS

Do each of the following statements apply to amino acid compounds (A) or lipids (L)?

1. protein hormones _____
2. can be formed by modifying cholesterol ____
3. hormones of the sex glands and the adrenal cortex ____
4. hormones not produced by the sex glands or adrenal cortex _____
5. prostaglandins ____

4. Explain how hormones are regulated.

EXERCISE 12-4.

INSTRUCTIONS

Define negative and positive feedback.

____

____

5. Identify the glands of the endocrine system on a diagram.

EXERCISE 12-5: The Endocrine Glands (Text Fig. 12-2)

INSTRUCTIONS

1. Write the name of each labeled part on the numbered lines in different colors.
2. Color the different structures on the diagram with the corresponding color. Some structures are present in more than one location on the diagram. Try to color all of a particular structure in the appropriate color. For instance, color both adrenal glands, although only one is indicated by a leader line.

1. ____________________
2. ____________________
3. ____________________
4. ____________________
5. ____________________
6. ____________________
7. ____________________
8. ____________________
9. ____________________

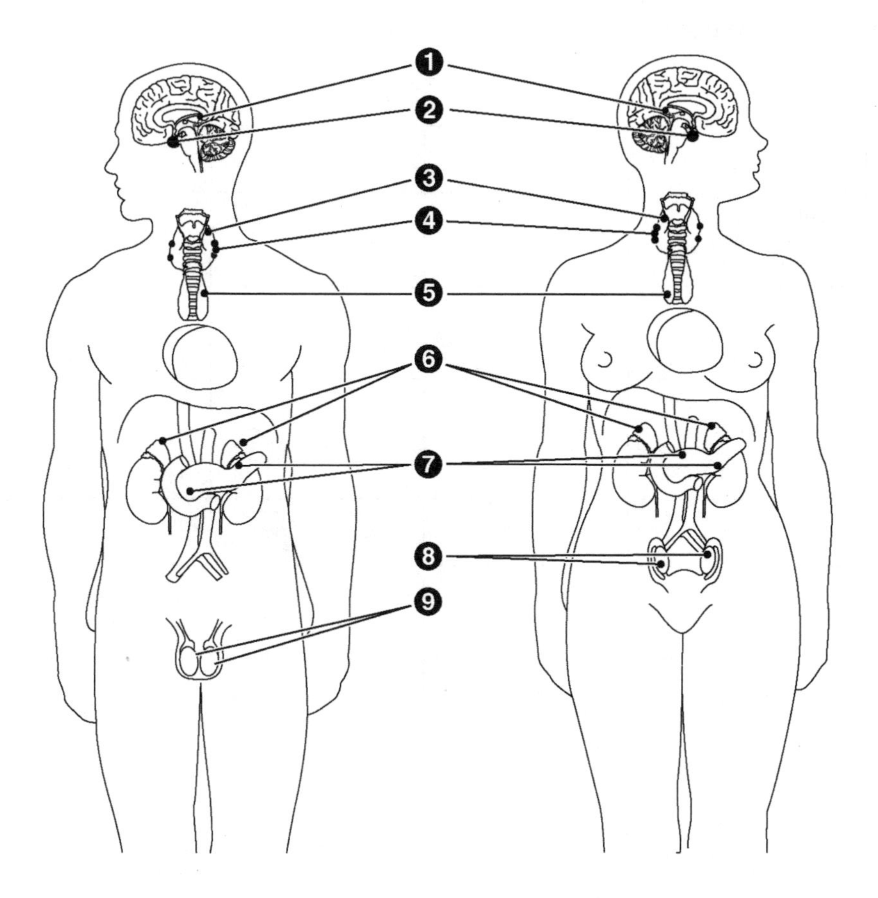

EXERCISE 12-6.

INSTRUCTIONS

Write the term in the appropriate blank.

parathyroid thymus pineal hypothalamus
thyroid adrenal pancreas

1. One of the tiny glands located behind the thyroid gland ________________
2. The largest of the endocrine glands, located in the neck ________________
3. The gland in the brain that is regulated by light ________________
4. An organ that contains islets ________________
5. The endocrine gland composed of a cortex and medulla, each with specific functions ________________

6. List the hormones produced by each endocrine gland and describe the effects of each on the body.

EXERCISE 12-7.

INSTRUCTIONS

Fill in the missing information in the chart below. This table does not cover all of the hormones and their actions discussed in the text. Do not fill in the three columns at the far right yet. These columns will be discussed under Outcomes 8 and 10.

Hormone	Gland	Effects	Hyposecretion	Hypersecretion	Medical Uses
ADH (antidiuretic hormone)					N/A
	Adrenal cortex	Nutrient metabolism during stress			
		Uterine contraction, milk ejection	N/A	N/A	
	Anterior pituitary	Promotes growth of all body tissues			
Parathyroid hormone					N/A
	Pancreas	Necessary for glucose uptake into cells			
Thyroid hormones: thyroxine (T4) and triiodothyronine (T3)					
	Adrenal cortex	Helps regulate water and electrolyte balance			N/A
		Stimulates glucose release from the liver	N/A	N/A	N/A
Melatonin			N/A	N/A	N/A

EXERCISE 12-8.

INSTRUCTIONS

Write the appropriate term in each blank.

epinephrine	antidiuretic hormone	ACTH	aldosterone
follicle-stimulating hormone	estrogen	calcitonin	prolactin

1. The anterior pituitary hormone that stimulates milk synthesis ________________
2. The main hormone of the adrenal medulla that, among other actions, raises blood pressure and increases the heart rate ________________
3. The anterior pituitary hormone that stimulates the adrenal cortex ________________
4. A hormone produced by the ovaries ________________
5. The hormone from the adrenal cortex that regulates sodium and potassium reabsorption in the kidney tubules ________________
6. A gonadotropic hormone ________________
7. A hormone synthesized in the hypothalamus ________________

EXERCISE 12-9.

INSTRUCTIONS

Write the appropriate term in each blank

insulin	glucagon	parathyroid hormone	testosterone
thymosin	cortisol		

1. A hormone that raises the blood calcium level ________________
2. A hormone that lowers the blood glucose level ________________
3. A pancreatic hormone that raises the blood glucose level ________________
4. An adrenal hormone that raises the blood glucose level ________________
5. A hormone that promotes T-cell development ________________

7. Describe how the hypothalamus controls the anterior and posterior pituitary.

EXERCISE 12-10: The Pituitary Gland (Text Fig. 12-3)

INSTRUCTIONS

1. Label the parts of the hypothalamo-pituitary system.
2. Color blood vessels red and nerves yellow.

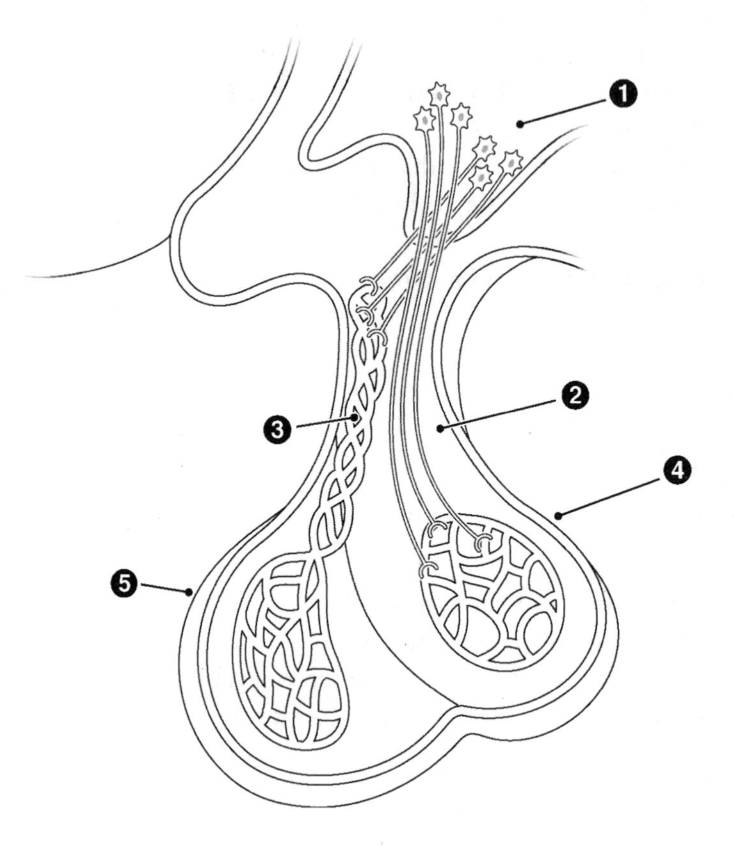

1. ____________________
2. ____________________
3. ____________________
4. ____________________
5. ____________________

EXERCISE 12-11.

Do the following characteristics apply to the anterior pituitary (AP) or the posterior pituitary (PP)?

1. Secretes hormones synthesized in the hypothalamus ____
2. Consists of neural tissue ____
3. Releases hormones under the regulation of hypothalamic releasing hormones ___

8. Describe the effects of hyposecretion and hypersecretion of the various hormones

EXERCISE 12-12.

INSTRUCTIONS

Add the disorders associated with each hormone to the table prepared in Exercise 12-7.

EXERCISE 12-13.

INSTRUCTIONS

Write the appropriate term in each blank.

myxedema	goiter	acromegaly	Graves disease
Addison disease	diabetes mellitus	hypoglycemia	

1. Any enlargement of the thyroid gland ____________
2. A disease in which insulin function is abnormally low ____________
3. A form of hyperthyroidism ____________
4. A form of hypothyroidism in adults ____________
5. A disease resulting from adrenal cortex hypoactivity ____________
6. Low blood sugar ____________

9. List tissues other than the endocrine glands that produce hormones.

EXERCISE 12-14.

INSTRUCTIONS

Fill in the missing information in the chart below.

Hormone	Site of Synthesis	Effects
Atrial natriuretic peptide		
	Kidney	Stimulates erythrocyte production

10. List some medical uses of hormones.

EXERCISE 12-15.

INSTRUCTIONS

Fill in the column "Medical Uses" in the table prepared for Exercise 12-7 for the following hormones: insulin, cortisol (a glucocorticoid), epinephrine, thyroid hormones, and oxytocin.

11. Explain how the endocrine system responds to stress.

EXERCISE 12-16.

INSTRUCTIONS

Explain why cortisol and epinephrine are released during stressful situations.

1. ______________________________

____

____

2. ______________________________

____

____

12. Show how word parts are used to build words related to the endocrine system.

EXERCISE 12-17.

INSTRUCTIONS

Complete the following table by writing the correct word part or meaning in the space provided. Write a word that contains each word part in the Example column.

Word Part	Meaning	Example
1. trop/o	________	________
2. ________	cortex	________
3. –poiesis	________	________
4. natri	________	________
5. ________	male	________
6. ________	enlargement	________
7. ren/o	________	________
8. acro-	________	________
9. oxy	________	________
10. nephr/o	________	________

Making the Connections

The following concept map deals with the relationship between the hypothalamus, pituitary gland, and some target organs. Each pair of terms is linked together by a connecting phrase into a sentence. The sentence should be read in the direction of the arrow. Complete the concept map by filling in the appropriate term or phrase. There is one right answer for each term. However, there are many correct answers for the connecting phrases (1, 4, 5, 9, 12).

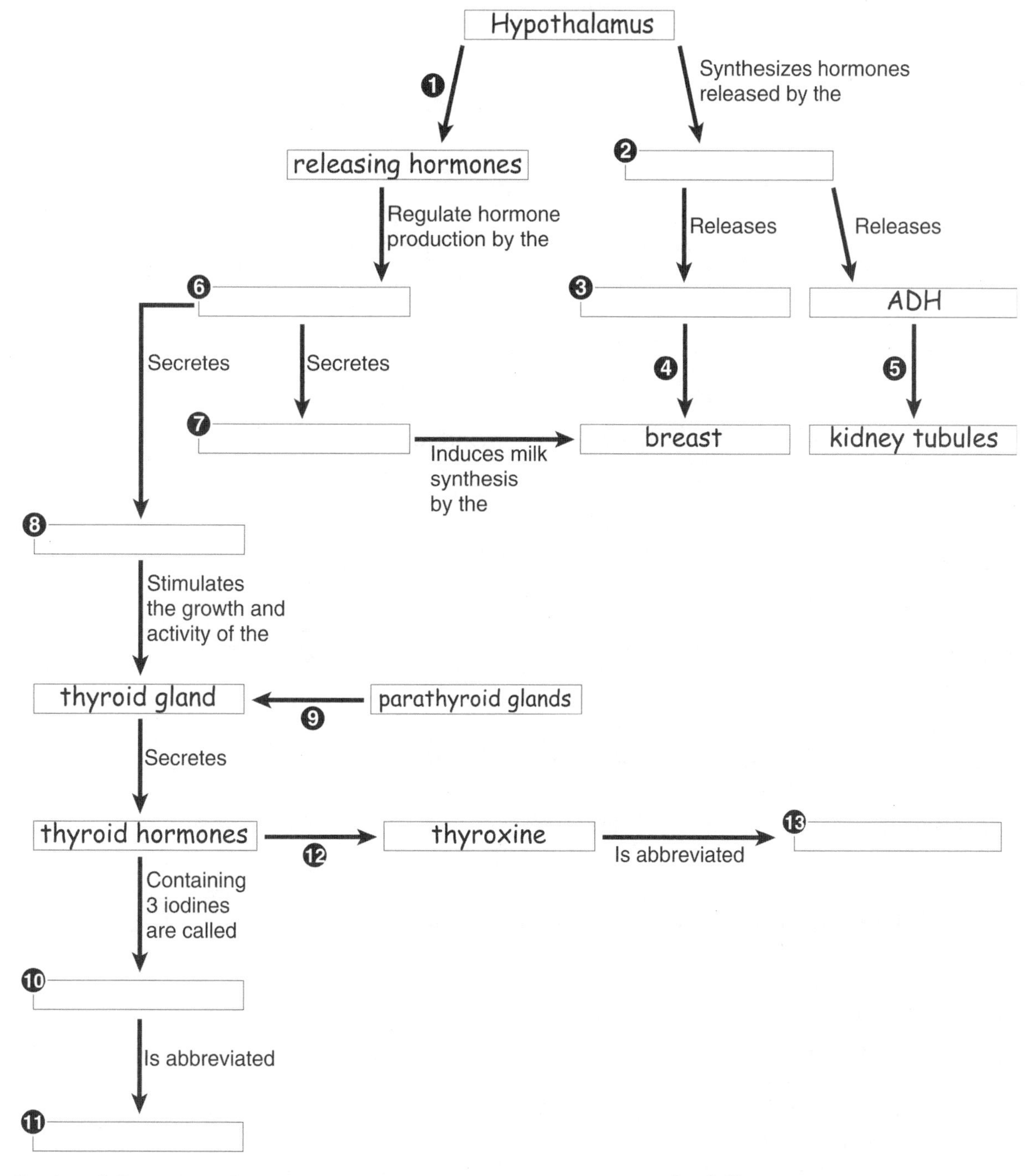

Optional Exercise: Construct your own concept map using the following terms: pancreas, insulin, glucagon, adrenal gland, medulla, cortex, epinephrine, cortisol, aldosterone, raises blood sugar, lowers blood sugar, Cushing syndrome, diabetes mellitus, and Addison disease. You can also add other appropriate terms (for instance, target sites or hormone actions).

Testing Your Knowledge

Building Understanding

I. Multiple Choice

1. An excess of growth hormone in an adult results in 1. _______
   a. acromegaly
   b. dwarfism
   c. tetany
   d. goiter
2. An androgen is a(n) 2. _______
   a. female sex hormone
   b. glucocorticoid
   c. male sex hormone
   d. atrial hormone
3. Which of the following hormones is NOT produced by the thyroid gland? 3. _______
   a. calcitonin
   b. thyroxine
   c. triiodothyronine
   d. thymosin
4. The hormone that causes milk ejection from the breasts is 4. _______
   a. oxytocin
   b. prolactin
   c. progesterone
   d. estrogen
5. Cushing syndrome is associated with an excess of 5. _______
   a. cortisol
   b. aldosterone
   c. insulin
   d. androgens
6. Diabetes insipidus results from a deficiency of 6. _______
   a. ACTH
   b. insulin
   c. ADH
   d. melatonin
7. Which of the following hormones is derived from cholesterol? 7. _______
   a. progesterone
   b. thyroid hormone
   c. growth hormone
   d. luteinizing hormone
8. Addison disease results from underactivity of the 8. _______
   a. adrenal medulla
   b. anterior pituitary
   c. parathyroid
   d. adrenal cortex

9. The pituitary hormone that regulates the activity of the thyroid gland is 9. _______
   a. TSH
   b. GH
   c. ACTH
   d. MSH
10. Erythropoietin is synthesized in the 10. _______
   a. kidneys
   b. skin
   c. heart
   d. placenta

II. Completion Exercise

1. An abnormal increase in production of the hormone epinephrine may result from a tumor of the _______________.
2. Releasing hormones are sent from the hypothalamus to the anterior pituitary by way of a special circulatory pathway called a(n) _______________.
3. When the blood glucose level decreases to less than average, the islet cells of the pancreas release less insulin. The result is an increase in blood glucose. This is an example of the regulatory mechanism called _______________.
4. The hypothalamus stimulates the anterior pituitary to produce ACTH, which in turn stimulates hormone production by the _______________.
5. The element needed for the production of thyroxine is _______________.
6. Local hormones that have a variety of effects, including the promotion of inflammation and the production of uterine contractions, are the _______________.
7. A hormone secreted from the posterior pituitary that is involved in water balance is _______________.
8. The primary target tissue for prolactin is the _______________.

Understanding Concepts

I. True/False

For each question, write *T* for true or *F* for false in the blank to the left of each number. If a statement is false, correct it by replacing the underlined term and write the correct statement in the blank below the question.

_______ 1. ACTH acts on the <u>adrenal medulla</u>.

____

_______ 2. Cortisol and the pancreatic hormone <u>insulin</u> both raise blood sugar.

____

_______ 3. <u>Diabetes insipidus</u> results from hyposecretion of antidiuretic hormone.

____

_______ 4. The ovaries and testes produce steroid hormones.

____

_______ 5. Cortisol is produced by the adrenal cortex.

____

_______ 6. Islet cells are found in the adrenal gland.

____

_______ 7. ADH and oxytocin are secreted by the anterior lobe of the pituitary.

____

_______ 8. Atrial natriuretic peptide (ANP) is produced by the kidneys.

____

II. Practical Applications

Write the appropriate word or phrase in the space provided.

1. Ms. H, age 8, was brought to the clinic because she had been losing weight, was very thirsty, and urinated frequently. A blood chemistry panel taken when she had not eaten for 8 hours showed an extremely high blood glucose reading of 186 mg/dL as well as other abnormal readings. A pancreatic disorder was suspected, leading to a tentative diagnosis of _______________.
2. Mr. G, age 38, had been taking high doses of oral steroid drugs for asthma for 6 years. He noted fat deposits between his shoulders and around his abdomen, muscle loss in his arms and legs, and thin skin that was prone to bruising. These symptoms are common with excess of a hormone produced normally by the _______________.
3. Mr. G was told that he was manifesting the symptoms of an endocrine disease. This disease is called _______________.
4. Mr. L, age 42, reported to the hospital emergency room with complaints of shortness of breath and heart palpitations. The initial assessment by the nurse included the following findings: rapid heart rate, nervousness with tremor of the hands, skin warm and flushed, sweating, rapid respiration, and protruding eyes. Laboratory tests confirmed the diagnosis of overactivity of the _______________.
5. After surgery for his endocrine problem, Mr. L had tetany, or contractions of the muscles of the hands and face. This was caused by the incidental surgical removal of the glands that control the release of calcium into the blood. The glands that maintain adequate blood calcium levels are the _______________.
6. Ms. M has just been stung by a bee. She is extremely allergic to bees. Her sister gives her a life-saving injection of an adrenal hormone. This hormone is called _______________.
7. Ms. Q was supposed to have her baby 10 days ago. Her relatives are anxiously awaiting the birth of her child, so she asks her obstetrician if he can do something to hasten the birth of her child. The obstetrician agrees to induce labor using a hormone called _______________.

8. Mr. S is preparing for his final law exams and is feeling very stressed. His partner is studying for a physiology exam and mentions that he probably has elevated levels of an anterior pituitary hormone that acts on the adrenal cortex. This hormone is called ______________.
9. Ms. J, age 35, is preparing for a weight-lifting competition. She wants to build muscle tissue very rapidly, so a friend recommends that she tries injections of a male steroid known to stimulate tissue building. This steroid is mainly produced by the ______________.

III. Short Essays

1. Explain why hormones, although they circulate throughout the body, exercise their effects only on specific target cells.

____

____

____

2. List two differences between the endocrine system and the nervous system.

____

____

____

3. List three complications associated with untreated diabetes mellitus.

____

____

____

4. Name three organs other than endocrine glands that produce hormones, and name a hormone produced in each organ.

____

____

____

5. Compare the anterior and the posterior lobes of the pituitary.

____

____

____

Conceptual Thinking

1. Ms. J and Ms. K both have a goiter. However, Ms. J has hyperthyroidism and Ms. K has hypothyroidism. List the specific diseases suffered by the two women, and explain why each patient has a goiter.

____

____

____

2. Young Ms. K suffers from asthma. She uses an inhaler containing epinephrine to treat her attacks. However, lately she has been suffering from a very rapid heart beat. Her physician advises her to use her inhaler less frequently. Why?

____

____

____

Expanding Your Horizons

In the past, the only source of protein hormones for therapeutic use was cadavers. Hormone supplies were very limited and only used in severe cases of hormone deficiency. Hormones can now be synthesized in unlimited quantities in a test tube. Some hormones, including growth hormone and anabolic steroids, are now abused by some athletes. They are also used by the elderly to treat some age-related deficits, such as reduced bone mass and impaired muscle strength, even though their safety has not been proven. Hormone use and abuse can be investigated using websites (http://www.hormone.org/ and http://www.nlm.nih.gov/medlineplus/). Hormone abuse in athletes is discussed in a *Scientific American* article.

Resources

1. Zorpette G. The athlete's body: the chemical games. *Scientific American* Presents: Building the Elite Athlete 2000;11:16–23.

UNIT V

Circulation and Body Defense

CHAPTER 13

The Blood

Overview

The blood maintains the constancy of the internal environment through its functions of transportation, regulation, and protection. Blood is composed of two portions: the liquid portion, or plasma, and the formed elements, consisting of the cells and cellular products. The plasma is 91% water and 9% proteins, carbohydrates, lipids, electrolytes, and waste products. The formed elements are composed of the erythrocytes, which carry oxygen to the tissues; the leukocytes, which defend the body against invaders; and the platelets, which are involved in the process of blood coagulation (clotting). The forerunners of the blood cells are called hematopoietic stem cells. These are formed in the red bone marrow, where they then develop into the various types of blood cells.

Blood coagulation is a protective mechanism that prevents blood loss when a blood vessel is ruptured by an injury. The steps in the prevention of blood loss (hemostasis) include constriction of the blood vessels, formation of a platelet plug, and formation of a clot, a complex series of reactions involving many different factors.

If the quantity of blood in the body is severely reduced because of hemorrhage or disease, the cells suffer from lack of oxygen and nutrients. In such instances, a transfusion may be given after typing and matching the blood of the donor with that of the recipient. Donor red cells with different surface antigens (proteins) than the recipient's red cells will react with antibodies in the recipient's blood, causing harmful agglutination reactions and destruction of the donated cells. Blood is most commonly tested for the ABO system involving antigens A and B. Blood can be packaged and stored in blood banks for use when transfusions are needed. Whenever possible, blood components such as cells, plasma, plasma fractions, or platelets are used. This practice is more efficient and can reduce the chances of incompatibility and transmission of disease.

The Rh factor, another red blood cell protein, also is important in transfusions. If blood containing the Rh factor (Rh positive) is given to a person whose blood lacks that factor (Rh negative), the recipient will produce antibodies to the foreign Rh factor. If an Rh-negative mother is thus sensitized by an Rh-positive fetus, her antibodies may damage fetal red cells in a later pregnancy, resulting in hemolytic disease of the newborn (erythroblastosis fetalis).

Anemia is a common blood disorder. It may result from loss or destruction of red blood cells or from impaired production of red blood cells or hemoglobin. Other abnormalities are leukemia, a neoplastic disease of white cells; and clotting disorders.

Scientists have devised numerous studies to measure the composition of blood. These include the hematocrit, hemoglobin measurements, cell counts, blood chemistry tests, and coagulation studies. These techniques can diagnose blood diseases, some infectious diseases, and some metabolic diseases. Modern laboratories are equipped with automated counters, which rapidly and accurately count blood cells, and with automated analyzers, which measure enzymes, electrolytes, and other constituents of blood serum.

Addressing the Learning Outcomes

1. List the functions of the blood.

EXERCISE 13-1.

INSTRUCTIONS

List three functions of blood, and provide an example for each.

1. ______________________________
2. ______________________________
3. ______________________________

2. List the main ingredients in plasma.

EXERCISE 13-2.

INSTRUCTIONS

Which of the following substances are found in plasma? Circle all that apply.

A. erythrocytes
B. water
C. proteins
D. platelets
E. nutrients
F. electrolytes

3. Describe the formation of blood cells.

EXERCISE 13-3.

INSTRUCTIONS

Name the type of stem cell that can develop into all types of blood cells.

____

4. Name and describe the three types of formed elements in the blood and give the function of each.

EXERCISE 13-4: Composition of Whole Blood (Text Fig. 13-1)

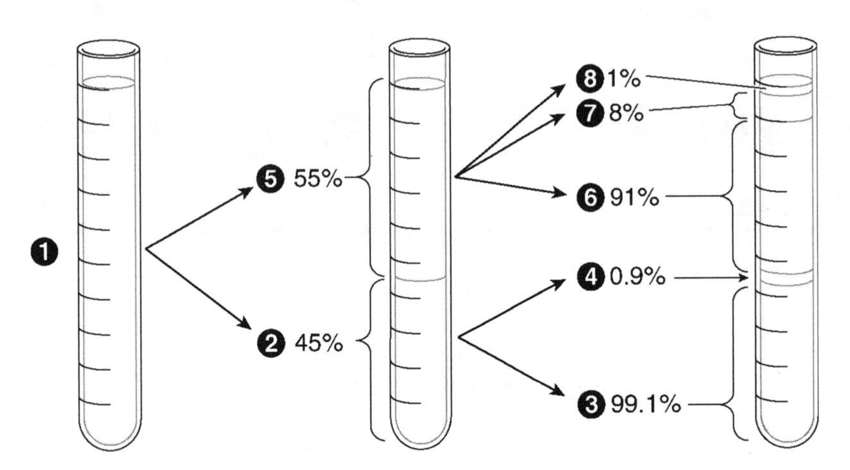

INSTRUCTIONS

1. Write the names of the different blood components on the appropriate numbered lines in different colors. Use the color red for parts 1, 2, and 3.
2. Color the blood components on the diagram with the corresponding color.

1. ____________________
2. ____________________
3. ____________________
4. ____________________
5. ____________________
6. ____________________
7. ____________________
8. ____________________

5. Characterize the five types of leukocytes.

EXERCISE 13-5: Leukocytes (Text Fig. 13-4)

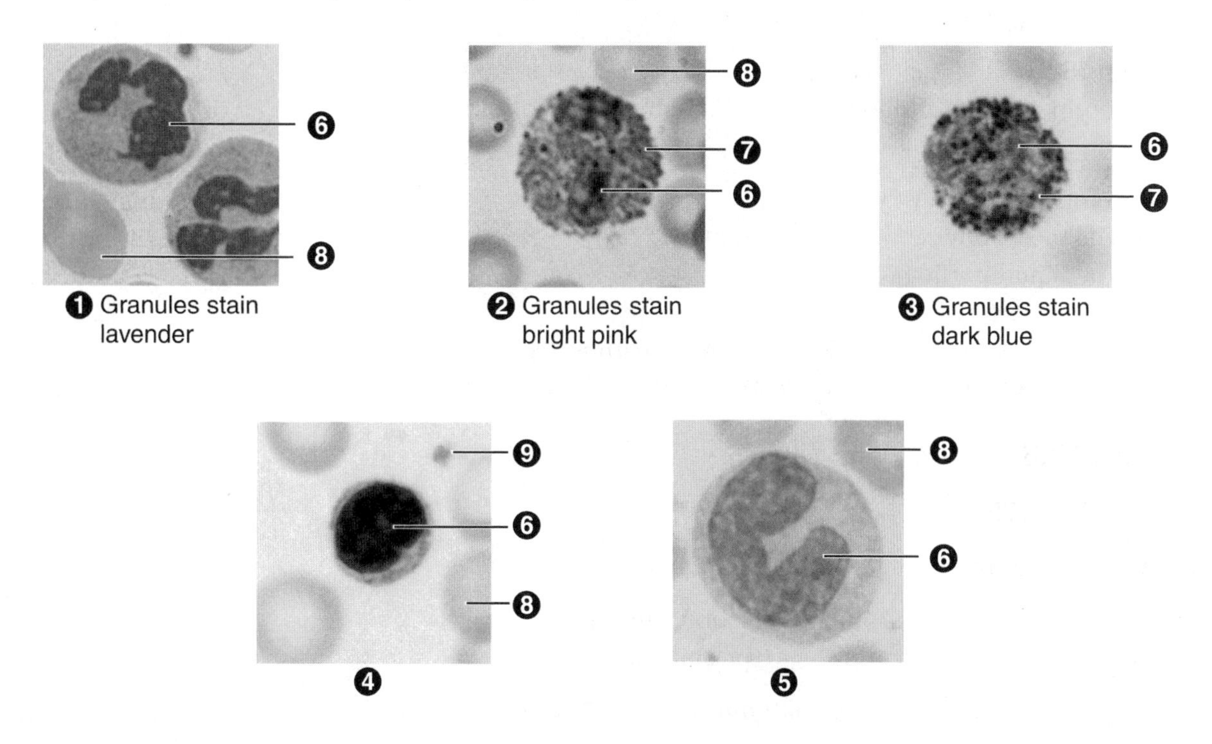

INSTRUCTIONS

1. Write the names of the five types of leukocytes on lines 1 to 5.
2. Write the names of the labeled cell parts and other blood cells on lines 6 to 9.

1. ____________________
2. ____________________
3. ____________________
4. ____________________
5. ____________________
6. ____________________
7. ____________________
8. ____________________
9. ____________________

EXERCISE 13-6.

INSTRUCTIONS

Write the appropriate term in each blank.

erythrocyte	platelet	leukocyte	serum
albumin	antibodies	complement	plasma

1. A red blood cell ________________
2. Another name for thrombocyte ________________
3. The most abundant protein(s) in blood ________________
4. The liquid portion of blood ________________
5. A white blood cell ________________
6. Enzymes that assist antibodies to battle pathogens ________________
7. The watery fluid that remains after a blood clot is removed ________________

EXERCISE 13-7.

INSTRUCTIONS

Write the appropriate term in each blank.

neutrophil	macrophage	monocyte	pus
eosinophil	basophil	plasma cell	

1. The most abundant type of white blood cell in whole blood ________________
2. A mature monocyte ________________
3. A lymphocyte that produces antibodies ________________
4. A leukocyte that stains with acidic dyes ________________
5. The largest blood leukocyte ________________
6. A substance that often accumulates when leukocytes are actively destroying bacteria ________________

6. Define *hemostasis* and cite three steps in hemostasis.

EXERCISE 13-8.

What is the difference between hemostasis and homeostasis?

____

____

____

EXERCISE 13-9.

INSTRUCTIONS

Write the appropriate term in each blank.

vasoconstriction coagulation hemorrhage platelet plug

1. A collection of cell fragments that temporarily repairs a vessel injury ____________
2. The process of blood clot formation ____________
3. Contraction of smooth muscles in the blood vessel wall ____________
4. Another term for profuse bleeding ____________

7. Briefly describe the steps in blood clotting.

EXERCISE 13-10: Formation of a Blood Clot (Text Fig. 13-8)

INSTRUCTIONS

Write the correct term in each of the numbered boxes.

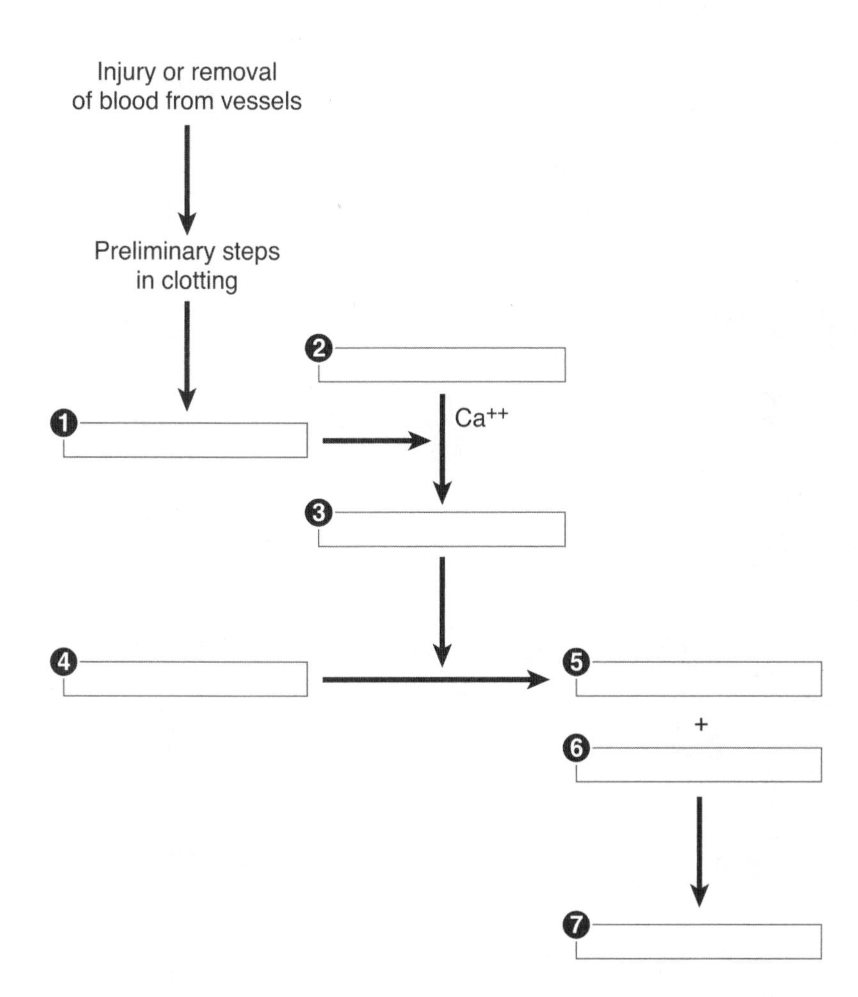

8. Define *blood type* and explain the relation between blood type and transfusions.

EXERCISE 13-11: Blood Typing (Text Fig. 13-9)

INSTRUCTIONS

Based on the agglutination reactions, write the name of each blood type on the numbered lines.

1. ____________________
2. ____________________
3. ____________________
4. ____________________

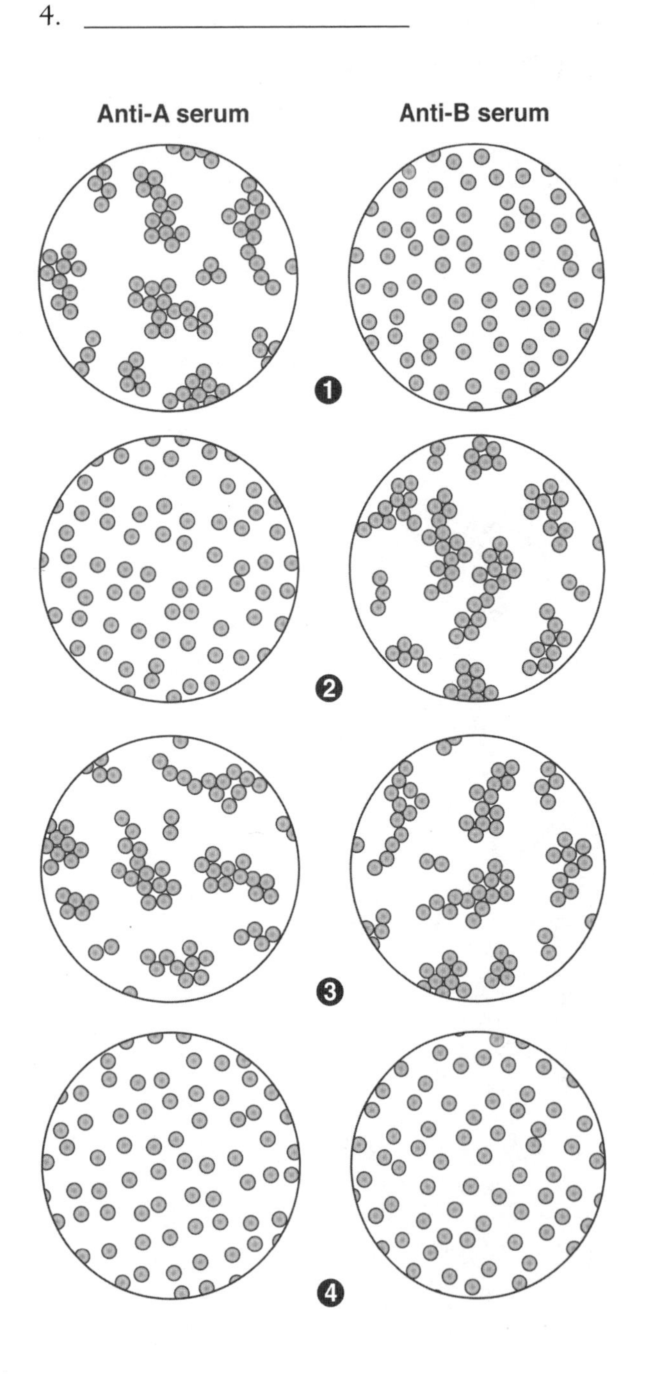

9. List the possible reasons for transfusions of whole blood and blood components.

EXERCISE 13-12.

INSTRUCTIONS

Write the appropriate term in each blank.

hematocrit	Rh factor	autologous	AB antigen	agglutination
hemapheresis	plasmapheresis	transfusion	antigen	

1. The blood antigen involved in hemolytic disease of the newborn, which results from a blood incompatibility between a mother and fetus ________________
2. The procedure for removing plasma and returning formed elements to the donor ________________
3. The procedure for removing specific components and returning the remainder of the blood to the donor ________________
4. Blood donated by an individual for use by the same individual ________________
5. The volume percentage of red cells in whole blood ________________
6. The administration of blood or blood components from one person to another person ________________
7. A general term describing a protein on blood cells that causes incompatibility reactions ________________
8. The process by which cells become clumped when mixed with a specific antiserum ________________

10. Define *anemia* and list the causes of anemia.

EXERCISE 13-13.

INSTRUCTIONS

Write the appropriate term in each blank.

hemolytic anemia	hemophilia	pernicious anemia
thalassemia	aplastic anemia	

1. A disease resulting from a lack of vitamin B_{12} ________________
2. A hereditary form of anemia associated with excess iron in the blood ________________
3. A disease in which red blood cells are excessively destroyed ________________
4. A disease resulting in insufficient red cell production in the bone marrow ________________

11. Define *leukemia* and name the two types of leukemia (see Exercise 13-14).

12. Describe several forms of clotting disorders.

EXERCISE 13-14.

INSTRUCTIONS

Write the appropriate term in each blank.

lymphocytic leukemia	myelogenous leukemia	von Willebrand disease
thrombocytopenia	disseminated intravascular coagulation	hemophilia

1. A clotting disorder involving excessive coagulation ________________
2. A clotting disorder that can be treated with an ADH-like drug ________________
3. A disease resulting from abnormal proliferation of stem cells in bone marrow ________________
4. A cancer that arises in lymphoid tissue ________________
5. A clotting disorder that can be treated with factor VIII ________________

13. Specify the tests used to study blood.

EXERCISE 13-15.

INSTRUCTIONS

Which blood test would you use to diagnose each disease? Choose from the list below. Each test can be used only once.

red cell count	coagulation study	bone marrow biopsy
platelet count	blood smear	

1. hemophilia ________________
2. polycythemia vera ________________
3. thrombocytopenia ________________
4. myelogenous leukemia ________________
5. malaria ________________

14. Show how word parts are used to build words related to the blood.

EXERCISE 13-16.

INSTRUCTIONS

Complete the following table by writing the correct word part or meaning in the space provided. Write a word that contains each word part in the Example column.

Word Part	Meaning	Example
1. erythr/o	________	________________
2. __________	blood clot	________________
3. pro-	________	________________
4. morph/o	________	________________
5. __________	white, colorless	________________
6. __________	producing, originating	________________
7. hemat/o	________	________________
8. __________	lack of	________________
9. -emia	________	________________
10. __________	dissolving	________________

Making the Connections

The following concept map deals with the classification of blood cells. Each pair of terms is linked together by a connecting phrase into a sentence. The sentence should be read in the direction of the arrow. Complete the concept map by filling in the appropriate term or phrase. There is one right answer for each term. However, there are many correct answers for the connecting phrases (5, 6, 8, 10, 12, 14, and 16).

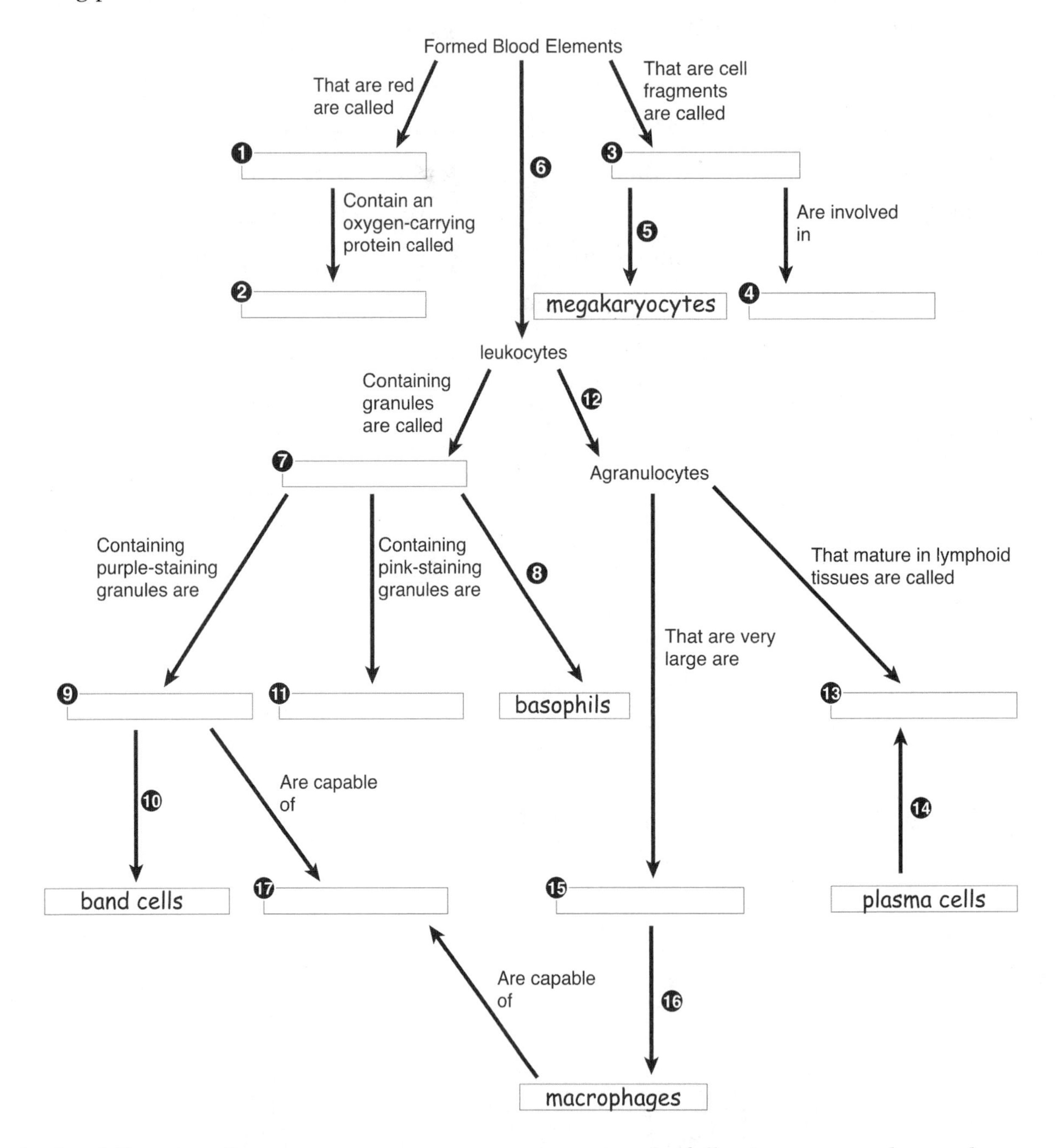

Optional Exercise: Construct your own concept map using the following terms and any others you would like to include: procoagulants, anticoagulants, platelet plug, hemostasis, vasoconstriction, blood clot, fibrinogen, fibrin, prothrombinase, thrombin, serum, hemophilia, thrombocytopenia, and disseminated intravascular coagulation.

Testing Your Knowledge

Building Understanding

I. Multiple Choice

1. Plasma can be given to anyone without danger of incompatibility because it lacks 1. _______
   a. serum
   b. red cells
   c. protein
   d. clotting factors
2. Polymorphs, PMNs, and segs are alternate names for 2. _______
   a. monocytes
   b. neutrophils
   c. basophils
   d. lymphocytes
3. Which of the following is NOT a type of white blood cell? 3. _______
   a. thrombocyte
   b. lymphocyte
   c. eosinophil
   d. monocyte
4. Blood clotting occurs in a complex series of steps. The substance that finally forms the clot is 4. _______
   a. albumin
   b. anticoagulant
   c. thromboplastin
   d. fibrin
5. Which of the following might result in an Rh incompatibility problem? 5. _______
   a. an Rh-positive mother and an Rh-negative fetus
   b. an Rh-negative mother and an Rh-positive fetus
   c. an Rh-negative mother and a type AB fetus
   d. an Rh-positive mother and an Rh-negative father
6. Intrinsic factor is 6. _______
   a. the first factor activated in blood clotting
   b. a substance needed for absorption of folic acid
   c. a type of hereditary bleeding disease
   d. a substance needed for absorption of vitamin B_{12}
7. Electrophoresis is the process by which 7. _______
   a. the volume proportion of red cells is determined
   b. normal and abnormal types of hemoglobin can be separated
   c. red blood cells are counted
   d. white blood cells are counted
8. An immature neutrophil is called a(n): 8. _______
   a. band cell
   b. monocyte
   c. eosinophil
   d. lymphocyte

9. Which of the following cells is NOT a granulocyte? 9. _______
   a. monocyte
   b. eosinophil
   c. neutrophil
   d. polymorph
10. Sickle cell anemia is a type of 10. _______
    a. nutritional anemia
    b. aplastic anemia
    c. hemolytic anemia
    d. pernicious anemia

II. Completion Exercise

1. The type of leukemia that originates in lymphoid tissues is ______________
2. A deficiency in the number of circulating platelets is called ______________
3. The gas that is necessary for life and that is transported to all parts of the body by the blood is ______________
4. Some monocytes enter the tissues, mature, and become active phagocytes. These cells are called ______________
5. One waste product of body metabolism is carried to the lungs to be exhaled. This gas is ______________
6. The hormone that stimulates red blood cell production is ______________
7. Blood cells are formed in the ______________
8. The most important function of certain lymphocytes is to engulf disease-producing organisms by the process of ______________
9. The chemical element that characterizes hemoglobin is ______________

Understanding Concepts

I. True/False

For each question, write *T* for true or *F* for false in the blank to the left of each number. If a statement is false, correct it by replacing the underlined term and write the correct statement in the blank below the question.

_______ 1. Eosinophils and basophils are <u>granular</u> leukocytes.

____

_______ 2. Blood cells from a person with type B blood will agglutinate with <u>type A</u> antiserum.

____

_______ 3. The form of anemia resulting from insufficient red cell production in the bone marrow is <u>myelogenous</u> anemia.

____

_______ 4. Erythropoietin, a hormone that stimulates production of red blood cells, is produced by the kidneys.

____

_______ 5. Type AB blood contains antibodies to both A and B antigens.

____

_______ 6. Monocytes are immature neutrophils.

____

_______ 7. Substances that induce blood clotting are called procoagulants.

____

_______ 8. The watery fluid that remains after a blood clot has been removed from the blood is plasma.

____

II. Practical Applications

Study each discussion. Then write the appropriate word or phrase in the space provided.

➤ Group A

1. A young girl named AL fell off her bike, sustaining a deep gash to her leg that bled copiously from a severed vessel. In describing this type of bleeding, the doctor in the emergency clinic used the word _______________.
2. While the physician attended to the wound, the technician drew blood for typing and other studies. AL's blood did not agglutinate with either anti-A or anti-B serum. Her blood was classified as group _______________.
3. Young AL required a blood transfusion. The only type of blood she could receive is _______________.
4. Further testing of AL's blood revealed that it lacked the Rh factor. She was, therefore, said to be _______________.
5. If AL were to be given a transfusion of Rh-positive blood, she might become sensitized to the Rh protein. In that event, her blood would produce counteracting substances called _______________.
6. The technician examined AL's blood under the microscope, and noticed some red blood cells were shaped like crescent moons. AL may have the disorder called _______________.

➤ Group B

1. Mr. KK, age 72, had an injury riding his motorized scooter down a steep hill. He suffered only minor scrapes, but he came to the hospital because his scrapes did not stop bleeding. The process by which blood loss is prevented or minimized is called _______________.
2. The physician discovered that Mr. KK was the great-great-great grandson of Queen Victoria. Like many of her offspring, Mr. KK suffers from a deficiency of a clotting factor. This hereditary disorder is called _______________.

3. Mr. KK must be treated with a rich source of clotting factors. The doctor gets a bag of frozen plasma, which has a white powdery substance at the bottom of the bag. The substance is called ________________.
4. Mr. KK also mentions that he has been very tired and pale lately. His cook recently retired, and he has been surviving on dill pickles and crackers. The physician suspects a deficiency in an element required to synthesize hemoglobin. This element is called ________________.
5. The physician orders a test to determine the proportion of red blood cells in his blood. The result of this test is called the ________________.
6. The test confirms that Mr. KK does not have enough red blood cells, due to a dietary deficiency. This disorder is called ________________.

➤ Group C

1. Ms. J, an elite cyclist, has come to the hospital complaining of a pounding headache. The physician, Dr. L., takes a sample of her blood, and notices that it is very thick. He decides to count her red blood cells, but the automatic cell counter is not functioning. Dr. L calls for a technician to perform a visual count using the microscope and a special slide called a(n) ________________.
2. The technician comes back with a result of 7.5 million cells per μL. This count is abnormally high, suggesting a diagnosis of ________________.
3. Dr. L is immediately suspicious. He asks Ms. J if she consumes any performance-enhancing drugs. Ms. J admits that she has been taking a hormone to increase blood cell synthesis. This hormone is called ________________.
4. Dr. L tells Ms. J that she is at risk for blood clot formation. He tells her to stop taking the hormone and to take aspirin for a few days to inhibit blood clotting. Drugs that inhibit clotting are called ________________.

III. Short Essays

1. Compare and contrast *leukopenia* and *leukocytosis*. Name one similarity and one difference.

____

____

____

2. What kind of information can be obtained from blood chemistry tests?

____

____

____

3. Briefly describe the final events in blood clot formation, naming the substances involved in each step.

4. Name one reason to transfuse an individual with:
   a. whole blood

   b. platelets

   c. plasma

   d. plasma protein

Conceptual Thinking

1. A dehydrated individual will have an elevated hematocrit. Explain why.

2. A man named JA has a history of frequent fevers. His skin is pale and his heart rate rapid. The doctor suspects that he has cancer. His white blood cell count was revealed to be 25,000 per μL of blood.
   a. Is this white blood cell count normal? What is the normal range?

   b. Based on his white blood cell count, what is a probable diagnosis of JA's condition?

   c. JA also has a hematocrit of 30%. Is this value normal? What is the normal range?

   d. What disorder does his hematocrit value suggest?

e. Consider the different functions of blood. Discuss some functions that might be impaired in JA's case.

____

____

____

3. Mr. R needs a blood transfusion. He has type AB blood. The doctor is considering a transfusion using A blood.
   a. Which antigens are present on his blood cells?

   ____

   b. Which antibodies will be present in his blood?

   ____

   c. Which antigens will be present on donor blood cells?

   ____

   d. Is this blood transfusion safe? Why or why not?

   ____

Expanding Your Horizons

The amount of oxygen carried by the blood is an important determinant of athletic prowess. Some elite athletes are always looking for ways to increase the oxygen in their blood. If you follow competitive cycling, you are probably familiar with the hormone erythropoietin (EPO). EPO came to the public's attention during the 1998 Tour de France, when it was found that EPO use is common among elite athletes. Learn about this and other methods that athletes have used to legally or illegally gain an advantage over their competitors. EPO is discussed in an article in *Scientific American*, and you can do a website search for "EPO sport." Information on other methods can be found by doing a website search for "blood doping."

Resources

1. Zorpette G. The athlete's body: the chemical games. *Scientific American* Presents: Building the Elite Athlete 2000;11:16–23.

CHAPTER 14

The Heart and Heart Disease

Overview

The ceaseless beat of the heart day and night throughout one's entire lifetime is such an obvious key to the presence of life that it is no surprise that this organ has been the subject of wonderment and poetry. When the heart stops pumping, life ceases. The cells must have oxygen, and it is the heart's pumping action that propels oxygenated blood to them.

In size, the heart is roughly the size of one's fist. It is located between the lungs, more than half to the left of the midline, with the **apex** (point) directed toward the left. Below is the **diaphragm**, the dome-shaped muscle that separates the thoracic cavity from the abdominopelvic cavity.

The heart consists of two sides separated by septa. The septa keep blood that is higher in oxygen entirely separate from blood that is lower in oxygen. The two sides pump in unison, the right side pumping blood to the lungs to be oxygenated, and the left side pumping blood to all other parts of the body.

Each side of the heart is divided into two parts or **chambers.** The upper chamber, or **atrium**, on each side is the receiving chamber for blood returning to the heart. The lower chamber, or **ventricle**, is the strong pumping chamber. Because the ventricles pump more forcefully, their walls are thicker than the walls of the atria. **Valves** between the chambers keep the blood flowing forward as the heart pumps. The muscle of the heart wall, the **myocardium**, has special features to enhance its pumping efficiency. The coronary circulation supplies blood directly to the myocardium.

The heartbeat originates within the heart at the **sinoatrial** (SA) **node**, often called the pacemaker. Electrical impulses from the pacemaker spread through special conducting fibers in the wall of the heart to induce contractions, first of the two atria and then of the two ventricles. After contraction, the heart relaxes and fills with blood. The relaxation phase is called **diastole**, and the contraction phase is called **systole**. Together, these two phases make up one **cardiac cycle**.

The heart rate is influenced by the nervous system and other circulating factors, such as hormones and drugs.

Heart diseases may be classified according to the cause or the area of the heart affected. Causes include congenital abnormalities, rheumatic fever, coronary artery disease, and heart failure.

Addressing the Learning Outcomes

1. Describe the three layers of the heart wall (see exercises for outcome 2).

2. Describe the structure of the pericardium and cite its functions.

EXERCISE 14-1: Layers of the Heart Wall and Pericardium (Text Fig. 14-4)

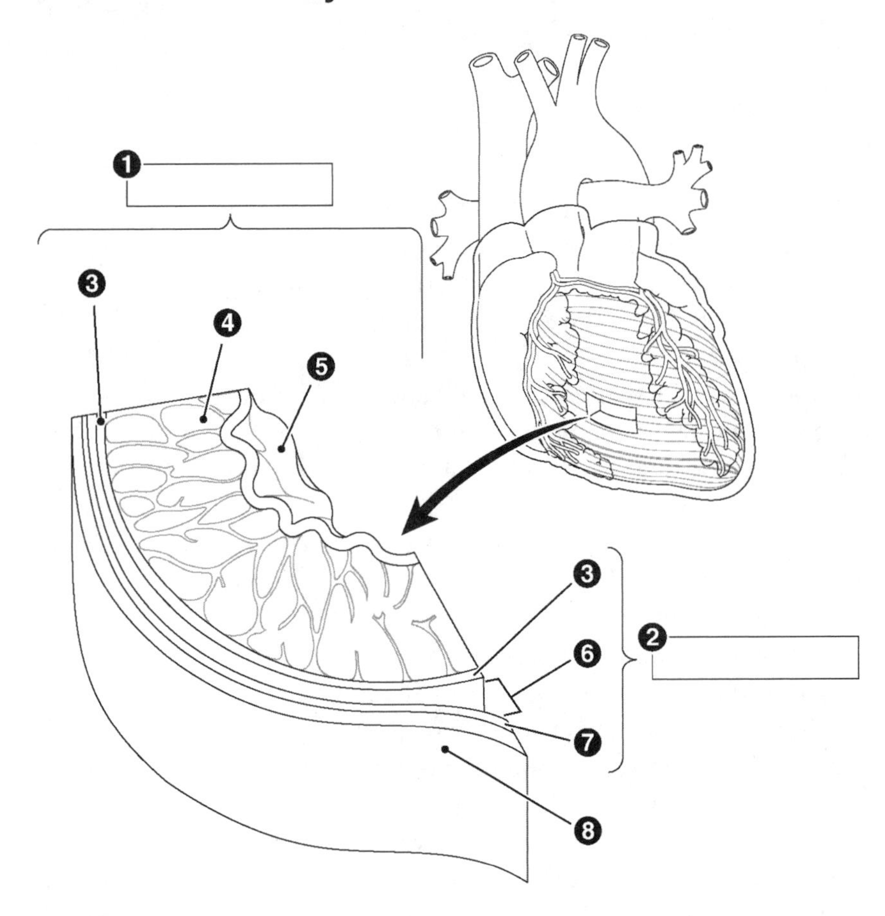

INSTRUCTIONS

1. Write the terms "heart wall" and "serous pericardium" in the corresponding boxes.
2. Write the names of the different structures on the numbered lines in different colors. Use black for structure 6, because it will not be colored.
3. Color the structures 3–8 on the diagram (except structure 6) with the corresponding colors.

3. ____________________ 6. ____________________
4. ____________________ 7. ____________________
5. ____________________ 8. ____________________

EXERCISE 14-2.

INSTRUCTIONS

Write the appropriate term in each blank.

base	apex	endocardium	fibrous pericardium
myocardium	serous pericardium	epicardium	

1. The pointed, inferior portion of the heart ________________
2. The membrane consisting of a visceral and a parietal layer ________________
3. A layer of epithelial cells in contact with blood within the heart ________________
4. The heart layer containing intercalated disks ________________
5. The outermost layer of the sac enclosing the heart ________________

3. Compare the functions of the right and left sides of the heart.

EXERCISE 14-3: The Heart Is a Double Pump (Text Fig. 14-4)

INSTRUCTIONS

1. Label the indicated parts.
2. Color the oxygenated blood red and the deoxygenated blood blue.
3. Use arrows to show the direction of blood flow.

1. ______________________
2. ______________________
3. ______________________
4. ______________________
5. ______________________
6. ______________________
7. ______________________
8. ______________________
9. ______________________
10. ______________________
11. ______________________
12. ______________________
13. ______________________
14. ______________________

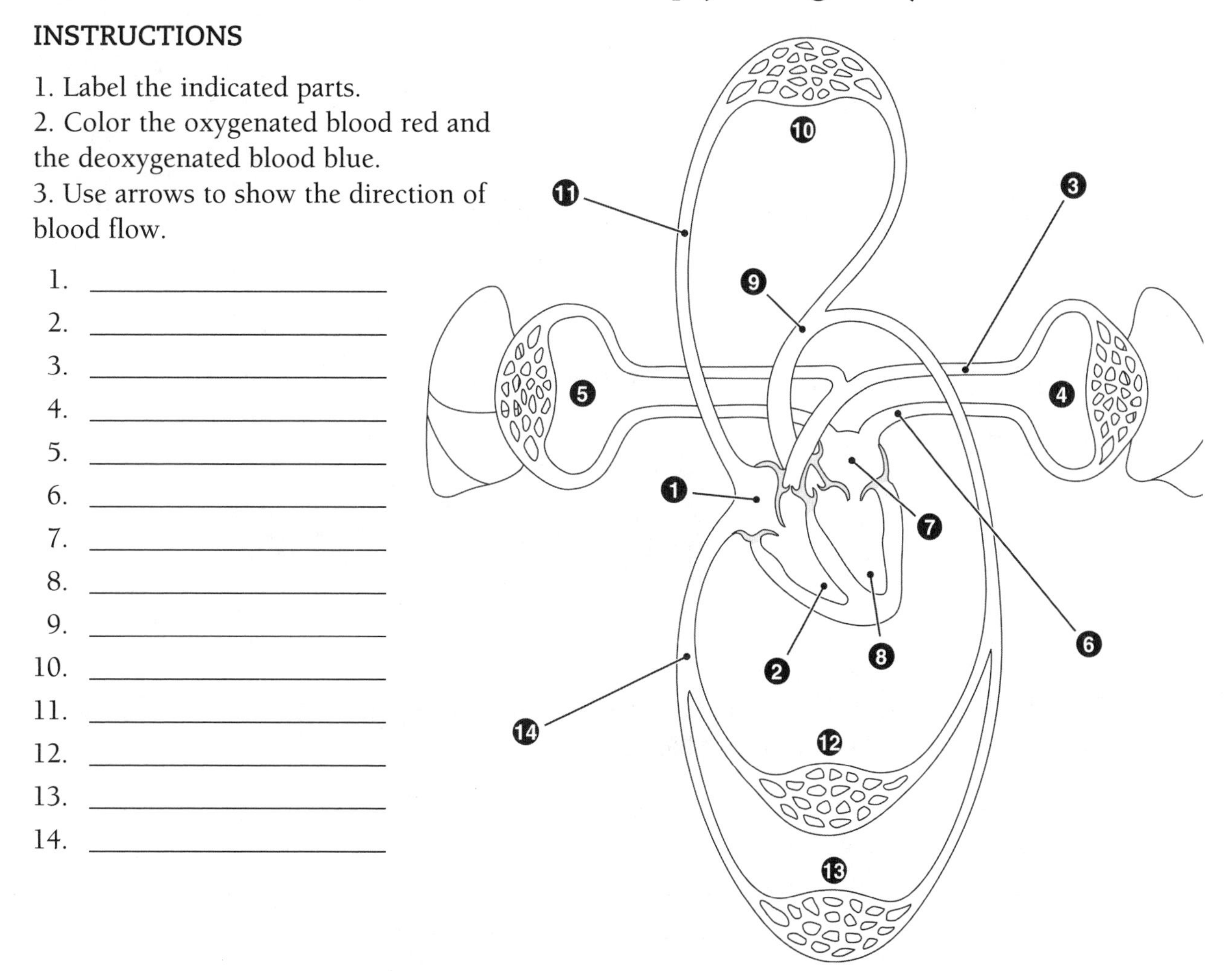

4. Name the four chambers of the heart and compare their functions (see exercises for Outcome 5).

5. Name the valves at the entrance and exit of each ventricle and cite the function of the valves.

EXERCISE 14-4: The Heart and Great Vessels (Text Fig. 14-5)

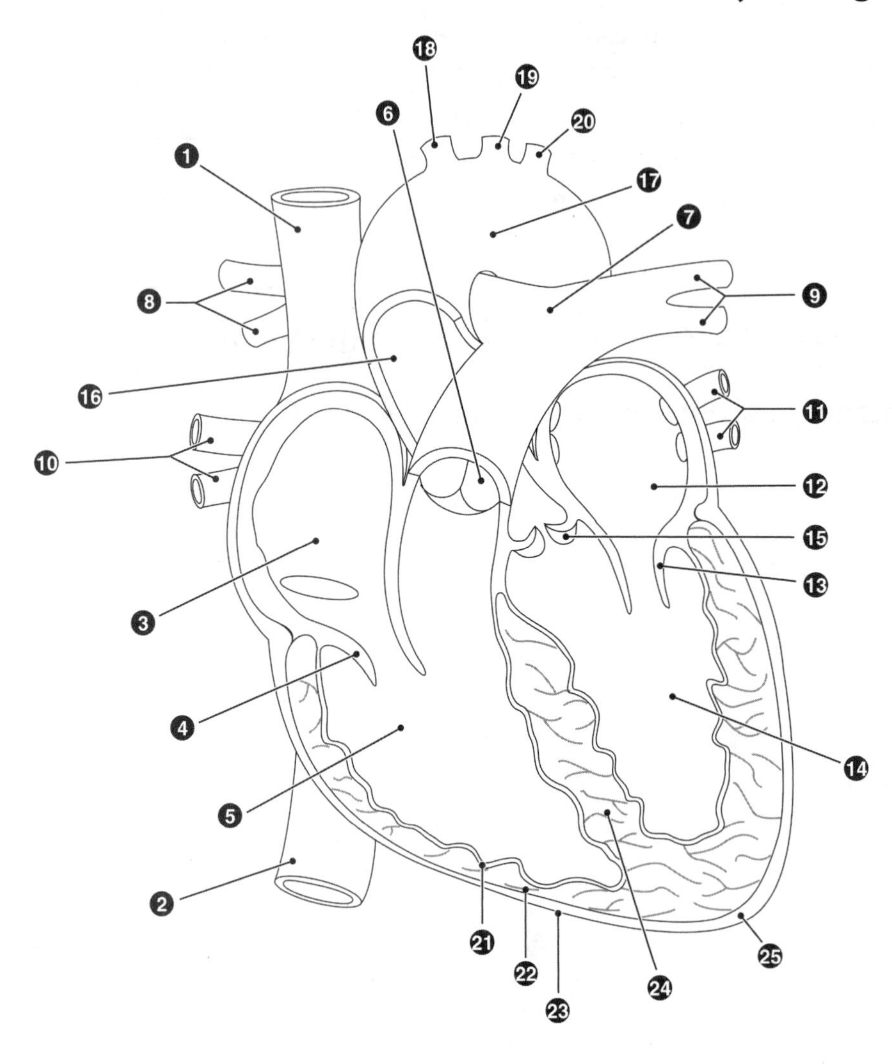

INSTRUCTIONS

1. Label the indicated parts.
2. Color the oxygenated blood red and the deoxygenated blood blue.
3. Use arrows to show the direction of blood flow.

1. ____________________
2. ____________________
3. ____________________
4. ____________________
5. ____________________
6. ____________________
7. ____________________
8. ____________________
9. ____________________
10. ____________________
11. ____________________
12. ____________________
13. ____________________
14. ____________________
15. ____________________
16. ____________________
17. ____________________
18. ____________________
19. ____________________
20. ____________________
21. ____________________
22. ____________________
23. ____________________
24. ____________________
25. ____________________

EXERCISE 14-5.

INSTRUCTIONS

Write the appropriate term in each blank.

ventricle	atrium	atrioventricular valve	pulmonary valve
aortic valve	interatrial septum	interventricular septum	

1. A lower chamber of the heart ____________
2. The valve that prevents blood from returning to the right ventricle ____________
3. An upper chamber of the heart ____________
4. The valve that prevents blood from returning to the left ventricle ____________
5. The partition that separates the two upper chambers of the heart ____________
6. One of two valves dividing the upper and lower chambers ____________

6. Briefly describe blood circulation through the myocardium.

EXERCISE 14-6.

INSTRUCTIONS

Fill in the blanks.

1. The blood vessels that supply the heart constitute the ____________ circulation.
2. The main arteries supplying blood to the heart muscle branch off from the aorta just superior to the ____________ valve.
3. Blood from capillaries in the heart muscle eventually enters a dilated vein called the ____________.
4. Blood from the heart muscle eventually drains into the ____________ atrium.

7. Briefly describe the cardiac cycle.

EXERCISE 14-7.

Do each of the following events occur in diastole (D), atrial systole (A), or ventricular systole (V)?

1. The ventricles are contracting. _____
2. The atria are contracting. _____
3. Blood is entering the aorta. _____
4. Neither the ventricles nor the atria are contracting. _____
5. The atrioventricular valves are closed. _____

EXERCISE 14-8.

INSTRUCTIONS

Write the appropriate term in each blank.

diastole atrial systole ventricular systole cardiac output stroke volume

1. The amount of blood ejected from a ventricle with each beat ____________
2. The stage of the cardiac cycle that directly follows the resting period ____________
3. The stage of the cardiac cycle which precedes the resting period ____________
4. The volume of blood pumped by each ventricle in 1 minute ____________
5. The resting period of the cardiac cycle ____________

8. Name and locate the components of the heart's conduction system.

EXERCISE 14-9: The Conduction System of the Heart (Text Fig. 14-11)

INSTRUCTIONS

1. Label the indicated parts.
2. Highlight the structures that conduct electrical impulses in yellow. Draw arrows to indicate the direction of impulse conduction.

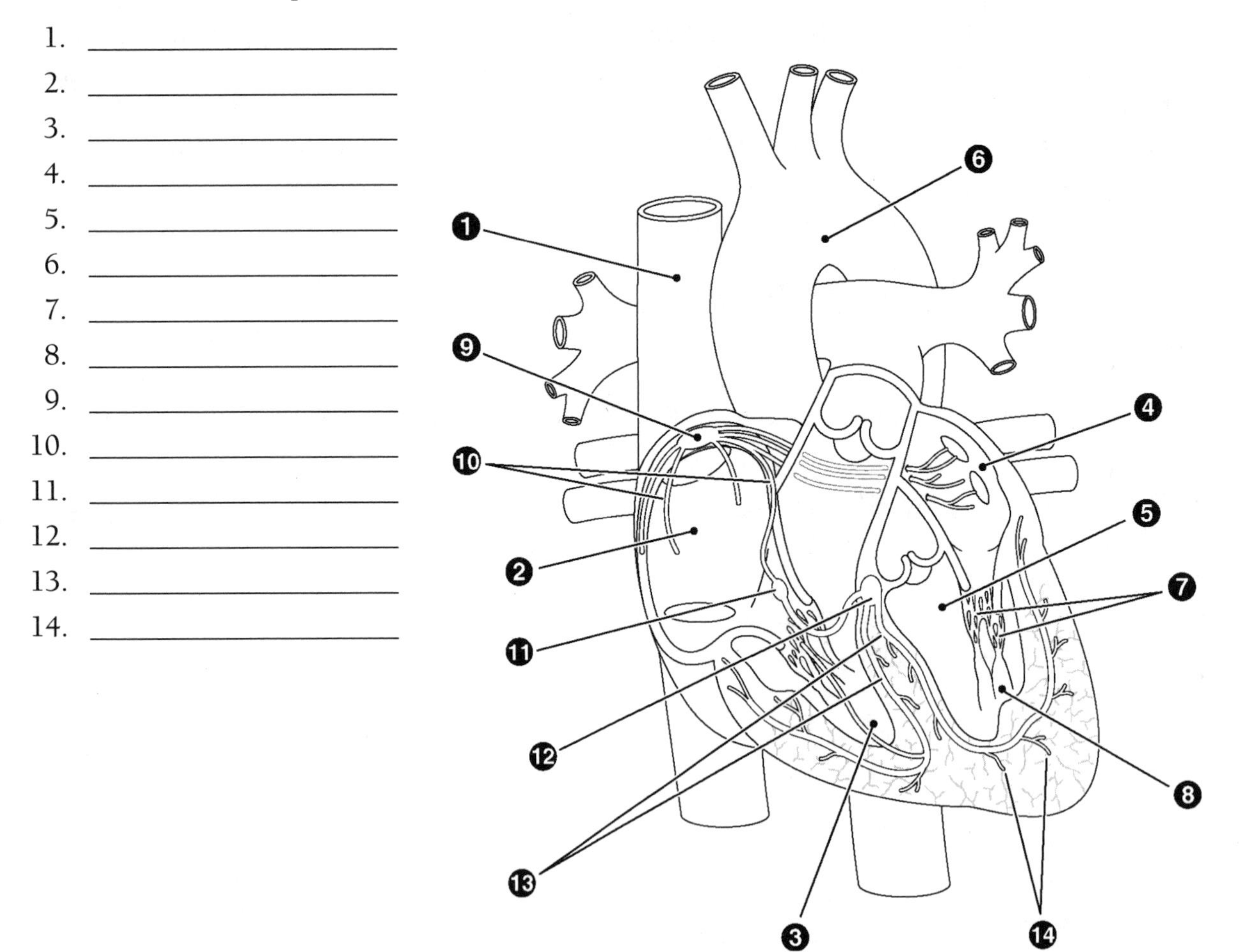

EXERCISE 14-10.

INSTRUCTIONS

Write the appropriate term in each blank.

bundle of His	Purkinje fibers	sinus rhythm
sinoatrial node	atrioventricular node	

1. The group of conduction fibers found in the ventricle walls ______________
2. The mass of conduction tissue located in the septum at the bottom of the right atrium ______________
3. A normal heart beat, originating from the normal heart pacemaker ______________
4. The group of conduction fibers carrying impulses from the AV node ______________
5. The name of the normal heart pacemaker, located in the upper wall of the right atrium ______________

9. Explain the effects of the autonomic nervous system on the heart rate.

EXERCISE 14-11.

Do each of the following characteristics refer to the parasympathetic nervous system (P) or the sympathetic nervous system (S)?

1. Regulates heart activity via the vagus nerve ______________
2. Uses ganglia located close to the spinal cord ______________
3. Decreases the heart rate ______________
4. Increases the force of each contraction ______________
5. Increases the heart rate ______________

10. List and define several terms that describe variations in heart rates.

EXERCISE 14-12.

INSTRUCTIONS

Write the appropriate term in each blank.

bradycardia   tachycardia   sinus arrhythmia   extrasystole

1. A beat that comes before the normal beat ______________
2. A normal variation in heart rate caused by changes in breathing rate ______________
3. A heart rate of less than 60 beats per minute ______________
4. A heart rate of greater than 100 beats per minute ______________

11. Explain what produces the two main heart sounds.

EXERCISE 14-13.

INSTRUCTIONS

Fill in the blanks.

1. The "lub" sound is caused by the closure of the ______________________ valves.
2. The "dub" sound is caused by the closure of the ______________________ valves.

12. Describe several common types of heart disease.

EXERCISE 14-14.

INSTRUCTIONS

Write the appropriate term in each blank below.

functional murmur	fibrillation	heart failure	plaque
organic murmur	flutter	infarct	ischemia

1. Rapid and uncoordinated heart muscle contractions ______________
2. Condition that results from a lack of blood supply to the tissues, as from narrowing of an artery ______________
3. A hard deposit of fatty material in a blood vessel ______________
4. Term for very rapid, coordinated heart contractions of up to 300 beats per minute ______________
5. A sound that may result from a heart defect, such as the abnormal closing of a heart valve ______________
6. An area of tissue damaged from a lack of blood supply ______________
7. A deterioration of the heart tissues, often resulting in fluid accumulation in the lungs ______________

EXERCISE 14-15.

INSTRUCTIONS

Write the appropriate term in each blank.

occlusion	atrial septal defect	patent ductus arteriosus	atherosclerosis	thrombus
endocarditis	myocarditis	pericarditis	angina pectoris	

1. Inflammation of the heart lining, often affecting the valves ______________
2. A blood clot formed within a vessel ______________
3. Persistence of an open foramen ovale after birth ______________
4. Inflammation of the heart muscle ______________
5. Complete closure, as of an artery ______________
6. Inflammation of the serous membrane on the heart surface ______________
7. Chest discomfort resulting from inadequate coronary blood flow ______________
8. A thickening and hardening of the vessels ______________

13. List five actions that can be taken to minimize the risk of heart disease.

EXERCISE 14-16.

INSTRUCTIONS

For each statement, write whether it is true (T) or false (F).

The risk of heart disease is increased if you:

1. smoke _____
2. have low blood pressure _____
3. have diabetes _____
4. exercise regularly ______
5. are a 30-year-old man (compared with a 30-year-old woman) _____
6. tend to deposit body fat on the thighs (rather than around the abdomen) _____

14. Briefly describe four methods for studying the heart.

EXERCISE 14-17.

INSTRUCTIONS

Write the appropriate term in each blank.

stethoscope	echocardiography	creatine kinase	catheterization
fluoroscope	electrocardiography	thrombus	

1. Technique that uses ultrasound to study the heart as it beats _______________
2. Levels of this substance are elevated when cardiac muscle tissue is damaged _______________
3. Technique that measures the electrical activity of the heart _______________
4. Instrument used to detect heart murmurs by their sound _______________
5. A procedure for measuring pressures within the heart chambers _______________

15. Describe several approaches to the treatment of heart disease.

EXERCISE 14-18.

INSTRUCTIONS

Write the appropriate term in each blank.

digitalis	anticoagulant	angioplasty
beta-blocker	nitroglycerin	artificial pacemaker

1. Aspirin is an example of this type of drug _______________
2. A device implanted under the skin that supplies impulses to regulate the heartbeat _______________
3. A drug that dilates the coronary vessels and is used to relieve angina pectoris _______________
4. A technique that uses a balloon to open a blocked blood vessel _______________
5. A drug derived from the foxglove plant that slows and strengthens heart contractions _______________

16. Show how word parts are used to build words related to the heart.

EXERCISE 14-19.

INSTRUCTIONS

Complete the following table by writing the correct word part or meaning in the space provided. Write a word that contains each word part in the Example column.

Word Part	Meaning	Example
1. ________	chest	________
2. sin/o	________	________
3. ________	vessel	________
4. cardi/o	________	________
5. scler/o	________	________
6. ________	lung	________
7. ________	slow	________
8. ________	rapid	________
9. isch-	________	________
10. cyan/o	________	________

Making the Connections

The following flow chart deals with the passage of blood through the heart. Beginning with the right atrium, outline the structures a blood cell would pass through in the correct order by filling in the boxes. Use the following terms: left atrium, left ventricle, right ventricle, right AV valve, left AV valve, pulmonary semilunar valve, aortic semilunar valve, superior/inferior vena cava, right lung, left lung, aorta, left pulmonary artery, right pulmonary artery, left pulmonary veins, right pulmonary veins, and body. You can write the names of structures that encounter blood high in oxygen in red and the names of structures that encounter blood low in oxygen in blue.

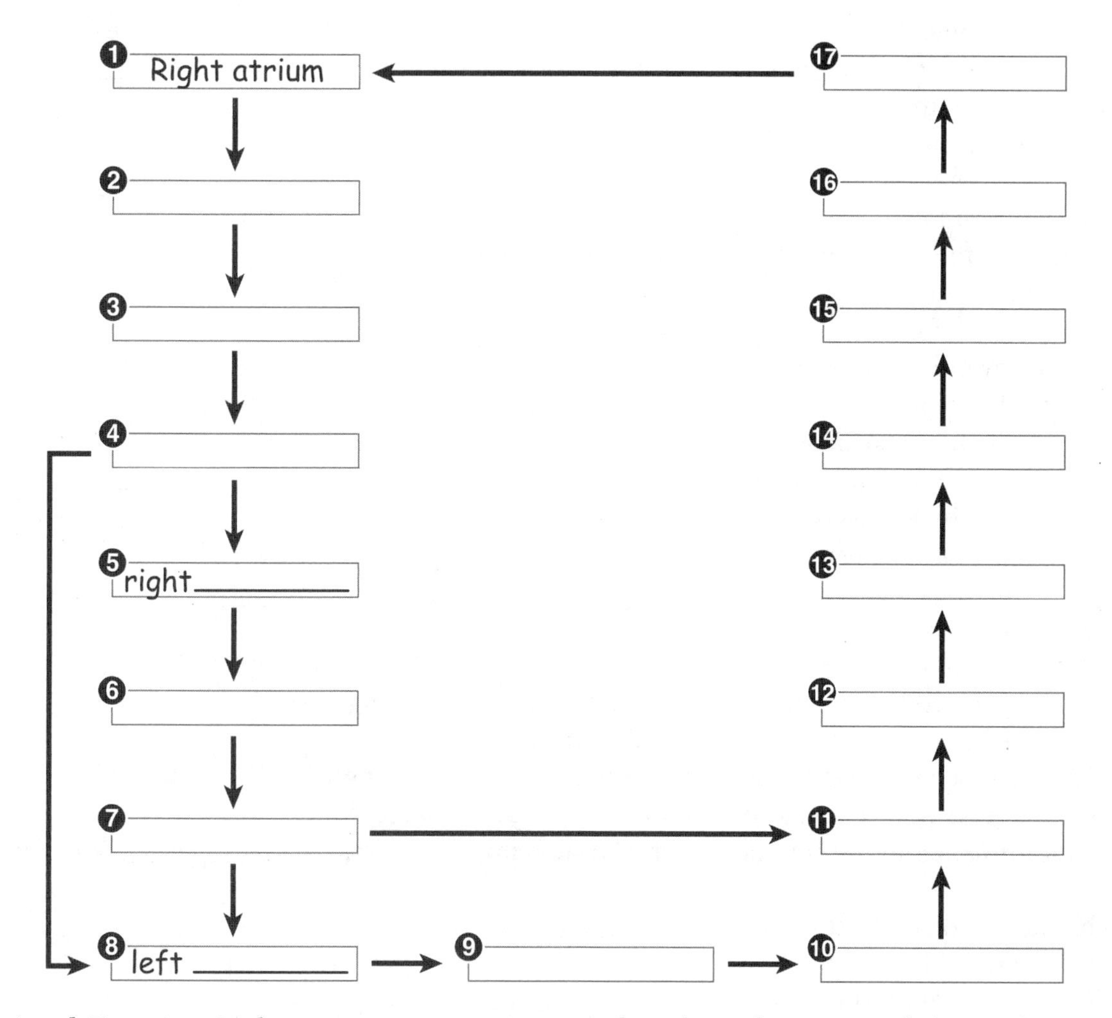

Optional Exercise: Make your own concept map, based on the events of the cardiac cycle. Choose your own terms to incorporate into your map, or use the following list: AV valves open, AV valves closed, blood flow from atria to ventricles, blood flow from ventricles to arteries, atrial diastole, atrial systole, ventricular diastole, and ventricular systole. There will be many links between the different terms.

Testing Your Knowledge

Building Understanding

I. Multiple Choice

Select the best answer and write the letter of your choice in the blank.

1. An average cardiac cycle lasts about — 1. _______
   a. 8 seconds
   b. 5 seconds
   c. 0.8 second
   d. 30 seconds
2. The volume of blood pumped by each ventricle in 1 minute is the — 2. _______
   a. stroke volume
   b. cardiac output
   c. heart rate
   d. ejection rate
3. Which of the following is NOT a part of the conduction system of the heart? — 3. _______
   a. bundle of His
   b. atrioventricular valve
   c. Purkinje fibers
   d. atrioventricular node
4. The function of a thrombolytic drug is to — 4. _______
   a. reduce hypertension
   b. regulate the heartbeat
   c. dilate the blood vessels
   d. dissolve blood clots
5. Activation of the parasympathetic nervous system would — 5. _______
   a. increase heart rate but not myocardial contraction strength
   b. increase heart rate and myocardial contraction strength
   c. decrease heart rate but not myocardial contraction strength
   d. decrease heart rate and myocardial contraction strength
6. Clot formation in the coronary arteries results in — 6. _______
   a. stroke
   b. myocardial infarction
   c. heart failure
   d. congenital heart disease
7. The vein that carries blood from the coronary circulation back into the right atrium is the — 7. _______
   a. right coronary artery
   b. interatrial septum
   c. pulmonary vein
   d. coronary sinus

8. The tetralogy of Fallot is 8. _______
   a. four specific defects associated with congenital heart disease
   b. the four cardiac valves
   c. four signs of an impending heart attack
   d. the four types of heart disease

II. Completion Exercise

Write the word or phrase that correctly completes each sentence.

1. The autonomic nerve that slows the heart beat is the ____________________.
2. The fibrous sac that surrounds the heart is the ____________________.
3. An abnormality in the rhythm of the hearbeat is a(n) ____________________.
4. The medical term for high blood pressure is ____________________.
5. One complete cycle of heart contraction and relaxation is called the ____________________.
6. The fibrous threads connecting the AV valves to muscles in the heart wall are called the ____________________.
7. The right atrioventricular valve is also known as the ____________________.
8. Normal heart sounds heard while the heart is working are called ____________________.
9. The small blood vessel connecting the pulmonary artery with the aorta that is found only in the fetus is the ____________________.

Understanding Concepts

I. True/False

For each question, write *T* for true or *F* for false in the blank to the left of each number. If a statement is false, correct it by replacing the underlined term and write the correct statement in the blank below the question.

_______ 1. Discomfort felt in the region of the heart as a result of coronary artery disease is called congestive heart failure.

____

_______ 2. The circulatory system of the fetus has certain adaptations for the purpose of bypassing the kidneys.

____

_______ 3. A heart rate of 150 beats/minute is described as tachycardia.

____

_______ 4. Blood in the heart chambers comes into contact with the epicardium.

____

_______ 5. The aorta is part of the pulmonary circuit.

____

_______ 6. Nerve impulses travel down the internodal pathways from the AV node

____

_______ 7. A heart rhythm originating at the SA node is termed a sinus rhythm.

____

_______ 8. Heart disease caused by antibodies to streptococci is called atherosclerosis.

____

II. Practical Applications

Study each discussion. Then write the appropriate word or phrase in the space provided.

➤ Group A

1. Ms. J, age 82, was participating in a lawn bowling tournament when she suddenly collapsed with chest pain. The paramedics were preparing her for transport to the hospital when they noted a sudden onset of pale skin and unconsciousness. The heart monitor showed a rapid, uncoordinated activity of the ventricles called _______________.
2. The paramedics administered an electric shock using an automated defibrillator with the aim of restoring the normal heart rhythm, which is called a(n) _______________.
3. In the hospital emergency room, Ms. J was given intravenous medications to dissolve any blood clots in her coronary circulation. These drugs are called _______________.
4. The drugs were not administered in time to prevent damage to the middle layer of the heart wall, which is known as the _______________.
5. The electrical activity of Ms. J's heart was analyzed using a(n) _______________.
6. The analysis showed that the conduction system was damaged. She was scheduled to have a device inserted to supply impulses needed to stimulate heart contractions. This device is called a(n) _______________.
7. Ms. J's long-term care included a daily dose of the anticoagulant drug acetylsalicylic acid, commonly known as _______________.

➤ Group B

1. Baby L has just been born. To her parents' dismay, her skin and mucous membranes are tinged with blue. The scientific term for this coloration is _______________.
2. The obstetrician listens to her heartbeat using an instrument called a(n) _______________.
3. The doctor notices an abnormality in the second heart sound (or "dup"), which is largely caused by the closure of the two _______________.
4. This abnormal sound probably reflects a structural problem with Baby L's heart and is thus termed a(n) _______________.
5. The baby is sent for a test that uses ultrasound waves to examine her heart structure. A hole is observed between the two lower chambers of her heart. This disorder is known as _______________.
6. In addition to the two defects already described, it was also observed that Baby L's aorta is not positioned properly and that her left ventricular wall was abnormally muscular. This common combination of four defects is called the _______________.
7. Baby L was diagnosed with a form of heart disease present at birth called _______________.

III. Short Essays

1. Although the heartbeat originates within the heart itself, it is influenced by factors in the internal environment. Describe some of these factors that can affect the heart.

___

___

___

2. What is the difference between thrombolytic drugs and anticoagulants? Describe their effects and their uses.

___

___

___

Conceptual Thinking

1. Atropine is a drug that inhibits activity of the parasympathetic nervous system. Discuss the effects of atropine on the heart. How does the parasympathetic nervous system affect the heart? Which aspects of heart function will be affected and which will be unaffected?

___

___

___

2. Mr. J is undertaking a gentle exercise program, primarily involving walking, to lose weight. His heart rate is 100 beats/minute and his stroke volume is 75 ml. What is his cardiac output, and what does cardiac output mean?

___

___

___

Expanding Your Horizons

How low can you go? You may have heard of the exploits of free divers, who dive to tremendous depths without the aid of SCUBA equipment. The world record for assisted free diving is held by Pipin Ferreras, who dove to 170 m (558 feet) with the aid of a sled. Free diving is not without its dangers. Pipin's wife (Audrey) died during a world record attempt. Free diving is facilitated by the dive reflex, which allows mammals to hold their breath for long periods of time underwater. Immersing one's face in cold water induces bradycardia and diverts blood away from the periphery. How would these modifications increase the ability to free dive? You can learn more about the underlying mechanisms, rationale, and dangers of the dive reflex by performing a website search for "dive reflex." Information is also available in the article listed below.

Resources

1. Hurwitz BE, Furedy JJ. The human dive reflex: an experimental, topographical, and physiological analysis. *Physiol Behav* 1986; 36:287–294.

CHAPTER 15

Blood Vessels and Blood Circulation

Overview

The blood vessels are classified as arteries, veins, or capillaries according to their function. **Arteries** carry blood away from the heart, **veins** return blood to the heart, and **capillaries** are the site of gas, nutrient, and waste exchange between the blood and tissues. Small arteries are called **arterioles**, and small veins are called **venules**. The walls of the arteries are thicker and more elastic than the walls of the veins in order to withstand higher pressure. All vessels are lined with a single layer of simple epithelium called **endothelium**. The smallest vessels, the capillaries, are made only of this single layer of cells. The exchange of fluid between the blood and interstitial spaces is influenced by **blood pressure**, which pushes fluid out of the capillary, and **osmotic pressure**, which draws fluid back in. Blood pressure is determined by the **cardiac output** and the **total peripheral resistance**.

The vessels carry blood through two circuits. The **pulmonary circuit** transports blood between the heart and the lungs for gas exchange. The **systemic circuit** distributes blood high in oxygen to all other body tissues and returns deoxygenated blood to the heart.

The walls of the vessels, especially the small arteries, contain smooth muscle that is under the control of the involuntary nervous system. The diameters of the vessels can be increased (**vasodilation**) or decreased (**vasoconstriction**) by the nervous system in order to alter blood pressure and blood distribution.

Several forces work together to drive blood back to the heart in the venous system. Contraction of skeletal muscles compresses the veins and pushes blood forward, **valves** in the veins keep blood from flowing backward, and changes in pressure that occur during breathing help to drive blood back to the heart. The **pulse rate** and **blood pressure** can provide information about an individual's cardiovascular health. Disorders of the circulatory system include hypertension, degenerative changes or obstructions that diminish blood flow in the vessels, and hemorrhage.

Addressing the Learning Outcomes

1. Differentiate among the five types of blood vessels with regard to structure and function.

EXERCISE 15-1: Sections of Small Blood Vessels (Text Fig. 15-2)

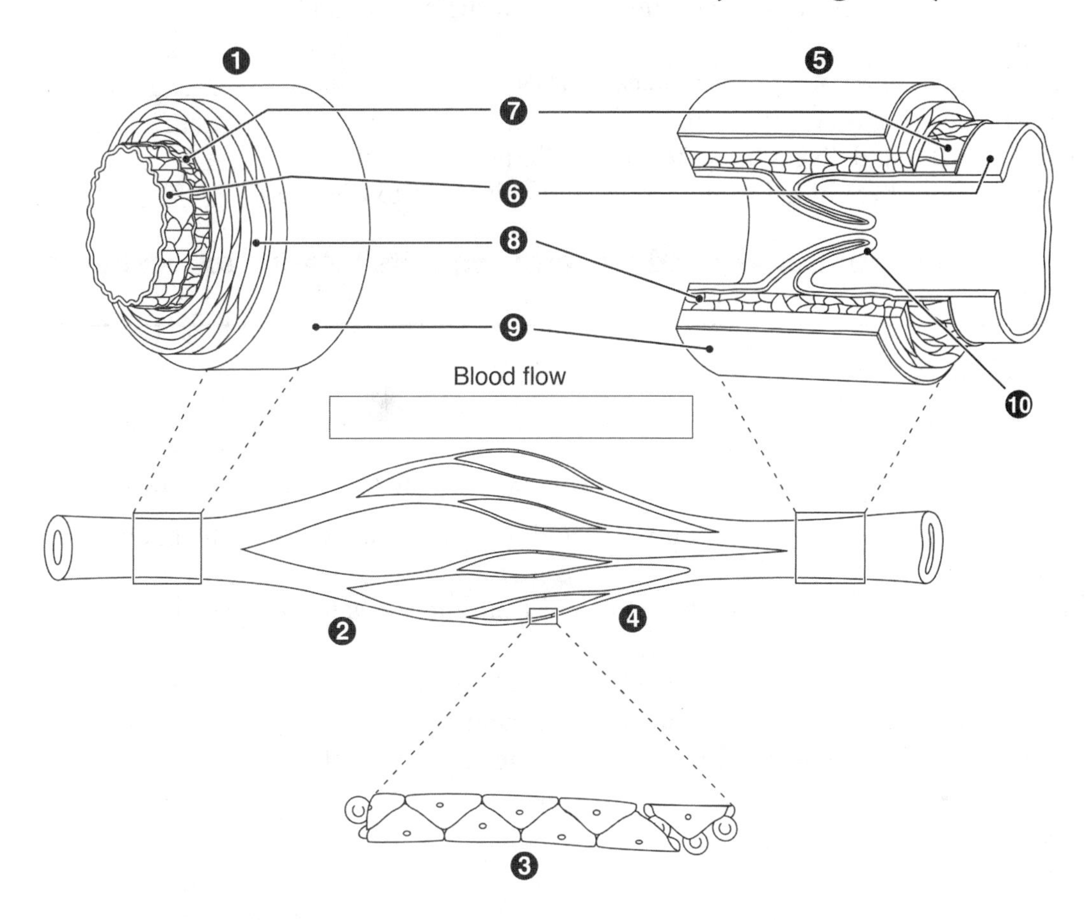

INSTRUCTIONS

1. Write the names of the different vessel types on lines 1 to 5.
2. Write the names of the different vascular layers on the appropriate numbered lines (6 to 10) in different colors. Use black for structure 10, because it will not be colored.
3. Color the structures on the diagram with the corresponding color (except for structures 1, 5, and 10 on the uppermost drawings).
4. Draw an arrow in the box to indicate the direction of blood flow.

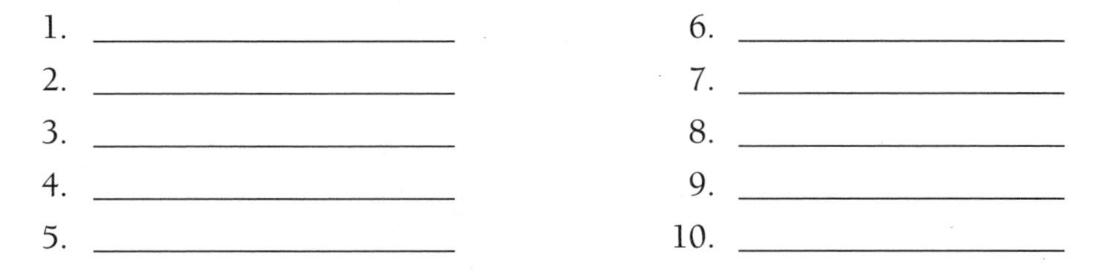

1. ______________________
2. ______________________
3. ______________________
4. ______________________
5. ______________________
6. ______________________
7. ______________________
8. ______________________
9. ______________________
10. ______________________

EXERCISE 15-2.

INSTRUCTIONS

Write the appropriate term in each blank.

artery capillary vein venule arteriole

1. A small vessel through which exchanges between the blood and the cells take place ________________
2. A vessel that receives blood from the capillaries ________________
3. A vessel that branches off the aorta ________________
4. A small vessel that delivers blood to the capillaries ________________
5. A vessel that receives blood from venules and delivers it to the heart ________________

2. Compare the pulmonary and systemic circuits relative to location and function.

EXERCISE 15-3.

INSTRUCTIONS

State whether each statement refers to the pulmonary circuit (P) or the systemic circuit (S).

1. The group of vessels that carries nutrients and oxygen to all tissues of the body except the lungs _____
2. The group of vessels that carries blood to and from the lungs for gas exchange _____
3. The group of vessels that includes the aorta _____
4. The group of vessels that delivers blood to the left atrium _____
5. The group of vessels that delivers blood to the right atrium ______
6. The group of vessels that receives blood from the right ventricle _____

3. Name the four sections of the aorta and list the main branches of each section.

EXERCISE 15-4: Aorta and Its Branches (Text Fig. 15-4)

INSTRUCTIONS

1. Write the names of the aortic sections on lines 1 to 4 in different colors, and color the appropriate structures on the diagram. Although the aorta is continuous, lines have been added to the diagram to indicate the boundaries of the different sections.
2. Write the names of the aortic branches on the appropriate lines 5 to 22 in different colors, and color the corresponding artery on the diagram. Use the same color for structures 8 and 9 and for structures 7 and 10. Use black for structure 15, because it will not be colored.

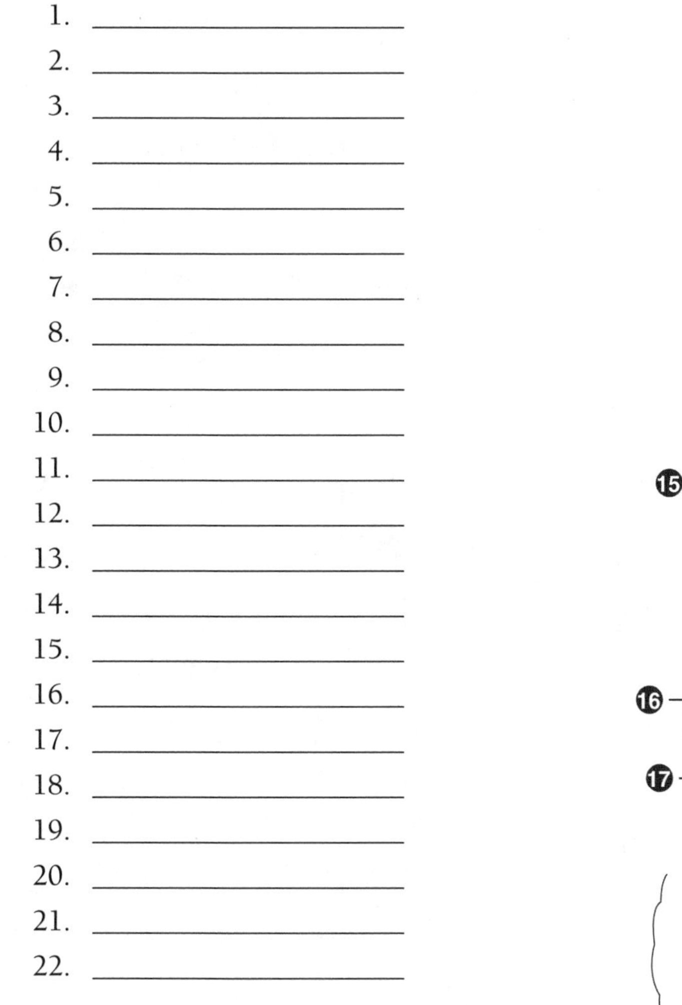

1. ____________
2. ____________
3. ____________
4. ____________
5. ____________
6. ____________
7. ____________
8. ____________
9. ____________
10. ____________
11. ____________
12. ____________
13. ____________
14. ____________
15. ____________
16. ____________
17. ____________
18. ____________
19. ____________
20. ____________
21. ____________
22. ____________

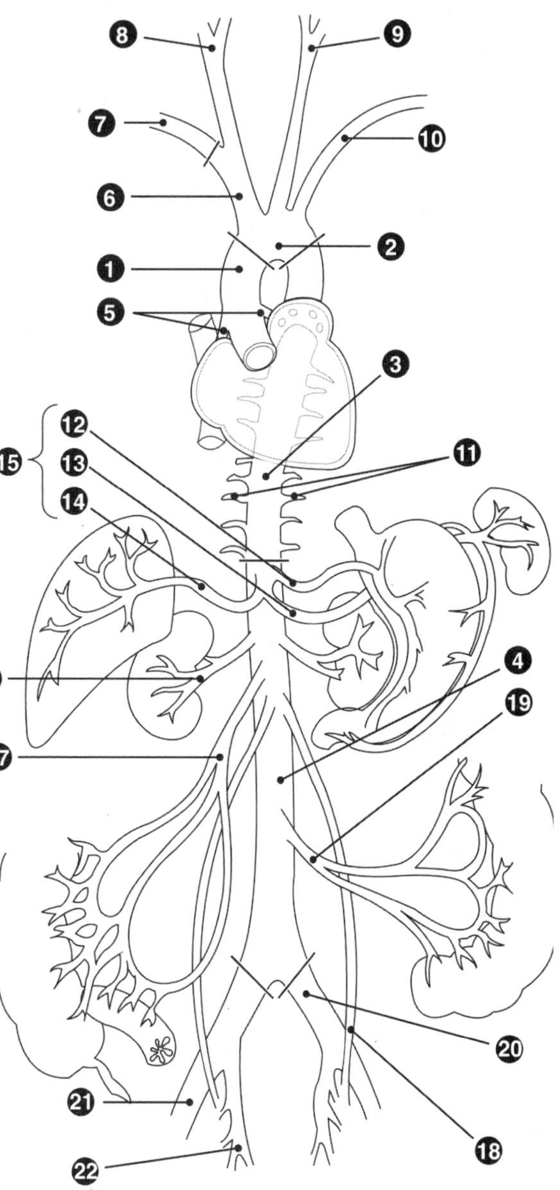

EXERCISE 15-5: Principal Systemic Arteries (Text Fig. 15-5)

INSTRUCTIONS

1. Write the names of the aortic segments and the principal systemic arteries on the appropriate lines in different, preferably darker, colors. Felt tip pens would work well for this exercise. Some structures were also labeled in Exercise 15-4 (for instance, the thoracic aorta); you may want to use the same color scheme for both exercises. Use black for structure 15, because it will not be colored.
2. Outline the arteries on the diagram with the corresponding color. If appropriate, color the left and right versions of each artery (for instance, you can color the right and left anterior tibial arteries). Some arteries change names along their length (for instance, the brachiocephalic artery). The boundary between different artery names is indicated by a perpendicular line.

1. ____________________
2. ____________________
3. ____________________
4. ____________________
5. ____________________
6. ____________________
7. ____________________
8. ____________________
9. ____________________
10. ____________________
11. ____________________
12. ____________________
13. ____________________
14. ____________________
15. ____________________
16. ____________________
17. ____________________
18. ____________________
19. ____________________
20. ____________________
21. ____________________
22. ____________________
23. ____________________
24. ____________________
25. ____________________
26. ____________________
27. ____________________
28. ____________________
29. ____________________

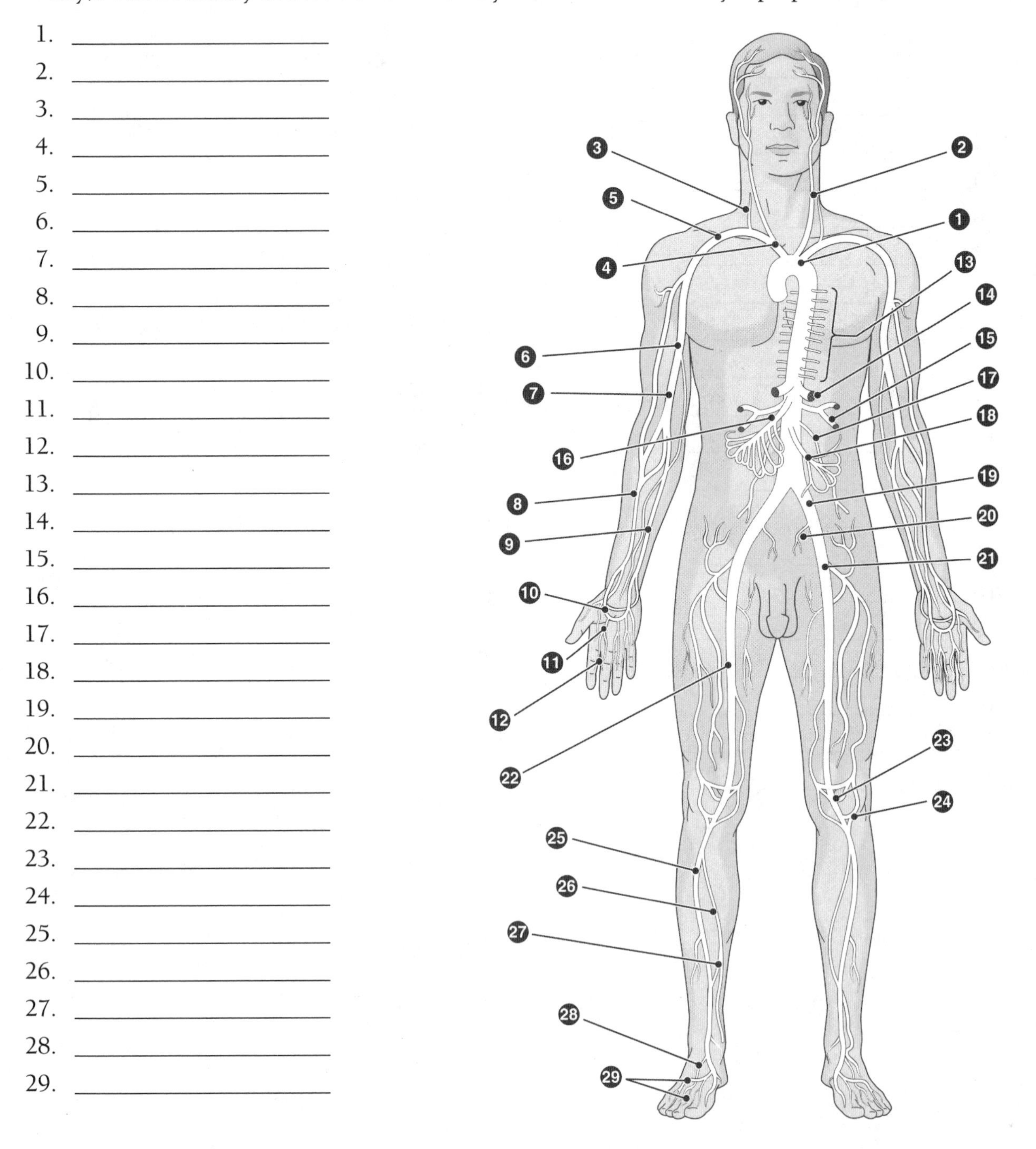

EXERCISE 15-6.

INSTRUCTIONS

Write the appropriate term in each blank.

coronary arteries, carotid arteries, lumbar arteries, common iliac arteries
phrenic arteries, intercostal arteries, ovarian arteries
suprarenal arteries, brachial arteries, renal arteries

1. Paired branches of the abdominal aorta that supply the diaphragm ______________
2. The vessels that branch off the ascending aorta and supply the heart muscle ______________
3. The large, paired branches of the abdominal aorta that supply blood to the kidneys ______________
4. The vessels formed by final division of the abdominal aorta ______________
5. The vessels that supply the head and neck on each side ______________
6. The paired arteries that branch into the radial and ulnar arteries ______________
7. A group of paired vessels that extend between the ribs ______________
8. Paired branches of the abdominal aorta that extend into the abdominal wall musculature ______________

EXERCISE 15-7.

INSTRUCTIONS

Write the appropriate term in each blank.

ascending aorta, aortic arch, thoracic aorta, abdominal aorta
brachiocephalic artery, celiac trunk, hepatic artery

1. The short artery that branches into the left gastric artery, the splenic artery, and the hepatic artery ______________
2. A large vessel found within the pericardial sac ______________
3. The portion of the aorta supplying the upper extremities, neck, and head ______________
4. The large vessel that branches into the right subclavian artery and the right common carotid artery ______________
5. The most inferior portion of the aorta ______________
6. The vessel supplying oxygenated blood to the liver ______________

4. Define *anastomosis*, cite its function, and give several examples.

EXERCISE 15-8: Principal Systemic Arteries of the Head (Text Fig. 15-5)

INSTRUCTIONS

1. Write the names of cranial arteries on the appropriate lines in different, preferably darker, colors. Felt tip pens would work well for this exercise.
2. Outline the arteries on the diagram with the corresponding color.

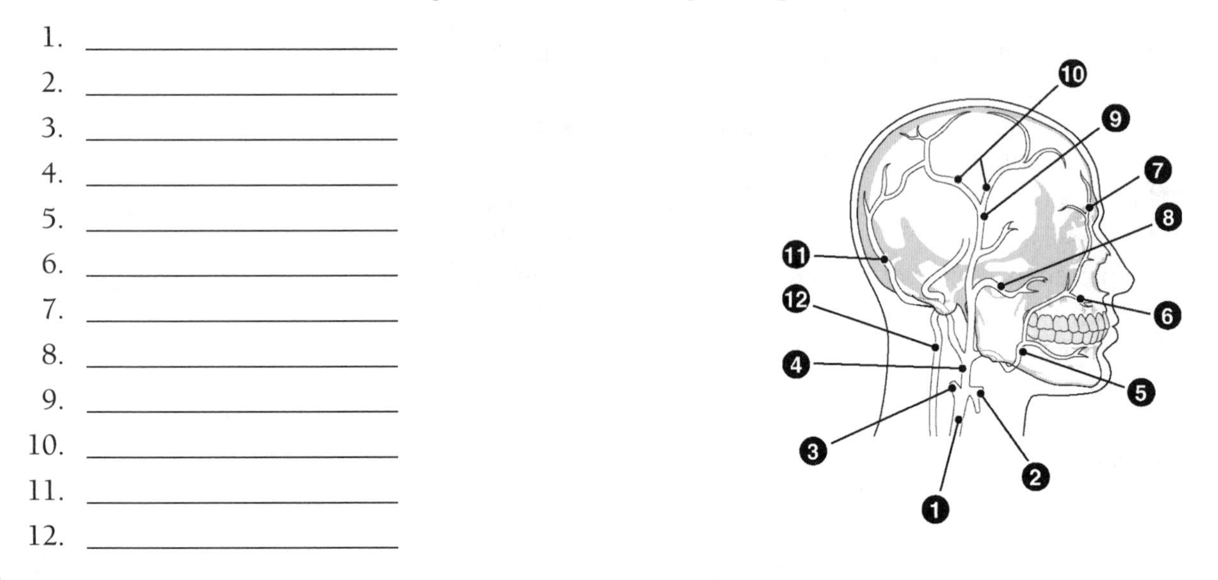

EXERCISE 15-9.

INSTRUCTIONS

Write the appropriate term in each blank.

mesenteric arch	arterial arch	circle of Willis
anastomosis	superficial palmar arch	basilar artery

1. A general term describing communication between two blood vessels ________________
2. An anastomosis between vessels supplying the intestines ________________
3. An anastomosis under the center of the brain formed by two internal carotid arteries and the basilar artery ________________
4. The vessel formed by union of the two vertebral arteries ________________
5. A vessel formed by the union of the radial and ulnar arteries ________________

5. Compare superficial and deep veins and give examples of each type.

EXERCISE 15-10.

INSTRUCTIONS

For each of the following veins, state whether they are deep (D) or superficial (S).

1. saphenous vein ____
2. basilic vein ____
3. brachial vein _____
4. femoral vein _____
5. jugular vein _____

6. Name the main vessels that drain into the superior and inferior venae cavae.

EXERCISE 15-11: Principal Systemic Veins (Text Fig. 15-8)

INSTRUCTIONS

Label each of the indicated veins.

1. ____________________
2. ____________________
3. ____________________
4. ____________________
5. ____________________
6. ____________________
7. ____________________
8. ____________________
9. ____________________
10. ____________________
11. ____________________
12. ____________________
13. ____________________
14. ____________________
15. ____________________
16. ____________________
17. ____________________
18. ____________________
19. ____________________
20. ____________________
21. ____________________
22. ____________________
23. ____________________
24. ____________________
25. ____________________
26. ____________________

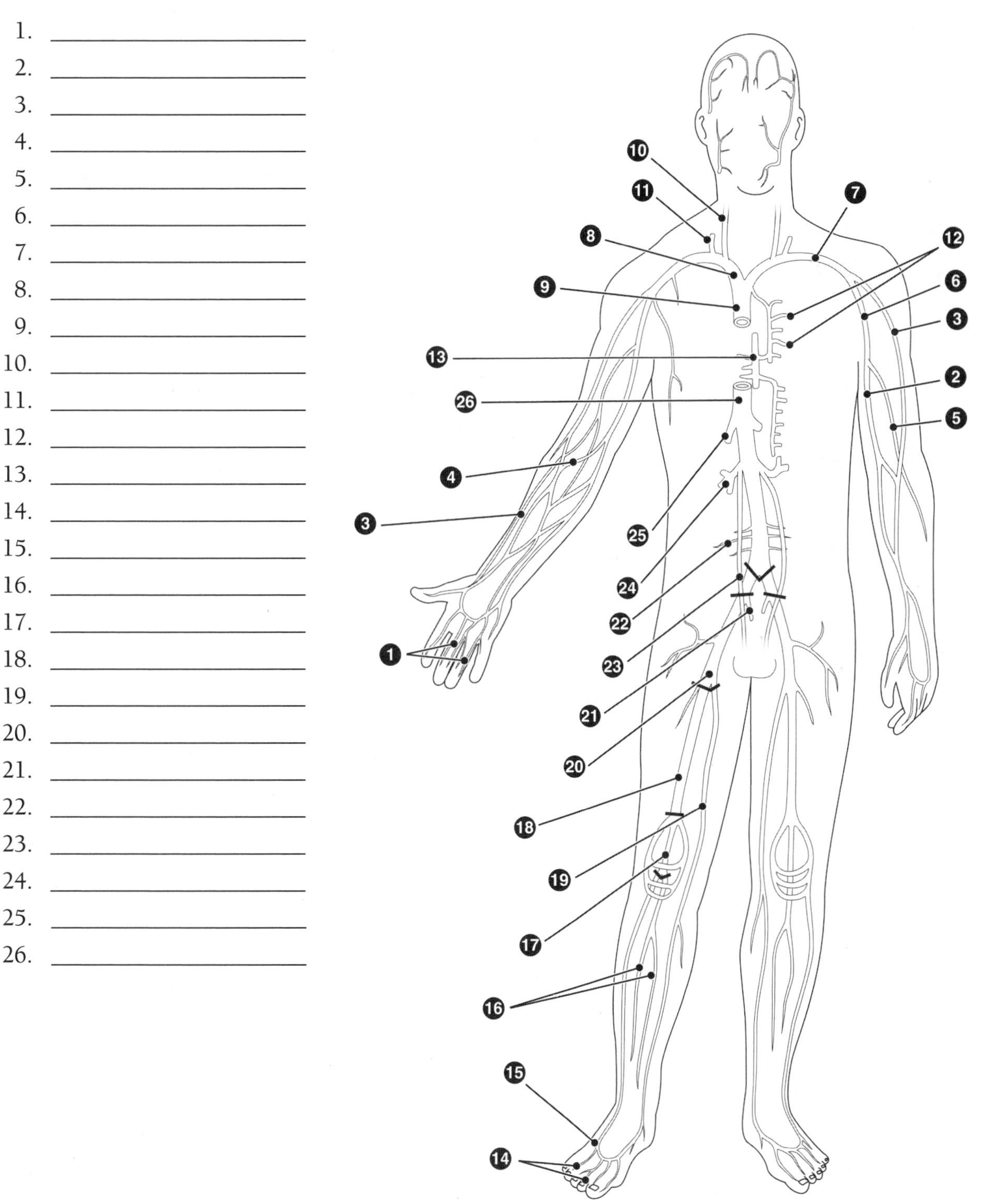

EXERCISE 15-12: Principal Systemic Veins of the Head (Text Fig. 15-8)

INSTRUCTIONS

Label each of the indicated veins.

1. ____________________
2. ____________________
3. ____________________
4. ____________________
5. ____________________
6. ____________________
7. ____________________

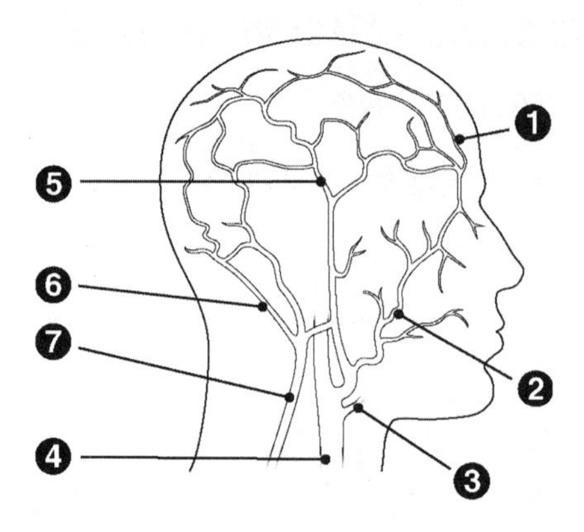

EXERCISE 15-13.

INSTRUCTIONS

Write the appropriate term in each blank.

cephalic vein	saphenous vein	jugular vein	gastric vein
femoral vein	lumbar vein	hepatic portal vein	common iliac vein

1. The vein that drains the area supplied by the carotid artery ____________
2. The longest vein ____________
3. A vessel that drains into the subclavian vein ____________
4. A deep vein of the thigh ____________
5. One of four pairs of veins that drain the dorsal part of the trunk ____________
6. A vein that sends blood to hepatic capillaries ____________
7. A vein that drains the stomach and empties into the hepatic portal vein ____________

7. Define *venous sinus* and give several examples of venous sinuses.

EXERCISE 15-14: Cranial Venous Sinuses (Text Fig. 15-9)

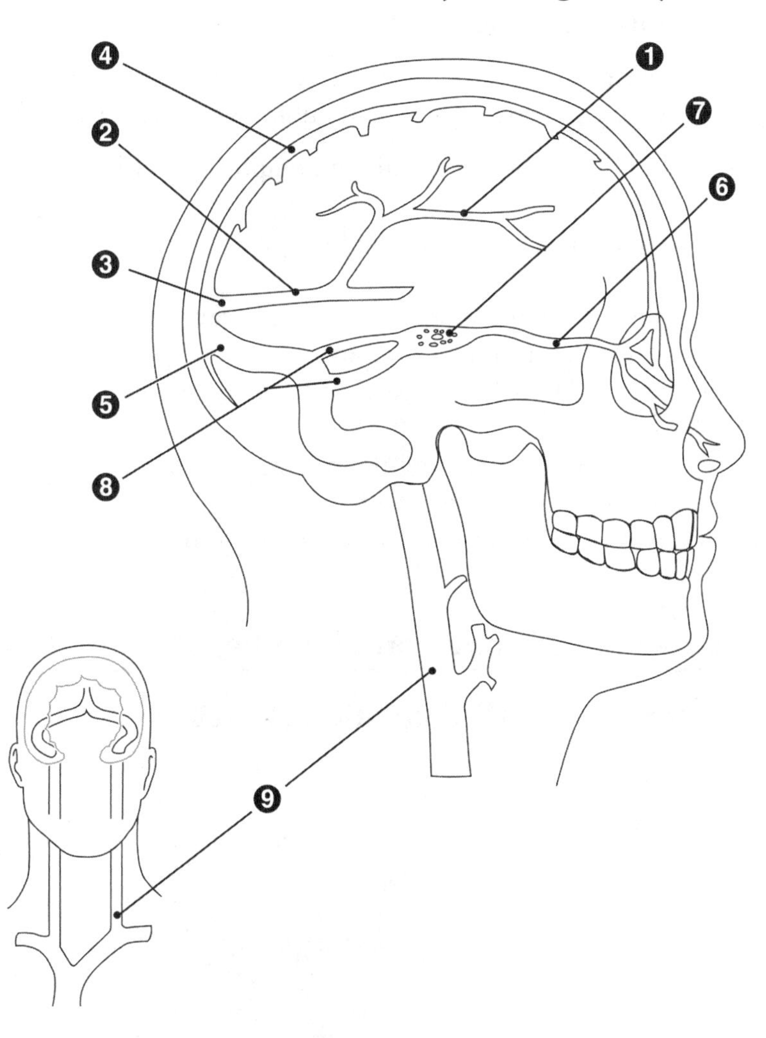

INSTRUCTIONS

1. Label each of the indicated veins and sinuses.
2. Draw arrows to indicate the direction of blood flow through the sinuses into the internal jugular vein.

1. ______________________
2. ______________________
3. ______________________
4. ______________________
5. ______________________
6. ______________________
7. ______________________
8. ______________________
9. ______________________

EXERCISE 15-15.

INSTRUCTIONS

Write the appropriate term in each blank.

coronary sinus	azygos vein	superior vena cava	inferior vena cava
transverse sinus	superior sagittal sinus	median cubital vein	cavernous sinus

1. A vessel that drains blood from the chest wall and empties into the superior vena cava ________________
2. The vein that receives blood draining from the head, neck, upper extremities, and chest ________________
3. A vein frequently used for removing blood for testing because of its location near the surface at the front of the elbow ________________
4. The large vein that drains blood from the parts of the body below the diaphragm ________________
5. The channel that drains blood from the ophthalmic vein of the eye
6. The channel that drains into the confluence of sinuses ________________
7. The channel that receives blood from most of the veins of the heart wall ________________

8. Describe the structure and function of the hepatic portal system.

EXERCISE 15-16: Hepatic Portal Circulation (Text Fig. 15-10)

INSTRUCTIONS

Label each of the indicated parts.

1. ________________________
2. ________________________
3. ________________________
4. ________________________
5. ________________________
6. ________________________
7. ________________________
8. ________________________
9. ________________________
10. ________________________
11. ________________________
12. ________________________
13. ________________________

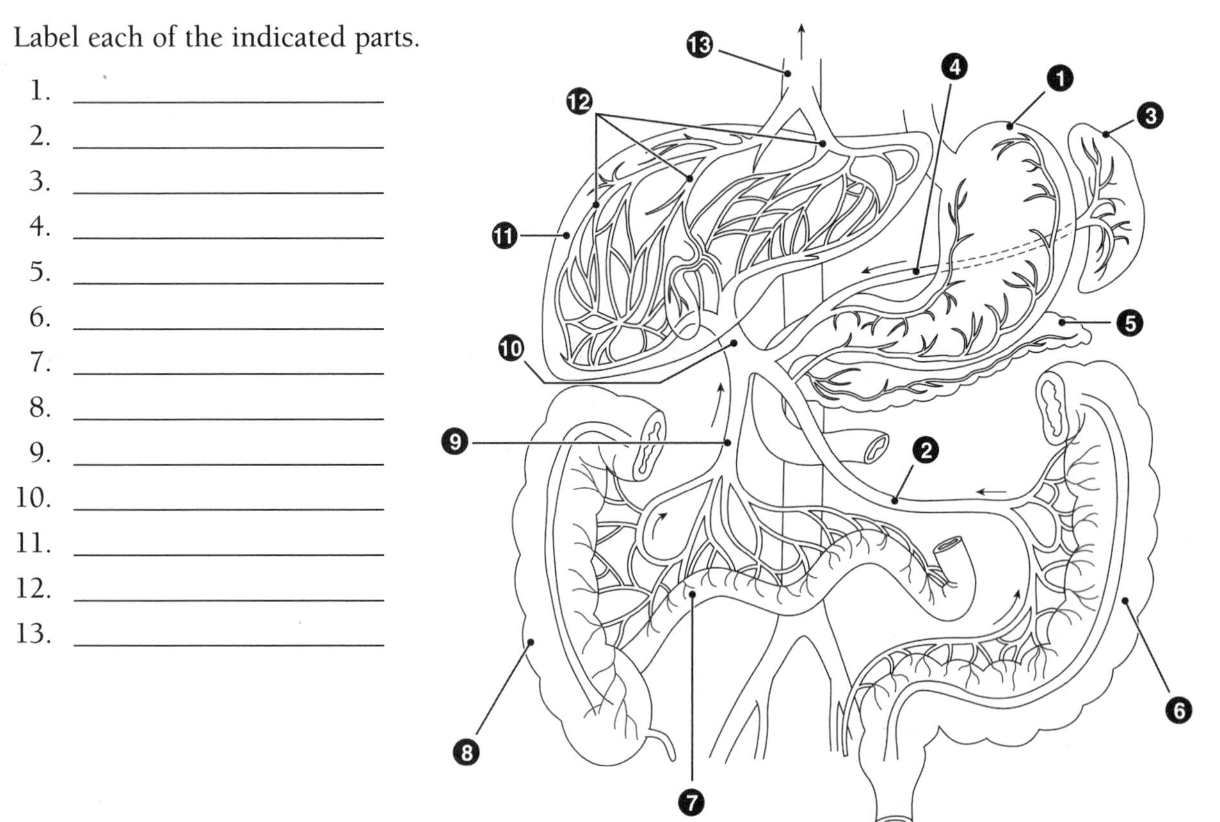

9. Explain the forces that affect exchange across the capillary wall.

EXERCISE 15-17.

INSTRUCTIONS

Write the appropriate term in each blank.

diffusion blood pressure osmotic pressure

1. A force that pushes water out of the capillary ________
2. A force that moves substances down their concentration gradients ________
3. A force that draws water into the capillary ________

10. Describe the factors that regulate blood flow.

EXERCISE 15-18.

INSTRUCTIONS

Write the appropriate term in each blank.

vasomotor center vasodilation vasoconstriction
valve precapillary sphincter

1. Structure that prevents blood from moving backward in the veins ________
2. A structure that regulates blood flow into an individual capillary ________
3. The change in blood vessel diameter caused by smooth muscle contraction ________
4. An increase in blood vessel diameter ________
5. A region of the medulla oblongata that controls blood vessel diameter ________

11. Define *pulse* and list factors that affect pulse rate.

EXERCISE 15-19.

INSTRUCTIONS

For each of the following situations, will the pulse rate most likely increase (I) or decrease (D)?

1. A newborn baby grows older _____
2. An adult falls asleep _____
3. Thyroid gland secretion increases _____
4. A teenager runs to catch a bus _____
5. A child gets a fever _____

12. List the factors that affect blood pressure.

EXERCISE 15-20.

INSTRUCTIONS

What is the effect of each of the following changes on blood pressure? In the blanks, write *I* for increase or *D* for decrease.

1. Decreased cardiac output _____
2. Increased blood thickness _____
3. Reduced volume of blood ejected from the heart per heartbeat _____
4. Decreased vessel elasticity (from atherosclerosis, for example) _____
5. Uncontrolled bleeding _____
6. Increased heart rate _____
7. Vasoconstriction _____

13. Explain how blood pressure is commonly measured.

EXERCISE 15-21.

INSTRUCTIONS

Write the appropriate term in each blank.

viscosity	systolic pressure	diastolic pressure	stethoscope
hypotension	sphygmomanometer	osmolarity	hypertension

1. Term for the blood pressure reading taken during ventricular relaxation _____
2. An instrument that is used to measure blood pressure _____
3. Term for blood pressure measured during heart muscle contraction _____
4. A term that describes the thickness of a solution _____
5. An abnormal increase in blood pressure _____
6. An abnormal decrease in blood pressure _____

14. List reasons why hypertension should be controlled.

EXERCISE 15-22.

INSTRUCTIONS

List four complications that can result from untreated hypertension.

1. _____
2. _____
3. _____
4. _____

15. List some disorders that involve the blood vessels.

EXERCISE 15-23.

INSTRUCTIONS

Write the appropriate term in each blank.

thrombosis endarterectomy phlebitis
catheterization hemorrhoid

1. A varicose vein in the rectum ____________
2. Removal of the thickened lining of diseased blood vessels ____________
3. Vein inflammation ____________
4. Blood clot formation within a blood vessel ____________

EXERCISE 15-24.

INSTRUCTIONS

Write the appropriate term in each blank.

aneurysm atherosclerosis gangrene
hemorrhage pulmonary embolism thrombosis

1. A bulging sac in the wall of an artery that results from weakness of the vessel wall ____________
2. An arterial disorder characterized by deposits of yellow, fatlike material ____________
3. Profuse bleeding ____________
4. A life-threatening condition resulting from a blood clot piece blocking a vessel in the lungs ____________
5. A condition resulting from the bacterial invasion of dead tissue ____________

16. List steps in first aid for hemorrhage.

EXERCISE 15-25.

INSTRUCTIONS

List the appropriate pressure point for each of the following situations. Choose from the following list: brachial artery, femoral artery, temporal artery, and subclavian artery. Each pressure point can be used only once.

1. A bleeding nose ____________
2. A serious leg laceration ____________
3. A cut to the shoulder ____________
4. A lacerated finger ____________

17. List four types of shock.

EXERCISE 15-26.

INSTRUCTIONS

Write the appropriate term in each blank.

anaphylactic shock   cardiogenic shock   septic shock   hypovolemic shock

1. Shock resulting from heart muscle damage ________________
2. Shock resulting from an overwhelming bacterial infection ________________
3. Shock resulting from an allergic reaction ________________
4. Shock resulting from hemorrhage ________________

18. Show how word parts are used to build words related to the blood vessels and circulation.

EXERCISE 5-27.

INSTRUCTIONS

Complete the following table by writing the correct word part or meaning in the space provided. Write a word that contains each word part in the Example column.

Word Part	Meaning	Example
1. __________	foot	____________________
2. phleb/o	_________	____________________
3. __________	mouth	____________________
4. hepat/o	_________	____________________
5. -ectomy	_________	____________________
6. __________	stomach	____________________
7. __________	arm	____________________
8. __________	intestine	____________________
9. sphygm/o	_________	____________________
10. celi/o	_________	____________________

Making the Connections

The following concept map deals with the measurement and regulation of blood pressure. Each pair of terms is linked together by a connecting phrase into a sentence. The sentence should be read in the direction of the arrow. Complete the concept map by filling in the appropriate term or phrase. There is one right answer for each term. However, there are many correct answers for the connecting phrases (3, 6, 8, and 11).

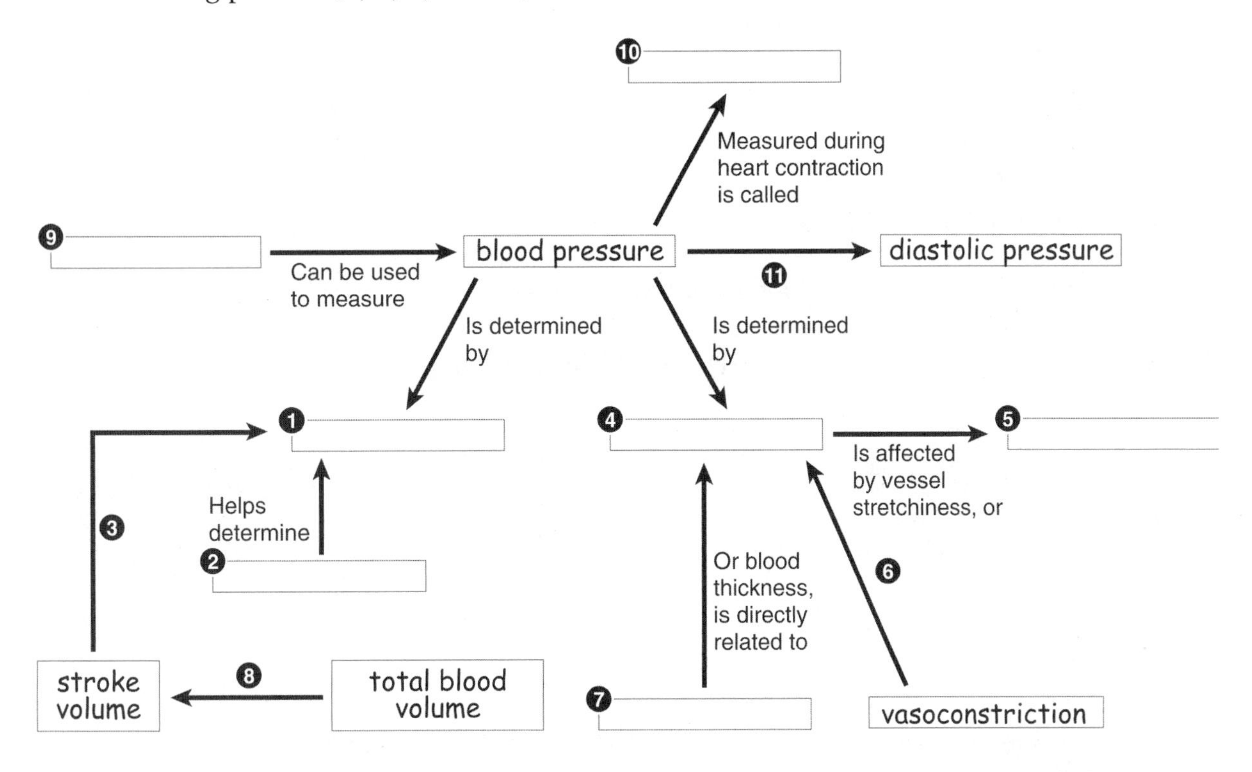

Optional Exercise: Make your own concept map/flow chart, based on the flow of blood from the left ventricle to the foot. It is not necessary to include connecting statements between the terms. You can also make flow charts based on the flow of blood from the leg to the heart, or to and from the arm and/or head. Choose your own terms to incorporate into your map, or use the following list: left ventricle, dorsalis pedis, femoral, anterior tibial, aortic arch, abdominal aorta, descending aorta, ascending aorta, popliteal, common iliac, and external iliac.

Testing Your Knowledge

Building Understanding

I. Multiple Choice

Select the best answer and write the letter of your choice in the blank.

1. Which of the following arteries is unpaired? 1. _______
   a. renal
   b. brachial
   c. brachiocephalic
   d. common carotid
2. The precapillary sphincter is a 2. _______
   a. ring of smooth muscle that regulates blood flow
   b. dilated vein in the liver
   c. tissue flap that prevents blood backflow in veins
   d. valve at the entrance to the iliac artery
3. Which of the following arteries carries blood low in oxygen? 3. _______
   a. pulmonary artery
   b. hepatic portal artery
   c. brachiocephalic artery
   d. superior vena cava
4. A varicose vein is also called a(n) 4. _______
   a. stent
   b. embolus
   c. thrombus
   d. varix
5. As blood flows through the tissues, a force that draws fluid back into the capillaries is 5. _______
   a. blood pressure
   b. osmotic pressure
   c. hypertension
   d. vasoconstriction
6. Which of the following veins is found in the lower extremity? 6. _______
   a. jugular
   b. brachial
   c. basilic
   d. popliteal
7. Which of the following arteries is NOT found in the circle of Willis? 7. _______
   a. anterior cerebral
   b. posterior communicating
   c. vertebral
   d. middle cerebral
8. Which of the following layers is found in arteries AND capillaries? 8. _______
   a. smooth muscle
   b. inner tunic
   c. outer tunic
   d. middle tunic

9. Blood pressure would be increased by 9. _______
   a. narrowing the blood vessels
   b. reducing the pulse rate
   c. increasing vasodilation
   d. decreasing blood viscosity
10. Which of the following is NOT a subdivision of the aorta? 10. _______
    a. thoracic aorta
    b. descending aorta
    c. pulmonary aorta
    d. abdominal aorta
11. Hemorrhage of the lower extremity can be stopped by compressing the 11. _______
    a. coronary artery
    b. femoral artery
    c. facial artery
    d. dorsalis pedis
12. The two veins that unite to form the inferior vena cava are the 12. _______
    a. gastric veins
    b. common iliac veins
    c. jugular veins
    d. mesenteric veins

II. Completion Exercise

Write the word or phrase that correctly completes each sentence.

1. One example of a portal system is the system that carries blood from the abdominal organs to the ____________________.
2. The inner, epithelial layer of blood vessels is called the ____________________.
3. A life-threatening condition in which there is inadequate blood flow to all body tissues is ____________________.
4. People whose work requires them to stand much of the time frequently suffer from varicosities of the long leg veins named the ____________________.
5. The aorta and the venae cavae are part of the group, or circuit, of blood vessels that make up the ____________________.
6. A large channel that drains deoxygenated blood is called a(n) ____________________.
7. The smallest subdivisions of arteries have thin walls in which there is little connective tissue and relatively more muscle. These vessels are ____________________.
8. A decrease in a blood vessel's diameter is called ____________________.
9. The circle of Willis is formed by a union of the internal carotid arteries and the basilar artery. Such a union of vessels is called a(n) ____________________.
10. The vessel that can be compressed against the lower jaw to stop bleeding in the mouth is the ____________________.
11. A condition in which vein inflammation contributes to abnormal clot formation is called ____________________.

Understanding Concepts

I. True/False

For each question, write *T* for true or *F* for false in the blank to the left of each number. If a statement is false, correct it by replacing the underlined term and write the correct statement in the blank below the question.

_______ 1. The anterior and posterior communicating arteries are part of the anastomosis supplying the brain.

____

_______ 2. Contraction of the smooth muscle in arterioles would decrease blood pressure.

____

_______ 3. Increased blood pressure would decrease the amount of fluid leaving the capillaries.

____

_______ 4. The external iliac artery continues in the thigh as the femoral artery.

____

_______ 5. The transverse sinuses receive most of the blood leaving the heart.

____

_______ 6. Sinusoids are found in the liver.

____

_______ 7. Blood flow into individual capillaries is regulated by precapillary sphincters.

____

_______ 8. Blood pressure is equal to the pulse rate × peripheral resistance.

____

_______ 9. Inadequate blood flow resulting from a severe allergic reaction is called anaphylactic shock.

____

_______ 10. Ms. L is bleeding from the ear. Pressure should be applied to the facial artery.

____

II. Practical Applications

Study each discussion. Then write the appropriate word or phrase in the space provided.

➤ Group A

1. Mr. Q, age 87, complained of pain and swelling in the area of his saphenous vein. The term for venous inflammation is ________________________.
2. Further study of Mr. Q's illness indicated that, associated with the inflammation, a blood clot had formed in one vein. This serious condition is called ________________________.
3. Mr. Q was put on bedrest with his legs elevated and was started on anticoagulant medications. These measures were taken to reduce the serious risk that the blood clot would break loose and travel to the lungs. This potentially fatal complication is called ________________________.
4. Mr. Q cut the palm of his hand while eating dinner. Due to the anticoagulant medication, the cut bled profusely. Profuse bleeding is called ________________________.
5. To attempt to stop the bleeding, the nurse applied pressure to the artery called the ________________________.

➤ Group B

1. Ms. L, aged 42, had her blood pressure examined during a routine physical. Her pressure reading was 165/120. Her diastolic pressure is thus ________________________.
2. Based on this reading, the physician diagnosed Ms. L with a condition called ________________________.
3. The physician was very alarmed by the finding and immediately prescribed a drug to reduce the production of an enzyme produced in the kidneys that causes blood pressure to increase. This enzyme is called ________________________.
4. Ms. L is at greater risk for a disease associated with plaque deposits in the arteries. This disease is called ________________________.
5. While at the doctor's office, Ms. L suddenly collapsed, complaining of severe, crushing chest pain. Further observation yielded these outcome signs: weak pulse of 120 beats per minute; blood pressure 76/40; gray, cold, clammy skin; and rapid, shallow respiration. Her symptoms were due to failure of the heart pump, a form of shock known as ________________________.

III. Short Essays

1. Explain the purpose of vascular anastomoses.

____

____

____

2. What is the function of the hepatic portal system, and what vessels contribute to this system?

____

____

____

3. List the vessels a drop of blood will encounter traveling from deep within the left thigh to the right atrium.

____

____

____

Conceptual Thinking

1. Mr. B, aged 54, is losing substantial amounts of blood due a bleeding ulcer. What will be the effect on his blood pressure? What are some physiological changes that could be made to correct this effect?

____

____

____

2. Ms. J has a blood pressure reading of 145/92 mm Hg. A. What is her systolic pressure? B. What is her diastolic pressure? C. Does she suffer from hypertension or hypotension? D. Should Ms. J be treated? If so, how?

____

____

____

Expanding Your Horizons

Eat, drink, and have happy arteries. Eat more fish. Eat less fish. Drink wine. Do not drink wine. We receive conflicting messages about the effect of different foodstuffs on arterial health. The following two articles discuss some of these claims.

Resources

1. Klatsky AL. Drink to your health? *Sci Am* 2003; 288:74–81.
2. Covington MB. Omega-3 fatty acids. *Am Fam Physician* 2004; 70:133–140.

CHAPTER

16 The Lymphatic System and Lymphoid Tissue

Overview

Lymph is the watery fluid that flows within the lymphatic system. It originates from the blood plasma and from the tissue fluid that is found in the minute spaces around and between the body cells. The fluid moves from the **lymphatic capillaries** through the **lymphatic vessels** and then to the **right lymphatic duct** and the **thoracic duct**. These large terminal ducts drain into the subclavian veins, adding the lymph to blood that is returning to the heart. Lymphatic capillaries resemble blood capillaries, but they begin blindly, and larger gaps between the cells make them more permeable than blood capillaries. The larger lymphatic vessels are thin-walled and delicate; like some veins, they have valves that prevent backflow of lymph.

The **lymph nodes**, which are the system's filters, are composed of lymphoid tissue. These nodes remove impurities and process **lymphocytes**, cells active in immunity. Chief among them are the cervical nodes in the neck, the axillary nodes in the armpit, the tracheobronchial nodes near the trachea and bronchial tubes, the mesenteric nodes between the peritoneal layers, and the inguinal nodes in the groin.

In addition to the nodes, there are several organs of lymphoid tissue with somewhat different functions. The **tonsils** filter tissue fluid; the **thymus** is essential for development of the immune system during early life. The **spleen** has numerous functions, including destruction of worn out red blood cells, serving as a reservoir for blood, and producing red blood cells before birth.

Another part of the body's protective system is the **reticuloendothelial system**, which consists of cells involved in the destruction of bacteria, cancer cells, and other possibly harmful substances.

Disorders of the lymphatic system include inflammation and enlargement of lymphoid tissue, neoplastic diseases, and elephantiasis, caused by a worm.

Addressing the Learning Outcomes

1. List the functions of the lymphatic system.

EXERCISE 16-1.

INSTRUCTIONS

List three functions of the lymphatic system in the blanks below.

1. ______________________________
2. ______________________________
3. ______________________________

EXERCISE 16-2: Lymphatic System in Relation to the Cardiovascular System (Text Fig. 16-1)

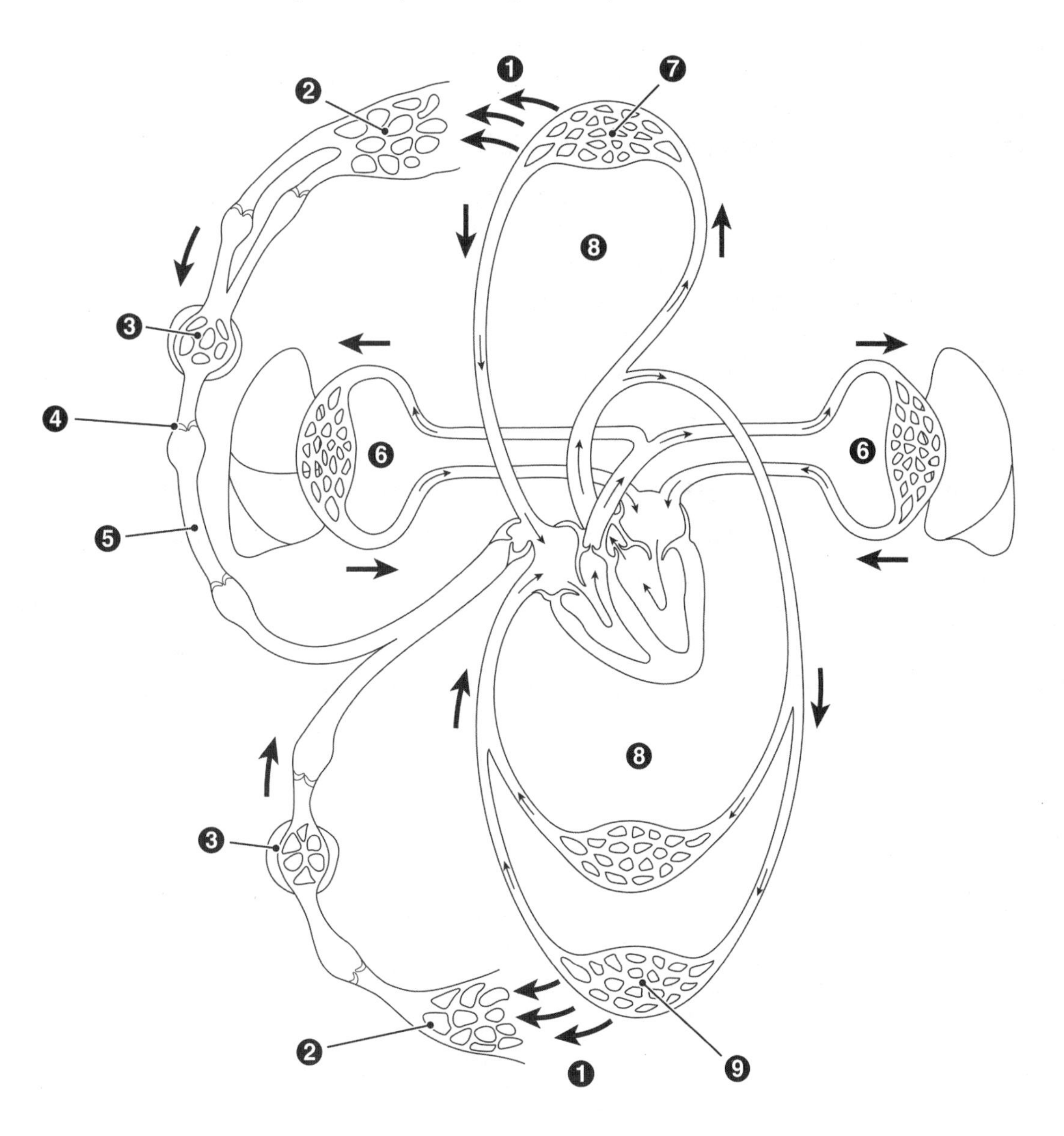

INSTRUCTIONS

1. Label the indicated parts.
2. Color the oxygenated blood red, the deoxygenated blood blue, and the lymph yellow.

1. ____________________
2. ____________________
3. ____________________
4. ____________________
5. ____________________
6. ____________________
7. ____________________
8. ____________________
9. ____________________

2. Explain how lymphatic capillaries differ from blood capillaries.

EXERCISE 16-3.

INSTRUCTIONS

Do the following statements apply to lymphatic capillaries (L), blood capillaries (B), or both (BOTH)? Write the appropriate abbreviation in each blank.

1. The vessel walls are constructed of a single layer of squamous epithelial cells. _____
2. The vessel walls are very permeable, permitting the passage of large proteins. _____
3. The cells in the vessel walls are called endothelial cells. _____
4. The gap between adjacent cells in the vessel wall is very small. _____
5. Lacteals are one example of this type of vessel. _____
6. The vessels form a bridge between two larger vessels. _____
7. The vessel transports erythrocytes. ______

3. Name the two main lymphatic ducts and describe the area drained by each.

EXERCISE 16-4.

INSTRUCTIONS

For each region, state whether its lymph will drain into the right lymphatic duct (R) or the thoracic duct (T).

1. left hand ____
2. right hand ____
3. right breast ____
4. left breast ____
5. left leg ____
6. right leg ____

4. List the major structures of the lymphatic system and give the locations and functions of each.

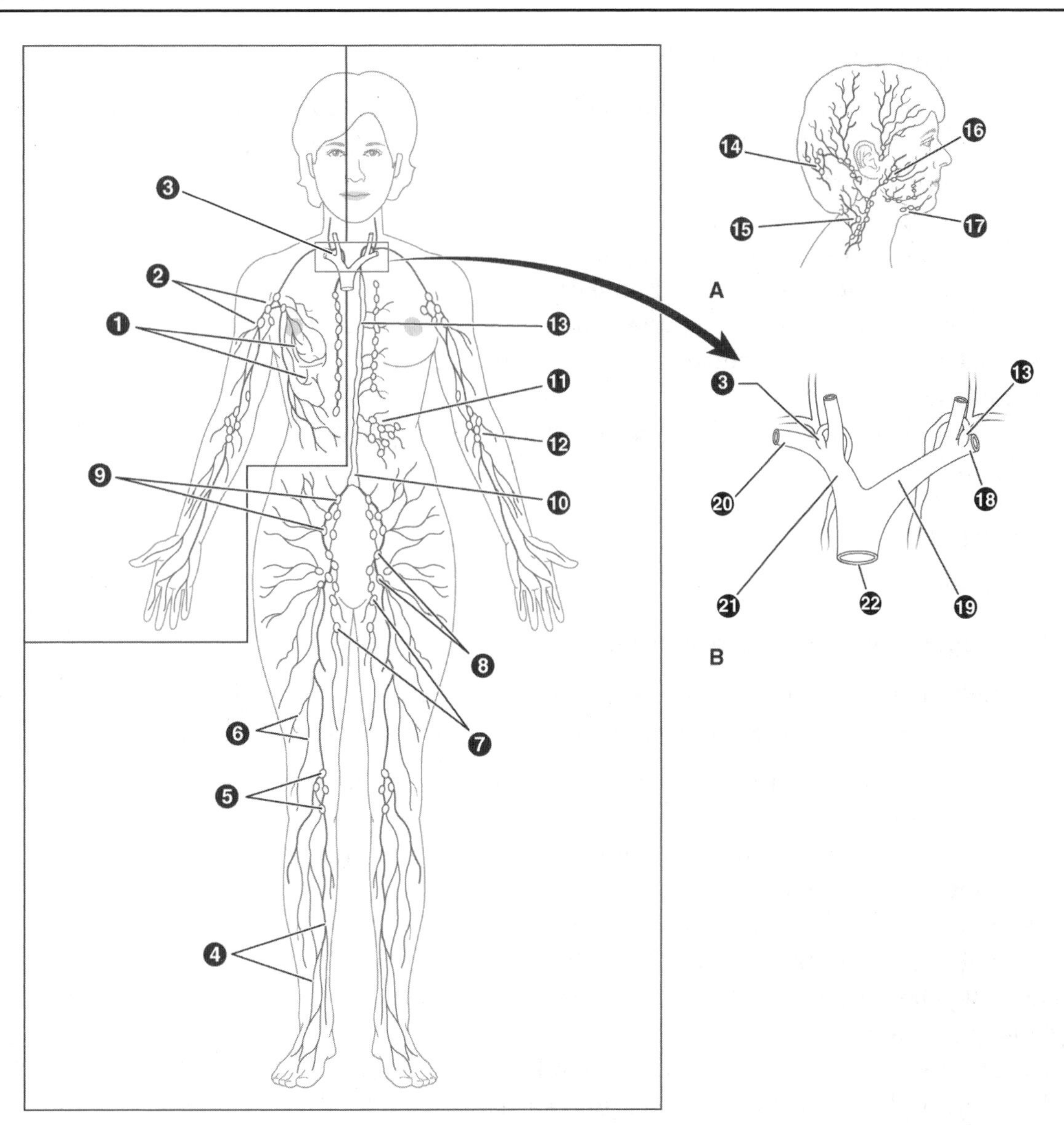

EXERCISE 16-5: Lymphatic System (Text Fig. 16-4)

INSTRUCTIONS

Label the indicated parts.

1. ____________
2. ____________
3. ____________
4. ____________
5. ____________
6. ____________
7. ____________
8. ____________
9. ____________
10. ____________
11. ____________
12. ____________
13. ____________
14. ____________
15. ____________
16. ____________
17. ____________
18. ____________
19. ____________
20. ____________
21. ____________
22. ____________

EXERCISE 16-6.

INSTRUCTIONS

Write the appropriate term in each blank.

lacteal	superficial	mesothelium	mesenteric nodes
deep	inguinal nodes	axillary nodes	cervical nodes

1. The nodes that filter lymph from the lower extremities and the external genitalia ________________
2. The lymph nodes located in the armpits ________________
3. Term for lymphatic vessels located under the skin ________________
4. The lymph nodes found between the two peritoneal layers ________________
5. A specialized vessel in the small intestine wall that absorbs digested fats ________________
6. The lymph nodes located in the neck that drain certain parts of the head and neck ________________

EXERCISE 16-7.

INSTRUCTIONS

Write the appropriate term in each blank.

lymphatic capillary	right lymphatic duct	chyle	lymph
valve	subclavian vein	thoracic duct	cisterna chyli

1. The temporary storage area formed by an enlargement of the first part of the thoracic duct ________________
2. The fluid formed when tissue fluid passes from the intercellular spaces into the lymphatic vessels ________________
3. The large lymphatic vessel that drains lymph from below the diaphragm and from the left side above the diaphragm ________________
4. The milky-appearing fluid that is a combination of fat globules and lymph ________________
5. Structure that prevents backflow of fluid in lymphatic vessels ________________
6. Blind-ended, thin-walled vessel that absorbs excess tissue fluid and proteins ________________

EXERCISE 16-8: Lymph Node (Text Fig. 16-5)

INSTRUCTIONS

Label the indicated parts.

1. ____________________
2. ____________________
3. ____________________
4. ____________________
5. ____________________
6. ____________________
7. ____________________
8. ____________________
9. ____________________
10. ____________________
11. ____________________
12. ____________________
13. ____________________

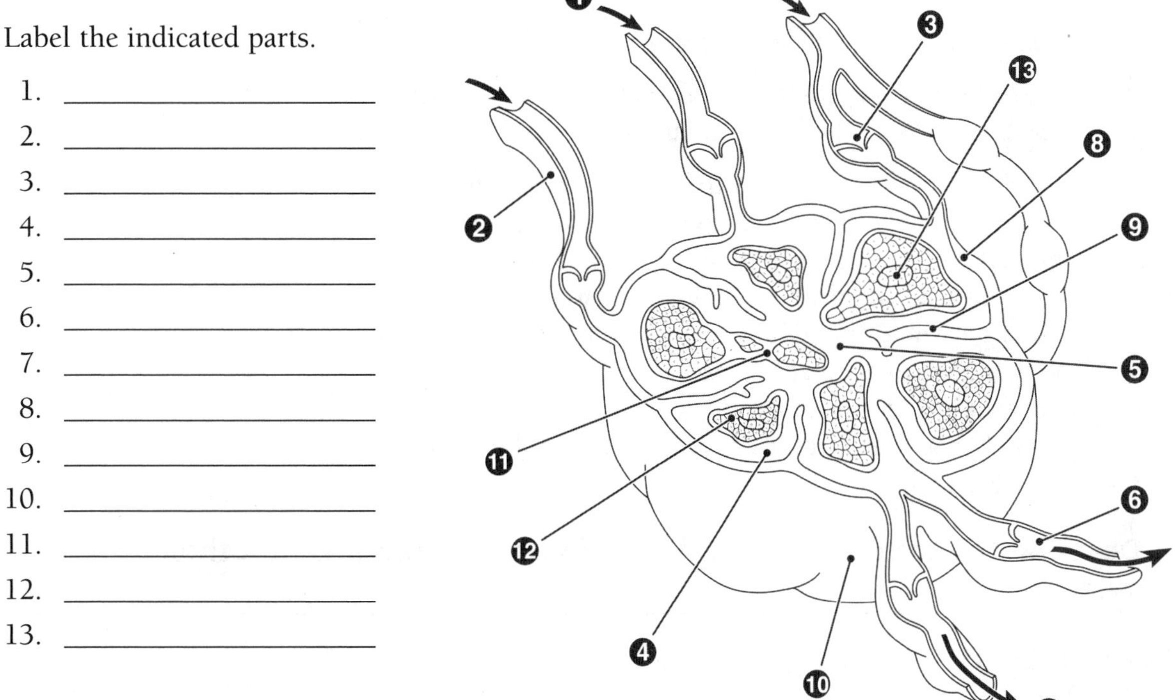

EXERCISE 16-9: Location of Lymphoid Tissue (Text Fig. 16-6)

INSTRUCTIONS

1. Write the names of the different lymphoid organs on the numbered lines in different colors.
2. Color the structures on the diagram with the corresponding colors.

1. ____________________
2. ____________________
3. ____________________
4. ____________________
5. ____________________
6. ____________________
7. ____________________
8. ____________________

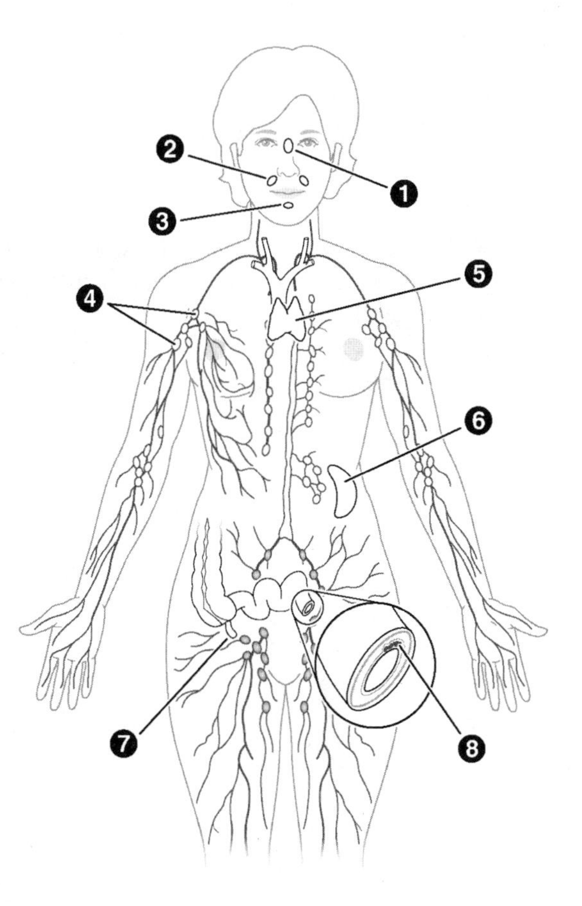

EXERCISE 16-10.

INSTRUCTIONS

Write the appropriate term in each blank.

hilum	germinal center	trabeculae	spleen
palatine tonsil	lingual tonsil	pharyngeal tonsil	thymus

1. An oval lymphoid body located at the side of the soft palate ____________
2. The organ that filters blood and is located in the upper left quadrant (left hypochondriac region) of the abdomen ____________
3. A mass of lymphoid tissue at the back of the tongue ____________
4. The mass of lymphoid tissue located in the pharynx behind the nose and commonly called adenoids ____________
5. The organ in which T cells mature ____________
6. The indented area of a lymph node where efferent lymphatic vessels exit the node ____________

5. Describe the composition and function of the reticuloendothelial system.

Exercise 16-11.

INSTRUCTIONS

Write the appropriate term in each blank.

Peyer patch	MALT	GALT	monocyte
Kupffer cell	reticuloendothelial system	immune system	

1. An alternate name for the tissue macrophage system ____________
2. Areas of lymphoid tissue found in mucous membranes ____________
3. Areas of lymphoid tissue specifically found throughout the gastrointestinal system ____________
4. The name given to a macrophage found in the liver ____________
5. An area of lymphoid tissue specifically found in the small intestine wall ____________
6. A large white blood cell circulating in the blood that gives rise to a macrophage ____________

6. Describe the major lymphatic system disorders.

EXERCISE 16-12.

INSTRUCTIONS

Write the appropriate term in each blank.

buboes	filariae	lymphangitis	lymphadenitis
lymphedema	Hodgkin disease	non-Hodgkin lymphoma	infectious mononucleosis

1. The small parasitic worms that cause elephantiasis ________________
2. A form of lymphoma diagnosed by the presence of Reed-Sternberg cells in a lymph node biopsy ________________
3. Abnormally large inguinal nodes, as may be found in certain infections ________________
4. An inflammatory disorder of lymph nodes ________________
5. Inflammation of lymphatic vessels ________________
6. A viral infection characterized by enlarged cervical lymph nodes ________________

7. Show how word parts are used to build words related to the lymphatic system.

EXERCISE 16-13.

INSTRUCTIONS

Complete the following table by writing the correct word part or meaning in the space provided. Write a word that contains each word part in the Example column.

Word Part	Meaning	Example
1. __________	gland	____________________
2. lingu/o	__________	____________________
3. __________	excessive enlargement	____________________
4. -oid	__________	____________________
5. -pathy	__________	____________________

Making the Connections

The following concept map deals with the structure and function of the lymphatic system. Each pair of terms is linked together by a connecting phrase into a sentence. The sentence should be read in the direction of the arrow. Complete the concept map by filling in the appropriate term or phrase. There is one right answer for each term. However, there are many correct answers for the connecting phrases (2, 5, 7, 9, 11, 12).

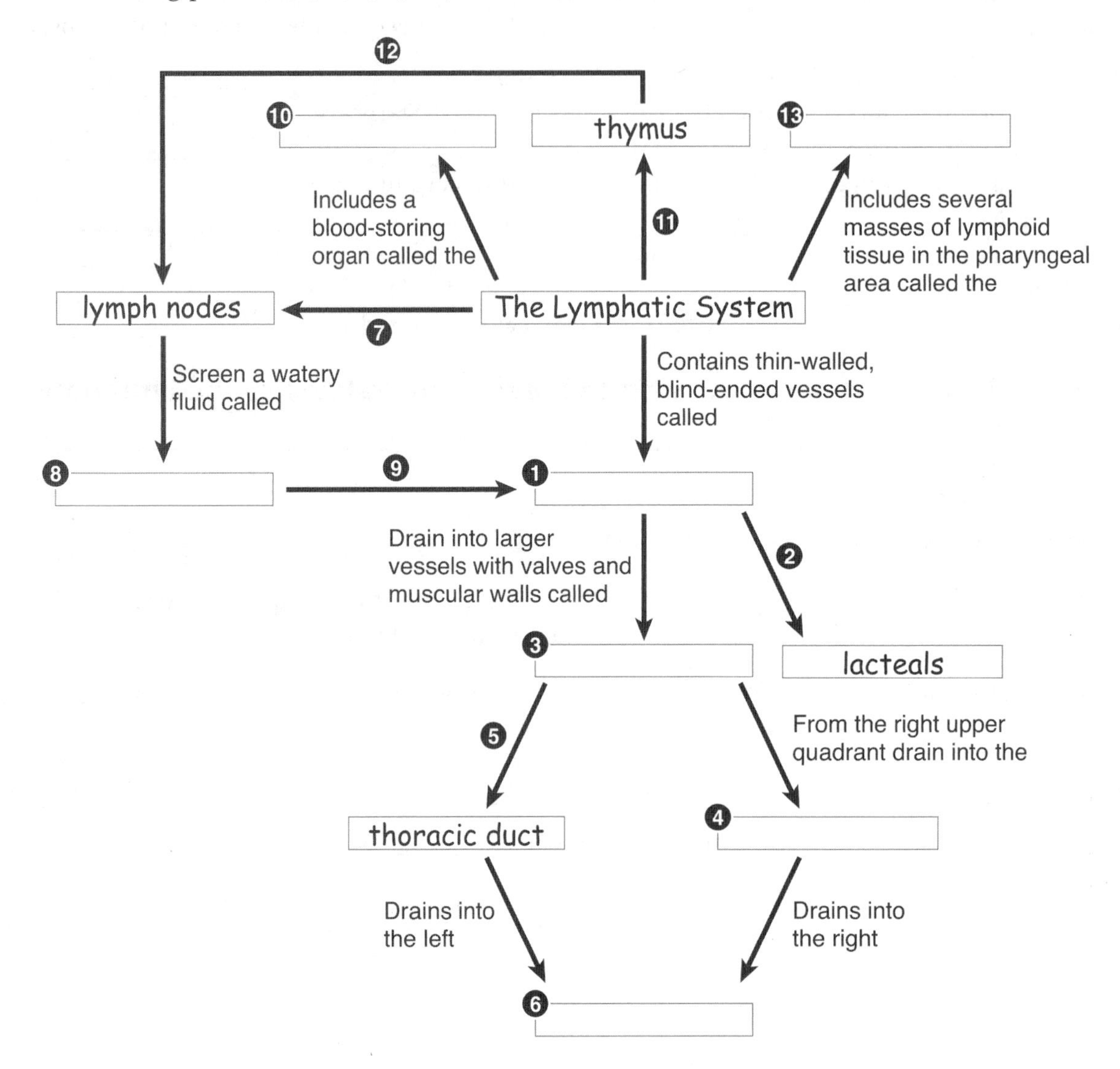

Optional Exercise: Make your own concept map/flow chart, based on the flow of lymph from a body part to the left atrium. It is not necessary to include connecting statements between the terms. For instance, you could map the flow of lymph from the right breast to the right atrium by linking the following terms: superior vena cava, right subclavian vein, right brachiocephalic vein, mammary vessels, axillary nodes, and right lymphatic duct.

Testing Your Knowledge

Building Understanding

I. Multiple Choice

Select the best answer and write the letter of your choice in the blank.

1. Which of the following is NOT a type of cell associated with the reticuloendothelial system? 1. _______
   a. monocytes
   b. macrophages
   c. dust cells
   d. red blood cells
2. Which of the following is NOT a function of the spleen? 2. _______
   a. destruction of old red blood cells
   b. blood filtration
   c. blood storage
   d. chyle drainage
3. The mesenteric nodes are found 3. _______
   a. in the groin region
   b. near the trachea
   c. between the two peritoneal layers
   d. in the armpits
4. An organ that shrinks in size after puberty is the 4. _______
   a. cisternal chyli
   b. thymus
   c. spleen
   d. liver
5. Unlike blood capillaries, lymphatic capillaries 5. _______
   a. contain a thin muscular layer
   b. are virtually impermeable to water and solutes
   c. are blind-ended
   d. do not contain any cells
6. Enlargement of the lymph nodes, as seen in Hodgkin disease, AIDS, and infectious mononucleosis, is termed 6. _______
   a. tonsillitis
   b. lymphadenopathy
   c. lymphangitis
   d. adenoidectomy
7. The enlarged portion of the thoracic duct is called the 7. _______
   a. cisterna chyli
   b. right lymphatic duct
   c. hilum
   d. lacteal

II. Completion Exercise

Write the word or phrase that correctly completes each sentence.

1. Thymosin is synthesized by the ____________________ gland.
2. Enlargement of the spleen is called ____________________.
3. The milky-appearing lymph that drains from the small intestine is called ____________________.
4. A benign or malignant tumor that occurs in lymphoid tissue is called a(n) ____________________.
5. The fluid that moves from tissue spaces into special collecting vessels for return to the blood is called ____________________.
6. Lymph from the right side of the body above the diaphragm joins the bloodstream when the right lymphatic duct empties into the ____________________.
7. The lymph nodes surrounding the breathing passageways may become black in individuals living in highly polluted areas. The nodes involved are the ____________________.
8. Nearly all of the lymph from the arm, shoulder, and breast passes through the lymph node known as the ____________________.
9. The spleen contains many cells that can engulf harmful bacteria and other foreign cells by a process called ____________________.

Understanding Concepts

I. True/False

For each question, write *T* for true or *F* for false in the blank to the left of each number. If a statement is false, correct it by replacing the <u>underlined</u> term and write the correct statement in the blank below the question.

_______ 1. Lymph filtered through the mesenteric nodes will drain into the <u>thoracic duct</u>.

____

_______ 2. An adenoidectomy involves the removal of the <u>lingual tonsils</u>.

____

_______ 3. Thymosin is produced by the <u>spleen</u>.

____

_______ 4. Kupffer cells are part of the <u>reticuloendothelial system</u>.

____

_______ 5. Cancer of the breast can cause lymphadenitis of the <u>axillary nodes</u>.

____

_______ 6. Lacteals are a type of <u>blood capillary</u>.

____

II. Practical Applications

Study each discussion. Then write the appropriate word or phrase in the space provided.

➤ Group A

1. Mr. W, age 21, detected a mass in his groin region. The physician suspected that one or more of the lymph nodes that drain the lower extremities was swollen. These nodes are called the ______________.
2. When these glands are enlarged, they are often referred to as ______________.
3. The physician informed Mr. W that he had a disease of the lymph nodes. The general term for these diseases is ______________.
4. A malignant tumor was detected by a lymph node biopsy. This tumor, like all tumors of lymphoid tissue, is called a(n) ______________.
5. Closer examination of the biopsy revealed the presence of Reed-Sternberg cells, leading to the diagnosis of ______________.

➤ Group B

1. Ms. L traveled in Asia for 2 years before starting nursing school. During the first week of classes, she noticed swelling in her lower right leg. Fearing the worst, she tearfully told her parents that her lymph vessels were blocked by small worms called ______________.
2. The physician confirmed that the swelling in Ms. L's leg was due to obstructed lymph flow. This type of swelling is called ______________.
3. Next, the lymph vessels in Ms. L's swollen lower leg were palpated. These vessels are called the ______________.
4. The physician noticed red streaks along the leg near a poorly healed cut. He concluded that Ms. L was suffering from inflammation of the lymphatic vessels, or ______________.
5. Ms. L was very relieved by the diagnosis. The doctor treated the infection aggressively to avoid the development of blood poisoning, or ______________.

III. Short Essays

1. Trace a lymph droplet from the interstitial fluid of the lower leg to the right atrium, based on the structures shown in Figure 16-4.

   ____

   ____

   ____

   ____

2. Compare and contrast between lymphangitis, lymphadenitis, and lymphadenopathy. What do these three disorders have in common? How do they differ?

   ____

   ____

   ____

Conceptual Thinking

1. Ms. Y, a healthy 24-year-old woman, is studying for her nursing finals. She has been sitting at her desk for 10 hours straight when she notices that her legs are swollen. Use your knowledge of the lymphatic system to explain the swelling, and suggest how she can prevent it in the future.

___

___

___

Expanding Your Horizons

Have you or one of your friends had the kissing disease? Infectious mononucleosis, or "mono," is an infection of lymphatic cells. It is easily transmitted between individuals living in close contact with one another (not necessarily by kissing) and is thus very common on college campuses. Learn more about this disease by reading the following articles.

Resources

1. Cozad J. Infectious mononucleoisis. *Nurse Pract* 1996; 21:14–18.
2. Bailey RE. Diagnosis and treatment of infectious mononucleosis. *Am Fam Physician* 1994; 49:879–888.

UNIT VI Energy: Supply and Use

CHAPTER 18

The Respiratory System

Overview

Oxygen is taken into the body and carbon dioxide is released by means of the organs and passageways of the **respiratory system.** This system contains the **nasal cavities, pharynx, larynx, trachea, bronchi,** and **lungs.**

Oxygen is obtained from the atmosphere and delivered to the cells by the process of **respiration.** The first phase of respiration is **pulmonary ventilation,** which is normally accomplished by breathing. During normal, quiet breathing, air enters the lungs (**inhalation**) because the diaphragm and intercostal muscles contract to expand the thoracic cavity. Air leaves the lungs (**exhalation**) when the muscles relax. Deeper breathing requires additional muscles during both inhalation and exhalation. The other two phases of respiration are: **external gas exchange** between the alveoli of the lungs and bloodstream; and **internal gas exchange** between the blood and tissues. In these exchanges, oxygen is delivered to the cells, and carbon dioxide is transported to the lungs for elimination.

Oxygen is transported to the tissues almost entirely by the **hemoglobin** in red blood cells. Some carbon dioxide is transported in the red blood cells as well, but most is converted into **bicarbonate ions** and **hydrogen ions.** Bicarbonate ions are carried in plasma, and the hydrogen ions (along with hydrogen ions from other sources) increase the acidity of the blood.

Breathing is primarily controlled by the **respiratory control centers** in the medulla and the pons of the brainstem. These centers are influenced by **chemoreceptors** located outside the medulla that respond to changes in the acidity of the cerebrospinal fluid. The acidity reflects the concentration of carbon dioxide.

Disorders of the respiratory tract include infection, allergy, chronic obstructive pulmonary disease (COPD), diseases and disorders of the pleura, and cancer.

Addressing the Learning Outcomes

1. Define *respiration* and describe the three phases of respiration.

EXERCISE 18-1.

INSTRUCTIONS

Write the appropriate term in each blank.

external gas exchange	internal gas exchange
cellular respiration	pulmonary ventilation

1. The exchange of air between the atmosphere and the alveoli ________________
2. The exchange of specific gases between the alveoli and the blood ________________
3. The exchange of specific gases between the blood and the cells ________________
4. The process by which cells use oxygen and nutrients to generate energy ________________

2. Name and describe all the structures of the respiratory system.

EXERCISE 18-2: Respiratory System (Text Fig. 18-2)

INSTRUCTIONS

1. Label the indicated parts.
2. Color all of the structures that encounter air green. Color structures containing oxygenated blood red. Color structures containing deoxygenated blood blue.

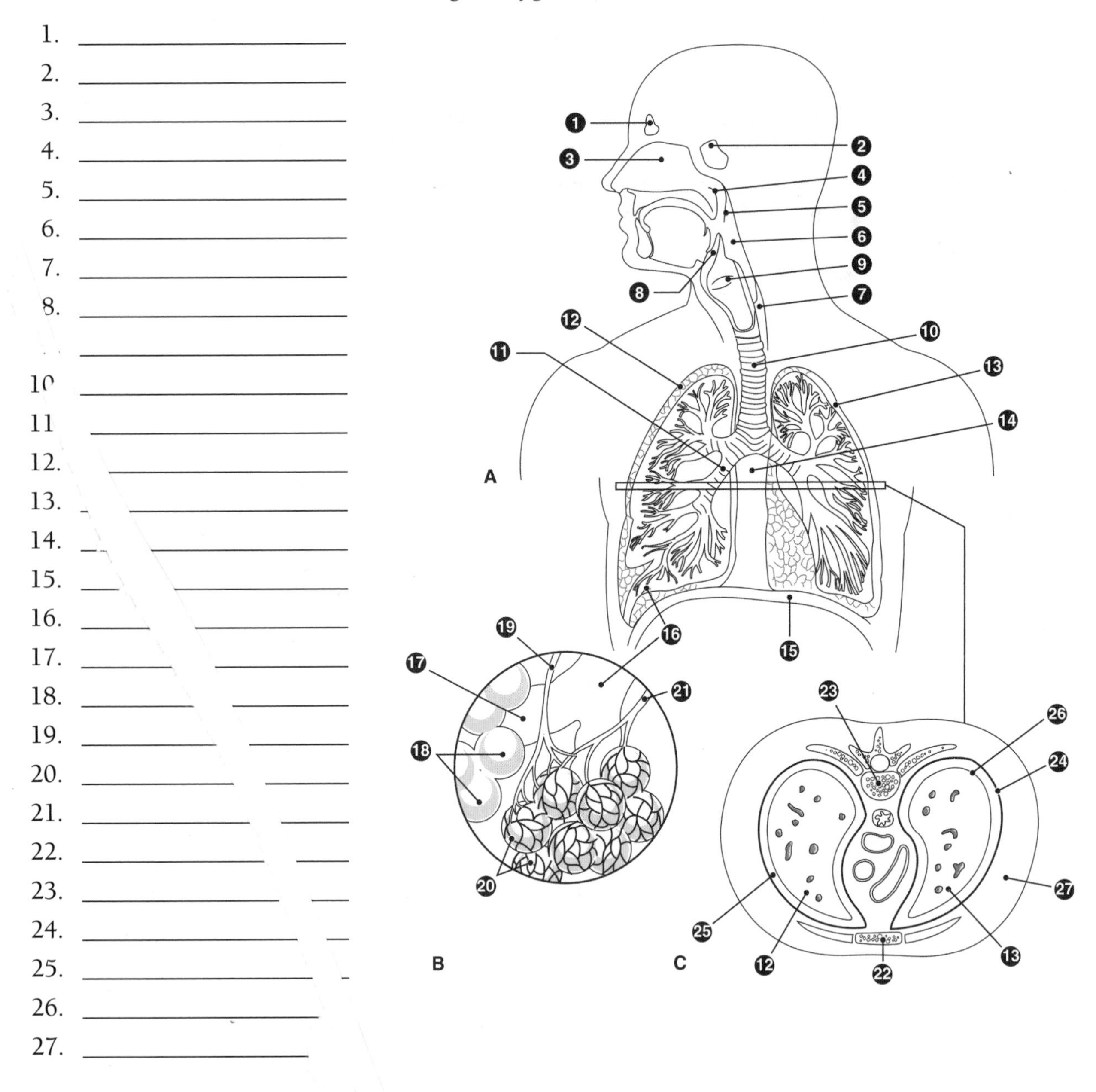

1. ____________________
2. ____________________
3. ____________________
4. ____________________
5. ____________________
6. ____________________
7. ____________________
8. ____________________
9. ____________________
10. ____________________
11. ____________________
12. ____________________
13. ____________________
14. ____________________
15. ____________________
16. ____________________
17. ____________________
18. ____________________
19. ____________________
20. ____________________
21. ____________________
22. ____________________
23. ____________________
24. ____________________
25. ____________________
26. ____________________
27. ____________________

EXERCISE 18-3: The Larynx (Text Fig. 18-3)

INSTRUCTIONS

1. Write the name of each labeled part on the numbered lines in different colors.
2. Color the different parts on the diagram with the corresponding color.

1. ____________________
2. ____________________
3. ____________________
4. ____________________
5. ____________________
6. ____________________
7. ____________________

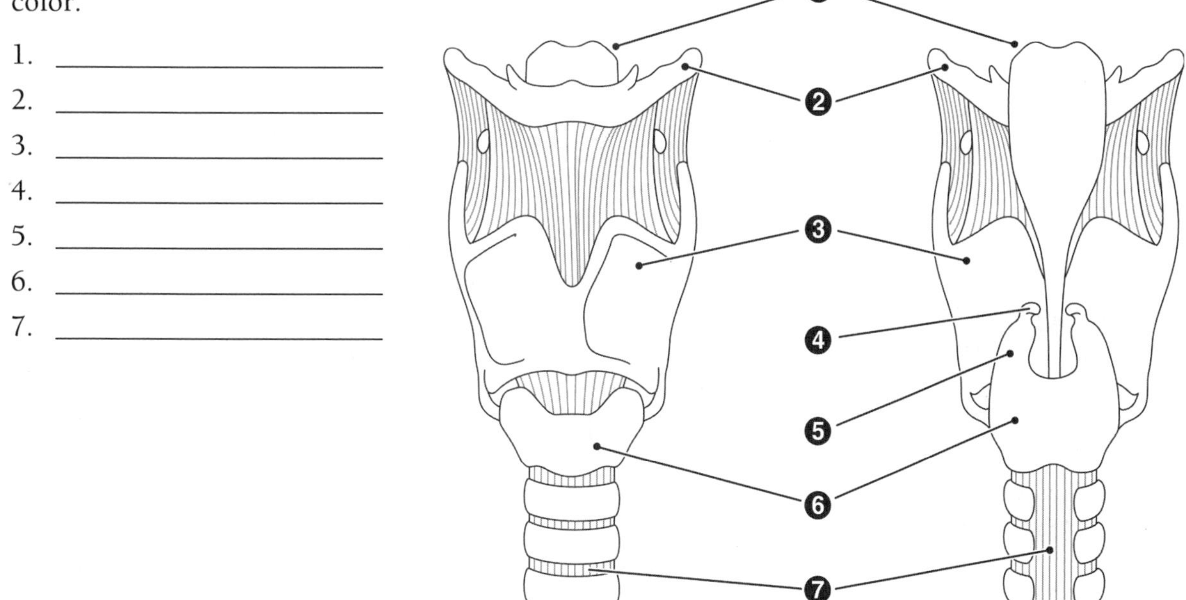

EXERCISE 18-4.

INSTRUCTIONS

Write the appropriate term in each blank.

nares	conchae	pharynx	glottis	epiglottis
larynx	hilum	bronchus	bronchiole	sinus

1. The openings of the nose ____________
2. The three projections arising from the lateral walls of each nasal cavity ____________
3. The scientific name for the voice box ____________
4. The leaf-shaped structure that helps to prevent the entrance of food into the trachea ____________
5. One of the two branches formed by division of the trachea ____________
6. The notch or depression where the bronchus, blood vessels, and nerves enter the lung ____________
7. The area below the nasal cavities that is common to both the digestive and respiratory systems ____________
8. A small air-conducting tube containing a smooth muscle layer but little or no cartilage ____________

3. Explain the mechanism for pulmonary ventilation.

EXERCISE 18-5: A Spirogram (Text Fig. 18-9)

INSTRUCTIONS

Write the names of the different lung volumes and capacities in the boxes on the diagram.

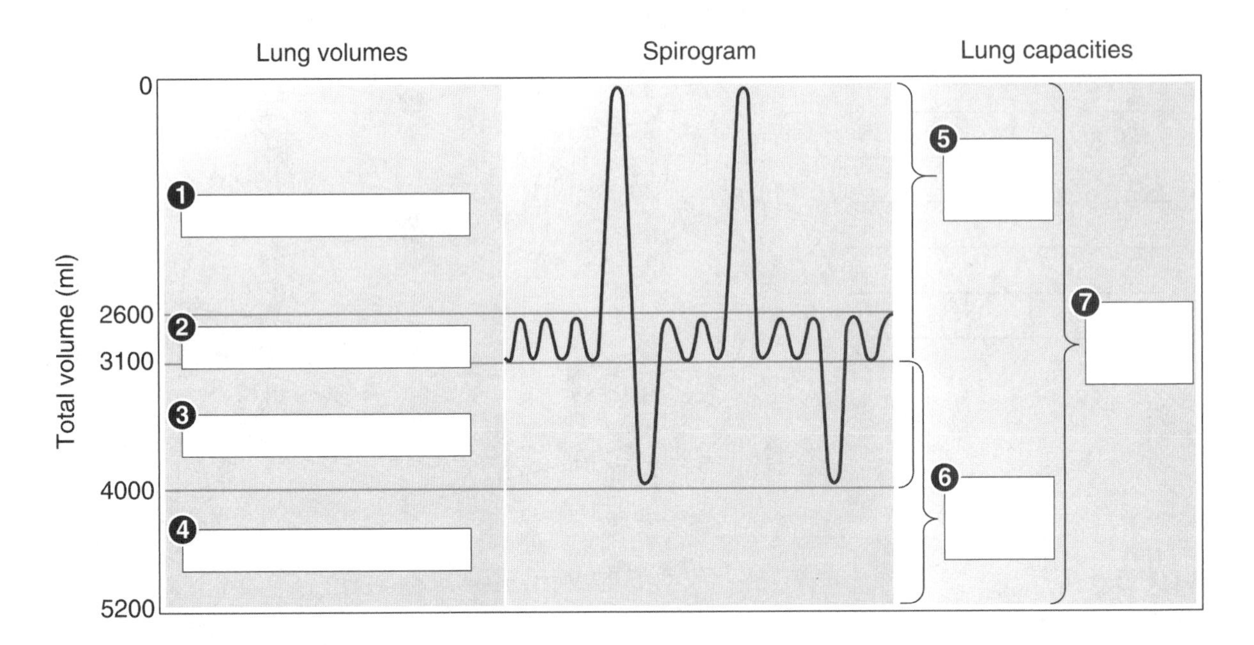

EXERCISE 18-6.

INSTRUCTIONS

Write the appropriate term in each blank.

alveoli	surfactant	pleura	inhalation	tidal volume
exhalation	intercostals	compliance	spirometer	vital capacity

1. The substance in the fluid lining the alveoli that prevents their collapse ____________
2. The phase of pulmonary ventilation in which air is expelled from the alveoli ____________
3. The phase of pulmonary ventilation in which the diaphragm contracts ____________
4. The serous membrane around each lung ____________
5. The only respiratory structures involved in external gas exchange ____________
6. The amount of air inhaled or exhaled during a relaxed breath ____________
7. The ease with which the lungs and thorax can be expanded ____________
8. The maximum volume of air that can be exhaled after maximum inspiration ____________

EXERCISE 18-7.

INSTRUCTIONS

In the blank following each statement, write T if is true and F it is false.

1. Inhalation is the active phase of breathing. ____
2. During quiet breathing, exhalation does not require any muscle contraction. ____
3. The diaphragm rises when it contracts. ____
4. The external intercostal muscles contract during active exhalation. ____
5. The external intercostal muscles contract during a large inhalation. ____
6. The diaphragm relaxes during exhalation. ____

4. List the ways in which oxygen and carbon dioxide are transported in the blood.

EXERCISE 18-8.

INSTRUCTIONS

Write the appropriate term in each blank.

bicarbonate ion	hemoglobin	carbonic anhydrase	carbon dioxide	15%
diffusion	hydrogen ion	oxygen	10%	75%

1. The process by which oxygen moves from the blood into tissues ____________
2. The gas converted into bicarbonate in the blood ____________
3. An important blood buffer produced from carbon dioxide ____________
4. The substance that carries most of the oxygen in the blood ____________
5. The gas that is more concentrated in the blood than in metabolically active tissues ____________
6. An ion that renders blood more acidic ____________
7. The proportion of total blood carbon dioxide dissolved in plasma ____________
8. The proportion of total blood carbon dioxide transported in the form of bicarbonate ____________
9. The proportion of total blood carbon dioxide carried on plasma proteins ____________

5. Describe nervous and chemical controls of respiration.

EXERCISE 18-9.

INSTRUCTIONS

Write the appropriate term in each blank.

hypercapnia	hydrogen ion	bicarbonate ion	phrenic nerve	vagus nerve
brainstem	aortic arch	carbon dioxide	oxygen	

1. The location of the central chemoreceptors ____________
2. A rise in the blood carbon dioxide concentration ____________
3. The location of a peripheral chemoreceptor ____________
4. The substance that acts directly on the central chemoreceptors to stimulate breathing ____________
5. The gas that stimulates breathing when its concentration increases ____________
6. The nerve that controls the diaphragm ____________

6. Give several examples of altered breathing patterns.

EXERCISE 18-10.

INSTRUCTIONS

For each of the following statements, write HYPO if it refers to hypoventilation and HYPER if it refers to hyperventilation.

1. The breathing pattern that causes hypocapnia ______________
2. The breathing pattern resulting from respiratory obstruction ______________
3. The breathing pattern that causes hypercapnia ______________
4. The breathing pattern that causes acidosis ______________
5. The breathing pattern that sometimes occurs during anxiety attacks ______________

EXERCISE 18-11.

INSTRUCTIONS

Write the appropriate term in each blank.

orthopnea	dyspnea	Cheyne-Stokes respiration	hyperpnea
hypopnea	tachypnea	apnea	Kussmaul respiration

1. Difficult or labored breathing ______________
2. An abnormal increase in the depth and rate of breathing ______________
3. A temporary cessation of breathing ______________
4. Difficult breathing that is relieved by sitting upright ______________
5. An abnormal decrease in the depth and rate of breathing ______________
6. Rapid breathing observed during exercise ______________
7. A variable respiratory rhythm observed in some critically ill patients ______________

7. List and define four conditions that result from inadequate breathing.

EXERCISE 18-12.

INSTRUCTIONS

In the blanks below, list and briefly describe four conditions resulting from inadequate breathing.

1. ____
2. ____
3. ____
4. ____

8. Describe several types of respiratory infection (see Exercise 18-13).

9. Describe some allergic responses that affect the respiratory system.

EXERCISE 18-13.

INSTRUCTIONS

Write the appropriate term in each blank.

croup	acute coryza	influenza	lobar pneumonia	TB
bronchopneumonia	effusion	asthma	hay fever	Pneumocystis pneumonia

1. Technical name for the common cold, based on the discharge of fluid from the nose ____________
2. A collection of fluid, as may occur in the pleural space ____________
3. A bacterial or viral infection that affects an entire lung at once ____________
4. An infectious lung disease characterized by the presence of small lung lesions ____________
5. An allergic reaction that affects the upper respiratory tract and eyes ____________
6. An allergic reaction that affects the bronchial tubes ____________
7. A condition in young children in which the airways are constricted as a result of a viral infection ____________
8. Infection of the alveoli occurring mainly in patients with suppressed immune systems ____________
9. A lung infection in which infected alveoli are scattered throughout the lung ____________

10. Name the diseases involved in chronic obstructive pulmonary disease (COPD) (see Exercise 18-14).

11. Describe some disorders that involve the pleurae.

EXERCISE 18-14.

INSTRUCTIONS

Write the appropriate term in each blank.

emphysema	SIDS	chronic bronchitis	atelectasis
pleurisy	hemothorax	pneumothorax	

1. Destruction of the alveoli of the lungs often related to heavy smoking ____________
2. Inflammation of the serous membrane covering the lungs ____________
3. The accumulation of blood in the pleural space ____________
4. The scientific term for a collapsed lung ____________
5. A type of COPD in which the airways are continually inflamed ____________
6. The scientific term for "crib death" ____________

12. Describe equipment used to treat respiratory disorders.

EXERCISE 18-15.

INSTRUCTIONS

Write the appropriate term in each blank.

bronchoscope	oxygen therapy	spirometer
suction apparatus	tracheostomy	tracheotomy

1. An operation to insert a metal or plastic tube into the trachea to serve as an airway for ventilation ______________
2. The incision in the trachea through which the tube is inserted ______________
3. An instrument used to inspect the bronchi and their branches ______________
4. An apparatus used to remove mucus from the respiratory tract ______________
5. An apparatus used to measure lung function ______________

13. Show how word parts are used to build words related to respiration.

EXERCISE 18-16.

INSTRUCTIONS

Complete the following table by writing the correct word part or meaning in the space provided. Write a word that contains each word part in the Example column.

Word Part	Meaning	Example
1. __________	incomplete	__________________
2. spir/o-	________	__________________
3. __________	nose	__________________
4. -centesis	________	__________________
5. -pnea	________	__________________
6. __________	carbon dioxide	__________________
7. __________	lung	__________________
8. __________	air, gas	__________________
9. orth/o-	________	__________________
10. or/o-	________	__________________

Making the Connections

The following concept map deals with the organization of the respiratory system. Each pair of terms is linked together by a connecting phrase into a sentence. The sentence should be read in the direction of the arrow. Complete the concept map by filling in the appropriate term or phrase. There is one right answer for each term. However, there are many correct answers for the connecting phrases (2, 9, and 12).

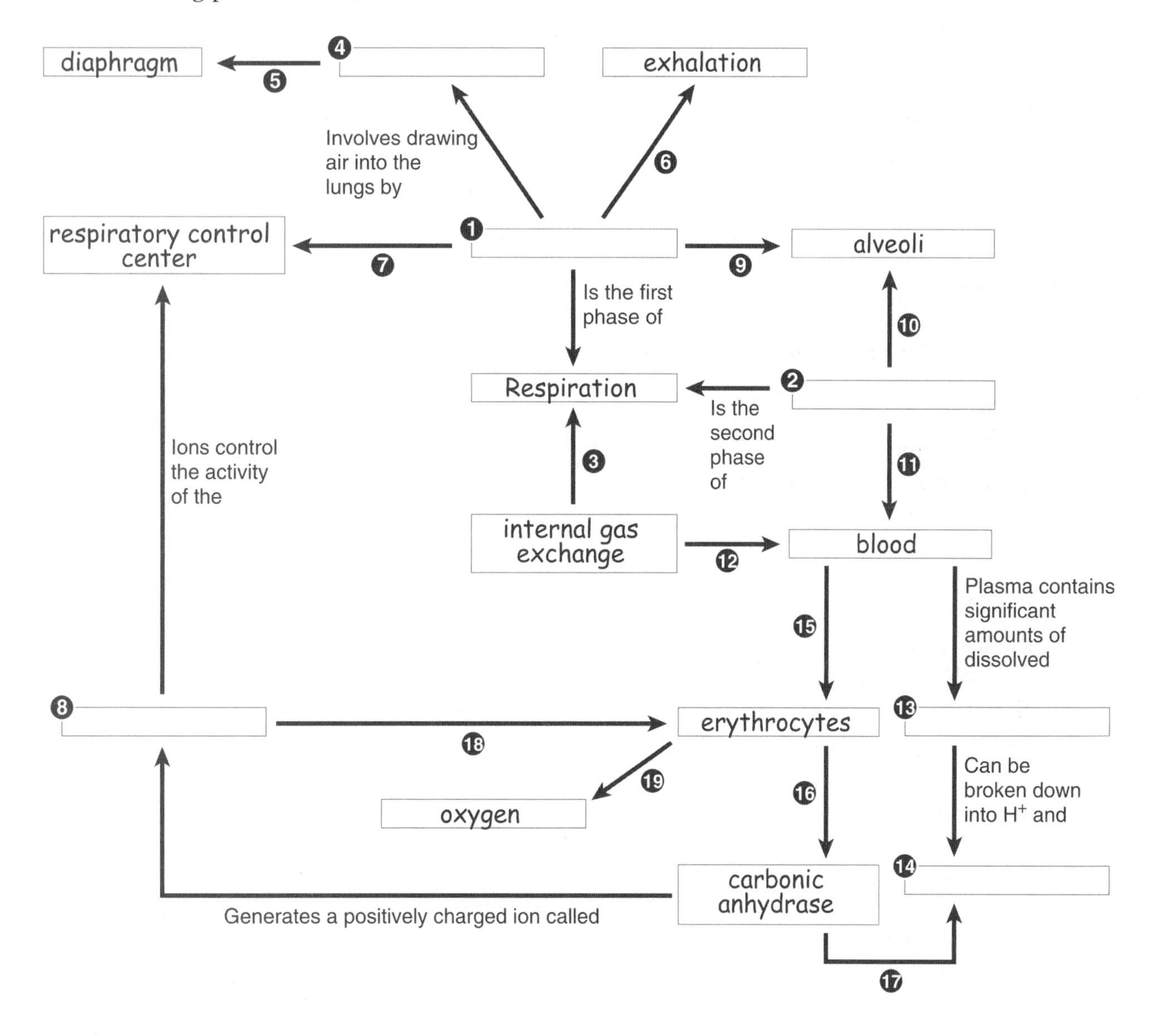

Optional Exercise: Make your own concept map/flow chart based on the structures an oxygen molecule will pass through from the atmosphere to a tissue. Choose your own terms to incorporate into your map, or use the following list: nostrils, atmosphere, nasopharynx, oropharynx, nasal cavities, laryngeal pharynx, trachea, larynx, bronchioles, blood cell, hemoglobin, plasma, tissue, bronchi, and alveoli.

Testing Your Knowledge

Building Understanding

I. Multiple Choice

Select the best answer and write the letter of your choice in the blank.

1. Carbon dioxide will diffuse out of blood during the phase of respiration called 1. _______
   a. internal exchange of gases
   b. external exchange of gases
   c. pulmonary ventilation
   d. none of the above
2. A structural defect of the partition in the nose is called a(n) 2. _______
   a. polyp
   b. epistaxis
   c. bifurcation
   d. deviated septum
3. A rhythmic abnormality in breathing that is seen in critically ill patients is termed 3. _______
   a. hypoxemia
   b. hyperpnea
   c. Cheyne-Stokes respiration
   d. hypocapnia
4. Allergic rhinitis is the medical term for 4. _______
   a. hives
   b. nosebleed
   c. hay fever
   d. asthma
5. Which of the following terms does NOT apply to the cells that line the conducting passages of the respiratory tract? 5. _______
   a. pseudostratified
   b. connective
   c. columnar
   d. ciliated
6. An increase in blood carbon dioxide levels would result in 6. _______
   a. fewer bicarbonate ions in the blood
   b. more hydrogen ions in the blood
   c. more alkaline blood
   d. hypocapnia
7. Which of the following is NOT an upper respiratory infection? 7. _______
   a. acute coryza
   b. RSV
   c. croup
   d. tuberculosis

8. The substance that reduces surface tension in the alveoli is 8. _______
   a. surfactant
   b. bicarbonate
   c. exudate
   d. effusion
9. The residual volume is 9. _______
   a. the amount of air that is always in the lungs, even after a maximal expiration
   b. the total amount of air in the lungs after a maximal inspiration
   c. the amount of air remaining in the lungs after a normal exhalation
   d. the amount of air that can be forced out of the lungs after a normal exhalation
10. Pneumothorax is the 10. _______
    a. presence of blood in the pleural space
    b. presence of pus in the pleural space
    c. removal of air from the pleural space
    d. presence of air in the pleural space

II. Completion Exercise

Write the word or phrase that correctly completes each sentence.

1. The space between the vocal cords is called the ____________________.
2. An abnormal decrease in the depth and rate of respiration is termed ____________________.
3. A lower than normal level of oxygen in tissues is called ____________________.
4. Heart disease and other disorders may cause the bluish color of the skin and visible mucous membranes characteristic of a condition called ____________________.
5. The space between the lungs is called the ____________________.
6. The type of epithelium that lines the nasal cavity is ____________________.
7. The term that is used for the pressure of each gas in a mixture of gases is ____________________.
8. Each heme region of a hemoglobin molecule contains an inorganic element called ____________________.
9. Certain diplococci, staphylococci, chlamydiae, and viruses may cause an inflammation of the lungs. This disease is called ____________________.
10. The scientific name for the organism that causes tuberculosis is ____________________.
11. The nerve that innervates the diaphragm is the ____________________.

Understanding Concepts

I. True/False

For each question, write *T* for true or *F* for false in the blank to the left of each number. If a statement is false, correct it by replacing the underlined term and write the correct statement in the blank below the question.

_______ 1. Hypercapnia results in greater blood acidity.

____

_______ 2. Most carbon dioxide in the blood is carried bound to hemoglobin.

____

_______ 3. The wall of an alveolus is made of stratified squamous epithelium.

____

_______ 4. During internal exchange of gases, oxygen moves down its concentration gradient out of blood.

____

_______ 5. The receptors that detect changes in blood gas concentrations are called mechanoreceptors.

____

_______ 6. The alveoli become filled with exudate in patients suffering from pneumonia.

____

_______ 7. As a result of a chronic lung disease, Ms. L's lungs do not expand very easily. Her lungs are said to be more compliant than normal.

____

_______ 8. Inhalation ALWAYS involves muscle contraction.

____

_______ 9. A tumor resulting from chronic sinusitis that obstructs air movement is called a bronchogenic carcinoma.

____

_______ 10. Hyperventilation results in an increase of carbon dioxide in the blood.

____

II. Practical Applications

Study each discussion. Then write the appropriate word or phrase in the space provided.

➤ Group A

1. Ms. L's complaints included shortness of breath, a chronic cough productive of thick mucus, and a "chest cold" of 2 months' duration. She was advised to quit smoking, a major cause of lung irritation in a group of chronic lung diseases collectively known as ________________.
2. Symptoms in Ms. L's case were due in part to the obstruction of groups of alveoli by mucous plugs. The condition in which the bronchial linings are constantly inflamed and secreting mucus is called ________________.
3. Ms. L was told that the mucus is not moving normally out of her airways because the toxins in cigarette smoke paralyze small cell extensions that beat to create an upward current. These extensions are called ________________.
4. Ms. L's respiratory function was evaluated by quantifying different lung volumes and capacities using a machine called a(n) ________________.
5. Evaluation of Ms. L's respiratory function showed a reduction in the amount of air that could be moved into and out of her lungs. The amount of air that can be expelled by maximum exhalation following maximum inhalation is termed the ________________.
6. The other abnormality in Ms. L's evaluation was also characteristic of her disease. There was an increase in the amount of air remaining in her lungs after a normal expiration. This amount is the ________________.

➤ Group B

1. Baby L was born at 32 weeks gestation. She is obviously struggling for breath, and her skin is bluish in color. This skin discoloration is called ________________.
2. The obstetrician fears that one or both of her lungs cannot inflate. The incomplete expansion of a lung is called ________________.
3. Baby L is placed on pressurized oxygen to inflate the lung. The physician tells the worried parents that her lungs have not yet matured sufficiently to produce the lung substance that reduces the surface tension in alveoli. This substance is called ________________.
4. A lack of this substance in a newborn results in a disorder called ________________.
5. Baby L is administered the substance she is lacking and is soon resting more comfortably. The neonatologist tells her parents that Baby L has an excellent prognosis, but because of her prematurity she will be at greater risk for "crib death," which is now known as ________________.

III. Short Essays

1. Are lungs passive or active players in pulmonary ventilation? Explain.

____

____

____

2. Name some parts of the respiratory tract where gas exchange does NOT occur.

____

____

____

Conceptual Thinking

A. Name the phase of respiration regulated by the respiratory control center.

____

B. Explain how the respiratory control center can alter this phase of respiration.

____

____

____

C. Name the chemical factor(s) that regulate(s) the activity of the control center.

____

____

____

Expanding Your Horizons

Everyone would agree that oxygen is a very useful molecule. Commercial enterprises, working on the premise that more is better, market water supplemented with extra oxygen. They claim that consumption of hyperoxygenated water increases alertness and exercise performance. Do these claims make sense? Do we obtain oxygen from the lungs or the digestive tract? A logical extension of this premise is that soda pop, supplemented with carbon dioxide, would increase carbon dioxide in the blood and increase the breathing rate! Is this true? You can read about another oxygen gimmick, oxygen bars, on the website of the Food and Drug Administration.

Resources

1. Bren L. Oxygen bars: Is a breath of fresh air worth it? 2002. Available at: http://www.fda.gov/fdac/features/2002/602_air.html.

CHAPTER 19

The Digestive System

Overview

The food we eat is made available to cells throughout the body by the complex processes of **digestion** and **absorption.** These are the functions of the **digestive system**, composed of the **digestive tract** and the **accessory organs**.

The digestive tract, consisting of the **mouth**, **pharynx**, **esophagus**, **stomach**, and small and large **intestines**, forms a continuous passageway in which ingested food is prepared for use by the body and waste products are collected to be expelled from the body. The accessory organs, the **salivary glands**, **liver**, **gallbladder**, and **pancreas**, manufacture and store various enzymes and other substances needed in digestion.

Digestion begins in the mouth with the digestion of starch. It continues in the stomach, where proteins are digested, and is completed in the small intestine. Most absorption of digested food also occurs in the small intestine through small projections of the lining called **villi.** The products of carbohydrate (monosaccharides) and protein (amino acids) digestion are absorbed into capillaries, but the products of fat digestion (glycerol and fatty acids) are absorbed into **lacteals.**

The process of digestion is controlled by both nervous and hormonal mechanisms, which regulate the activity of the digestive organs and the rate at which food moves through the digestive tract.

Addressing the Learning Outcomes

1. Name the three main functions of the digestive system.

EXERCISE 19-1.

INSTRUCTIONS

List the three main functions of the digestive system in the blanks below, in the order in which they occur.

1. ______________________
2. ______________________
3. ______________________

2. Describe the four layers of the digestive tract wall.

EXERCISE 19-2: The Wall of the Small Intestine (Text Fig. 19-1)

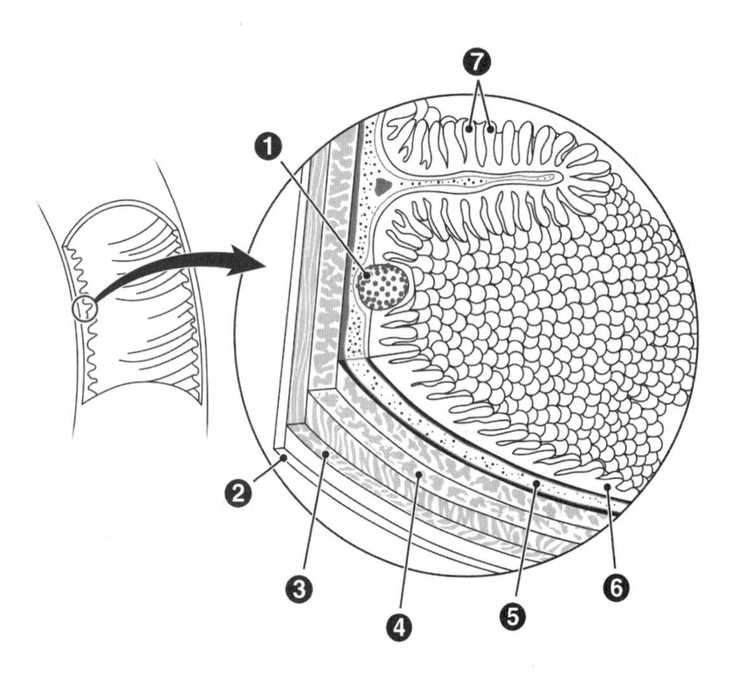

INSTRUCTIONS

1. Write the names of the intestinal layers and associated structures on the appropriate numbered lines in different colors. Use the same color for parts 6 and 7.
2. Color the layers and structures on the diagram with the corresponding color.

1. ______________________
2. ______________________
3. ______________________
4. ______________________
5. ______________________
6. ______________________
7. ______________________

EXERCISE 19-3.

INSTRUCTIONS

Write the appropriate term in each blank.

digestion	absorption	elimination	mucous membrane	submucosa
serosa	smooth muscle	squamous epithelium	simple columnar epithelium	

1. The transfer of nutrients into the bloodstream ____________
2. The visceral peritoneum attached to the surface of a digestive organ ____________
3. The layer of connective tissue beneath the mucous membrane in the wall of the digestive tract ____________
4. The layer of the digestive tract wall that is responsible for peristalsis ____________
5. The breakdown of food into small particles that can pass through intestinal cells ____________
6. The layer of the digestive tract wall that forms villi in the small intestine ____________
7. The type of epithelial tissue lining the esophagus ____________
8. The type of epithelial tissue lining the stomach ____________

3. Differentiate between the two layers of the peritoneum.

EXERCISE 19-4: Abdominal Cavity Showing Peritoneum (Text Fig. 19-3)

INSTRUCTIONS

1. Write the names of the abdominal organs on the appropriate numbered lines 1 to 9 in different colors. Use the same color for parts 3 and 4.
2. Color the organs on the diagram with the corresponding color.
3. Label the parts of the peritoneum (lines 10 to 15). Color the greater and lesser peritoneal cavities in contrasting colors.

1. ____________________
2. ____________________
3. ____________________
4. ____________________
5. ____________________
6. ____________________
7. ____________________
8. ____________________
9. ____________________
10. ____________________
11. ____________________
12. ____________________
13. ____________________
14. ____________________
15. ____________________

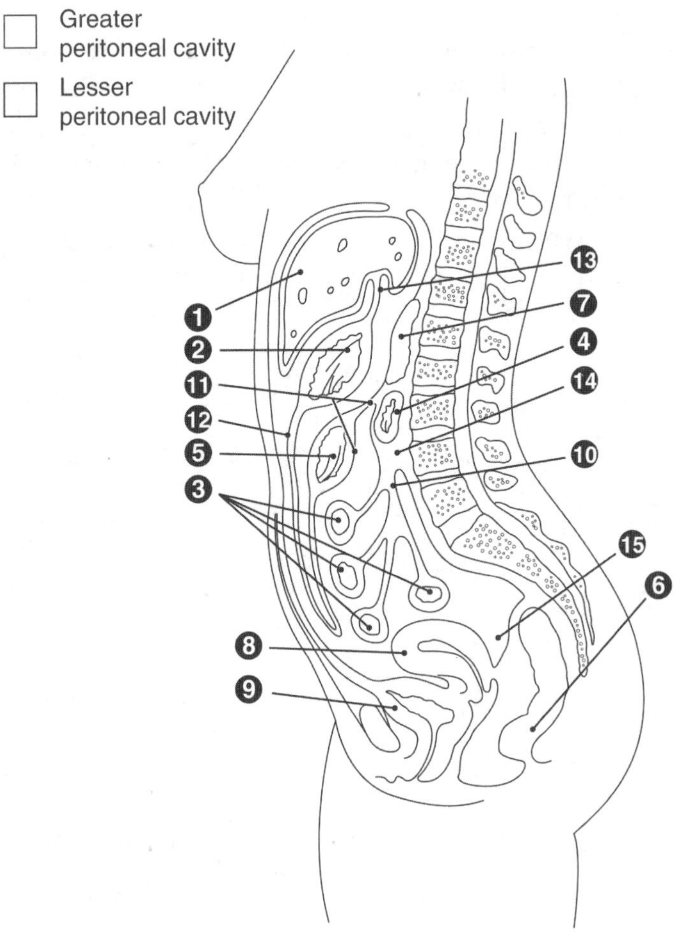

☐ Greater peritoneal cavity

☐ Lesser peritoneal cavity

EXERCISE 19-5.

INSTRUCTIONS

Write the appropriate term in each blank.

parietal peritoneum	visceral peritoneum	greater peritoneal cavity	lesser peritoneal cavity
mesocolon	greater omentum	lesser omentum	

1. The innermost layer of the serous membrane lining the abdominopelvic cavity ________________
2. The outer layer of the serous membrane lining the abdominopelvic cavity ________________
3. The subdivision of the peritoneum that contains fat and hangs over the front of the intestine ________________
4. The subdivision of the peritoneum extending between the stomach and liver ________________
5. The subdivision of the peritoneum that extends from the colon to the posterior abdominal wall ________________
6. The fluid-filled cavity that extends behind the stomach to the liver and the posterior portion of the diaphragm ________________

4. Name and locate the different types of teeth.

EXERCISE 19-6: Digestive System (Text Fig. 19-4)

INSTRUCTIONS

1. Trace the path of food through the digestive tract by labeling parts 1 to 12. You can color all of these structures orange.
2. Write the names of the accessory organs and ducts on the appropriate lines in different colors. Use black for structure 20, because it will not be colored.
3. Color the accessory organs on the diagram with the corresponding colors.

1. ____________________
2. ____________________
3. ____________________
4. ____________________
5. ____________________
6. ____________________
7. ____________________
8. ____________________
9. ____________________
10. ____________________
11. ____________________
12. ____________________
13. ____________________
14. ____________________
15. ____________________
16. ____________________
17. ____________________
18. ____________________
19. ____________________
20. ____________________

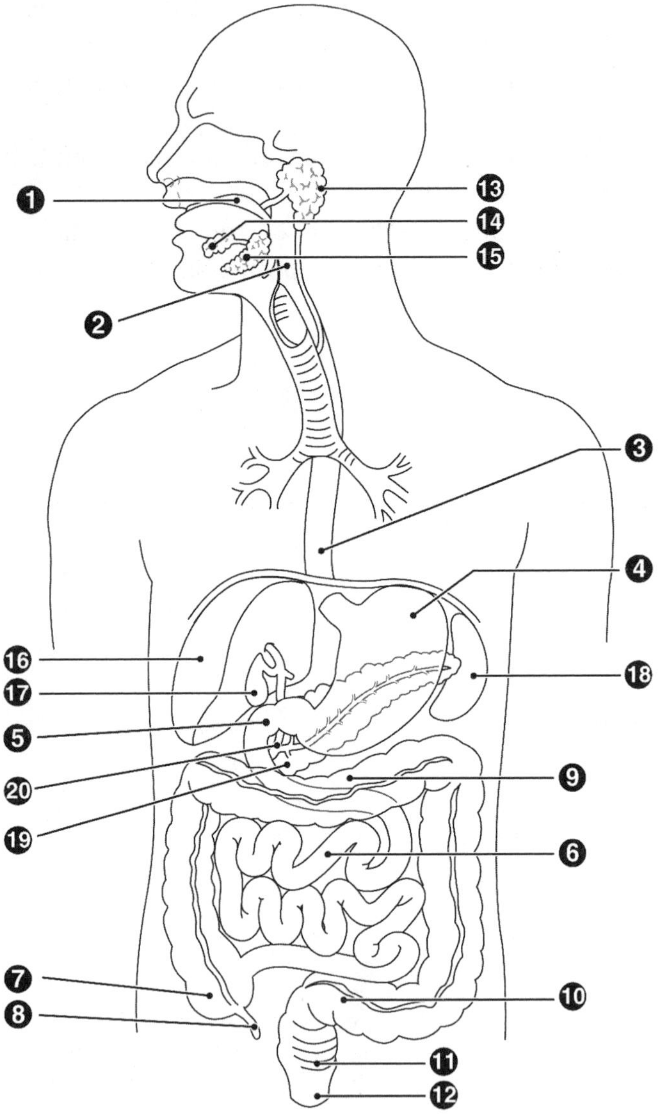

EXERCISE 19-7: The Mouth (Text Fig. 19-5)

INSTRUCTIONS

1. Write the names of the teeth on the appropriate lines 1 to 6 in different colors. Use the same color to label parts 5 and 6.
2. Color all of the teeth with the corresponding colors.
3. Label the other parts of the mouth.

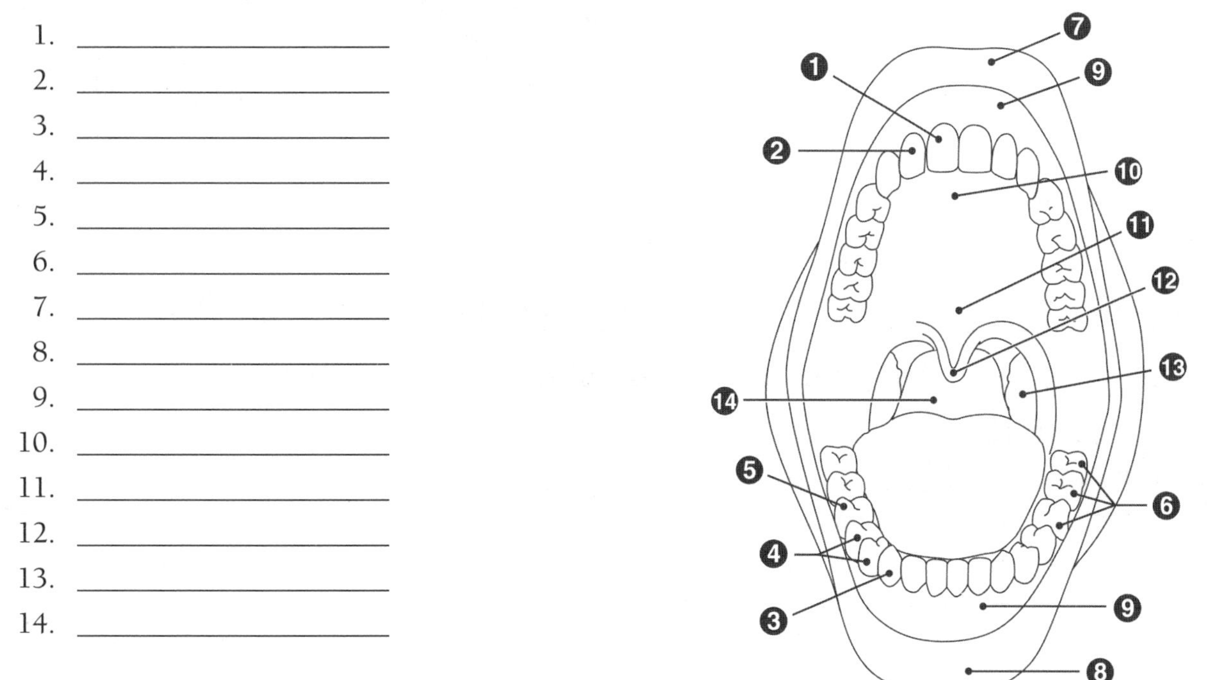

1. ____________
2. ____________
3. ____________
4. ____________
5. ____________
6. ____________
7. ____________
8. ____________
9. ____________
10. ____________
11. ____________
12. ____________
13. ____________
14. ____________

EXERCISE 19-8: Molar (Text Fig. 19-6)

INSTRUCTIONS

1. Write the names of the two divisions of a tooth in the numbered boxes.
2. Write the names of the parts of the tooth and gums on the appropriate numbered lines in different colors. Use the same color for parts 3 and 4 and for 6 and 7. Use a dark color for structure 8, because it will be outlined.
3. Label the other parts of the mouth.

1. ____________
2. ____________
3. ____________
4. ____________
5. ____________
6. ____________
7. ____________
8. ____________
9. ____________
10. ____________

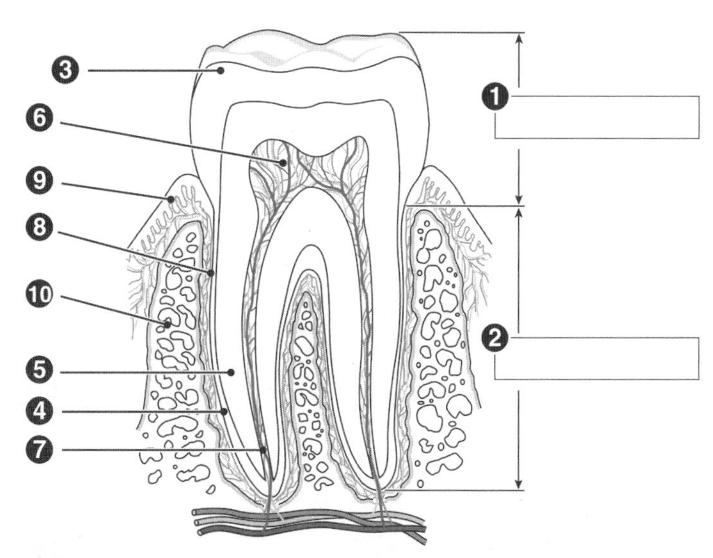

EXERCISE 19-9.

INSTRUCTIONS

Write the appropriate term in each blank.

deglutition	mastication	dentin	gingiva
deciduous	cuspids	incisors	

1. Term that describes the baby teeth, based on the fact that they are lost ____________
2. The process of chewing ____________
3. The act of swallowing ____________
4. The medical term for the gum ____________
5. The eight cutting teeth located in the front part of the oral cavity ____________
6. A calcified substance making up most of the tooth structure ____________

5. Name and describe the functions of the organs of the digestive tract.

EXERCISE 19-10: Longitudinal Section of the Stomach (Text Fig. 19-7)

INSTRUCTIONS

1. Label the parts of the stomach, esophagus, and duodenum (parts 1 to 7).
2. Label the layers of the stomach wall (parts 8 to 11). You can use colors to highlight the different muscular layers.
3. Label the two stomach curvatures (labels 12 and 13).

1. ____________
2. ____________
3. ____________
4. ____________
5. ____________
6. ____________
7. ____________
8. ____________
9. ____________
10. ____________
11. ____________
12. ____________
13. ____________

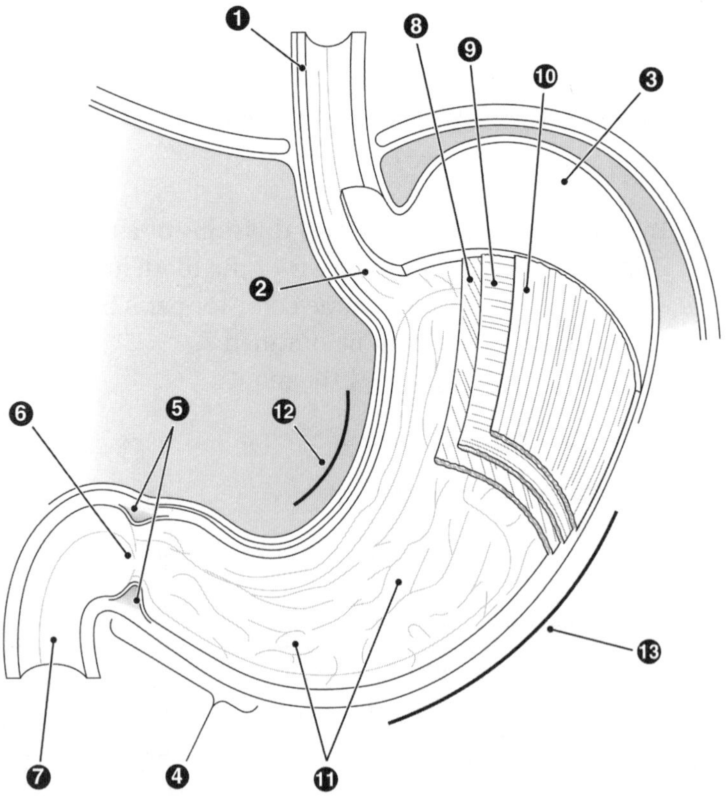

EXERCISE 19-11: The Small and Large Intestines (Text Fig. 19-8)

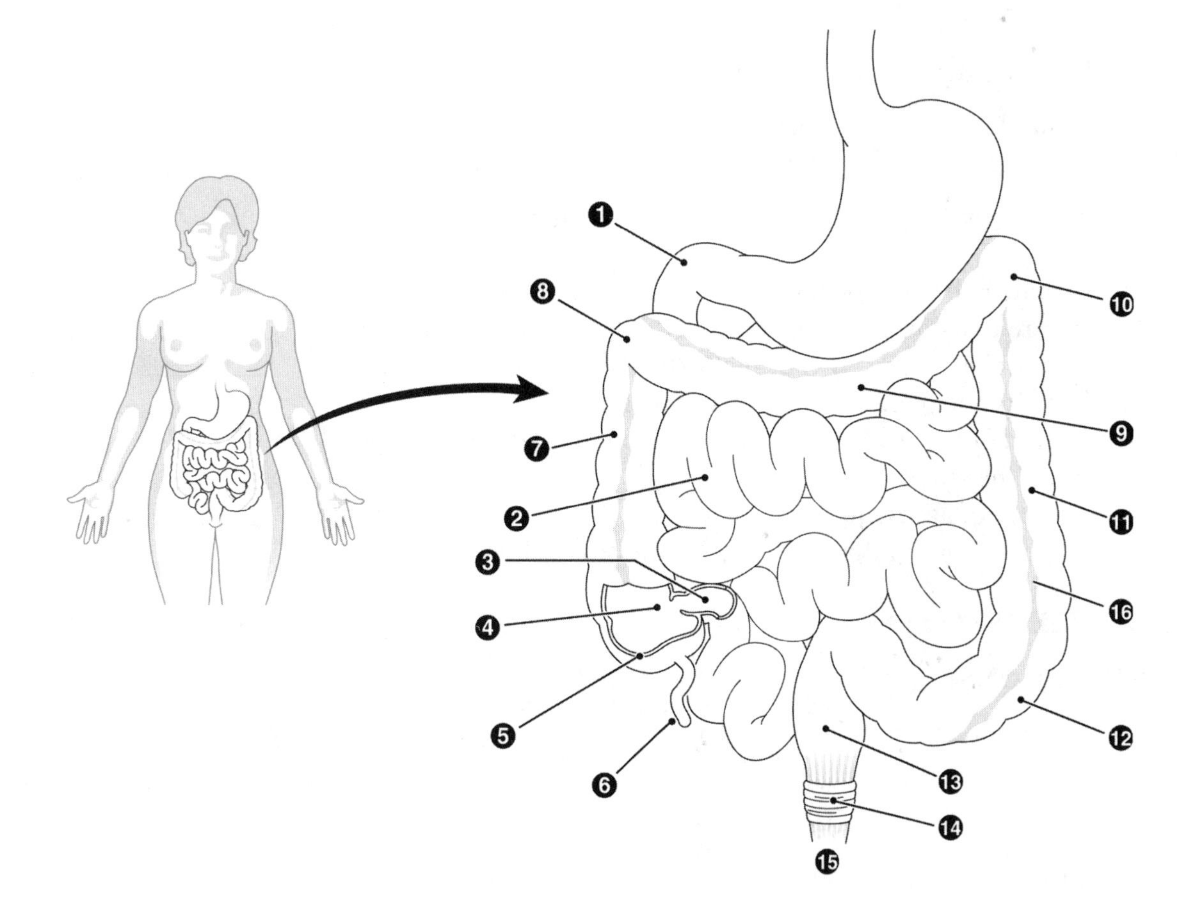

INSTRUCTIONS

Label the indicated parts.

1. ____________________
2. ____________________
3. ____________________
4. ____________________
5. ____________________
6. ____________________
7. ____________________
8. ____________________
9. ____________________
10. ____________________
11. ____________________
12. ____________________
13. ____________________
14. ____________________
15. ____________________
16. ____________________

EXERCISE 19-12.

INSTRUCTIONS

Write the appropriate term in each blank.

ileum	hard palate	soft palate	epiglottis	pyloric sphincter
LES	rugae	chyme	duodenum	jejunum

1. The valve between the distal end of the stomach and the small intestine ________________
2. The structure that guards the entrance into the stomach ________________
3. A structure that covers the opening of the larynx during swallowing ________________
4. The part of the oral cavity roof that extends to form the uvula ________________
5. The final, and longest, section of the small intestine ________________
6. The section of the small intestine that receives gastric juices and food from the stomach ________________
7. The mixture of gastric juices and food that enters the small intestine ________________
8. Folds in the stomach that are absent if the stomach is full ________________

EXERCISE 19-13.

INSTRUCTIONS

Write the appropriate term in each blank.

villi	lacteal	teniae coli	cecum	transverse colon
vermiform appendix	sigmoid colon	rectum	ileocecal valve	

1. The part of the large intestine just proximal to the anus ________________
2. The small blind tube attached to the first part of the large intestine ________________
3. The sphincter that prevents food moving from the large intestine into the small intestine ________________
4. Fingerlike extensions of the mucosa in the small intestine ________________
5. A blind-ended lymphatic vessel that absorbs fat ________________
6. Bands of longitudinal muscle in the large intestine ________________
7. The portion of the large intestine that extends across the abdomen ________________
8. The most proximal part of the large intestine ________________

6. Name and describe the functions of the accessory organs of digestion.

EXERCISE 19-14: Accessory Organs (Text Fig. 19-10)

INSTRUCTIONS

1. Write the names of the labeled parts on the appropriate lines in different colors.
2. Color the structures on the diagram.

1. ____________________
2. ____________________
3. ____________________
4. ____________________
5. ____________________
6. ____________________
7. ____________________
8. ____________________
9. ____________________
10. ____________________

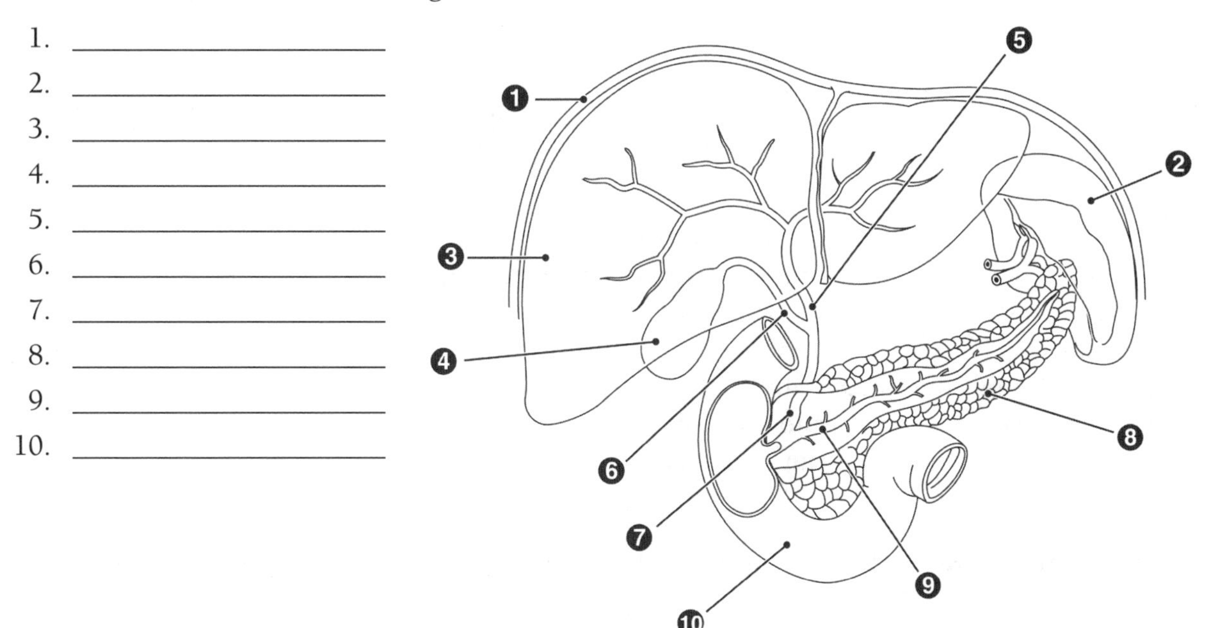

EXERCISE 19-15.

INSTRUCTIONS

Write the appropriate term in each blank.

parotid glands	submandibular glands	sublingual glands
gallbladder	liver	pancreas

1. The gland that secretes bicarbonate and digestive enzymes ____________
2. An organ that stores nutrients and releases them as needed into the bloodstream ____________
3. The accessory organ that stores bile ____________
4. The salivary glands that are inferior and anterior to the ear ____________
5. Glands found just under the tongue that secrete into the oral cavity ____________

7. Describe how bile functions in digestion (see Exercise 19-16).

8. Name and locate the ducts that carry bile from the liver into the digestive tract.

EXERCISE 19-16.

INSTRUCTIONS

Write the appropriate term in each blank.

common bile duct	bile	common hepatic duct	cystic duct	pancreatic duct
bicarbonate	bilirubin	glycogen	urea	

1. A substance that emulsifies fat ______________
2. The form in which glucose is stored in the liver ______________
3. A waste product resulting from the destruction of red blood cells ______________
4. A waste product synthesized by the liver as a result of protein metabolism ______________
5. The duct connecting the hepatic duct to the gallbladder ______________
6. The duct that delivers bile into the duodenum ______________
7. The duct that carries bile from both lobes of the liver to the common bile duct ______________

9. Explain the role of enzymes in digestion and give examples of enzymes.

EXERCISE 19-17.

INSTRUCTIONS

Write the appropriate term in each blank.

protein	hydrolysis	fat	nuclease	pepsin
sodium bicarbonate	lipase	maltose	trypsin	maltase

1. An enzyme that acts on a particular type of disaccharide ______________
2. A substance (NOT an enzyme) released into the small intestine that neutralizes the acidity in chyme ______________
3. A substance that digests DNA ______________
4. A pancreatic enzyme that splits proteins into amino acids ______________
5. An enzyme secreted into the stomach that initiates protein digestion ______________
6. The splitting of food molecules by the addition of water ______________
7. Lipase participates in the digestion of this nutrient ______________

10. Name the digestion products of fats, proteins, and carbohydrates.

EXERCISE 19-18.

INSTRUCTIONS

For each of the following statements, write if they apply to carbohydrates (C), proteins (P), or fats (F).

1. This nutrient type is digested into sugars. ____
2. This nutrient type is digested into amino acids. ____
3. This nutrient type is digested into glycerol and fatty acids. ____
4. This nutrient type is partially digested in the stomach in both children and adults. ____
5. This nutrient type can be broken down into disaccharides. ____

11. Define *absorption*.

EXERCISE 19-19.

INSTRUCTIONS

Write a definition of *absorption* in the space below.

____

12. Define *villi* and state how villi function in absorption.

EXERCISE 19-20.

INSTRUCTIONS

For each statement, state whether it is true (T) or false (F).

1. Digested carbohydrates are absorbed into lacteals. ____
2. Villi are small extensions of individual intestinal cells. ____
3. Villi are folds in the mucosa, each composed of many cells. ____
4. Digested fats are absorbed into lacteals. ____

13. Explain the use of feedback in regulating digestion and give several examples.

EXERCISE 19-21.

INSTRUCTIONS

Complete the following discussion by choosing one of the terms in brackets after each blank.

Digestion is regulated by a type of feedback that maintains homeostasis and is called (1) ____________________ [negative, positive] feedback, For instance, (2) ________________ [gastrin, secretin] is released from duodenal cells in response to (3) ____________________ [increased, decreased] acidity in the intestine. This hormone, in turn, stimulates (4) ____________________ [bile, bicarbonate] release from the pancreas, which neutralizes the change in acidity. Once the intestinal pH returns to normal, the release of this hormone is (5) ____________________ [increased, decreased].

14. List several hormones involved in regulating digestion.

EXERCISE 19-22.

INSTRUCTIONS

Write the appropriate term in each blank.

leptin	gastrin	gastric-inhibitory peptide
cholecystokinin (CCK)	secretin	

1. A hormone released from fat cells that inhibits appetite ________________
2. A duodenal hormone that stimulates insulin release ________________
3. A hormone that stimulates the secretion of gastric juice and increases stomach motility ________________
4. An intestinal hormone that causes the gallbladder to contract, releasing bile ________________

15. Describe common disorders of the digestive tract and the accessory organs.

EXERCISE 19-23.

INSTRUCTIONS

Write the appropriate term in each blank.

anorexia	bulimia	peritonitis	periodontitis	hiatal hernia
cirrhosis	GERD	emesis	Vincent disease	

1. Any infection of the gums and supporting bone ________________
2. Inflammation of the membrane lining the abdominal cavity ________________
3. A chronic condition in which stomach contents often flow into the esophagus ________________
4. A protrusion of the stomach through the diaphragm ________________
5. A specific type of gum infection caused by a spirochete or bacillus ________________
6. A chronic loss of appetite ________________
7. Vomiting ________________
8. An eating disorder in which affected individuals "binge and purge" ________________

EXERCISE 19-24.

INSTRUCTIONS

Write the appropriate term in each blank.

diverticulitis	obstipation	flatulence	Crohn disease	ulcerative colitis
hepatitis	gastroenteritis	cirrhosis	diverticulosis	

1. A general term referring to inflammation of the liver ________
2. A chronic disorder in which liver cells are replaced by scar tissue ________
3. A form of inflammatory bowel disease associated with inflammation of the large intestine and rectum ________
4. An autoimmune disease involving inflammation of the small intestine ________
5. Infection in the small pouches found in the intestinal wall ________
6. Extreme constipation ________
7. A condition characterized by large numbers of small pouches in the intestines ________
8. Inflammation of the stomach and intestine ________

16. Show how word parts are used to build words related to digestion.

EXERCISE 19-25.

INSTRUCTIONS

Complete the following table by writing the correct word part or meaning in the space provided. Write a word that contains each word part in the Example column.

Word Part	Meaning	Example
1. ________	starch	________
2. mes/o-	________	________
3. ________	intestine	________
4. chole	________	________
5. bil/i	________	________
6. ________	bladder, sac	________
7. ________	stomach	________
8. ________	away from	________
9. hepat/o	________	________
10. lingu/o	________	________

Making the Connections

The following concept map deals with the structure and regulation of the gastrointestinal system. Each pair of terms is linked together by a connecting phrase into a sentence. The sentence should be read in the direction of the arrow. Complete the concept map by filling in the appropriate term or phrase. There is one right answer for each term (1, 2, 4, 7, 15, 17, 18). However, there are many correct answers for the connecting phrases. Write the connecting phrases along the arrows if possible. If your phrases are too long, you may want to write them in the margins or on a separate sheet of paper.

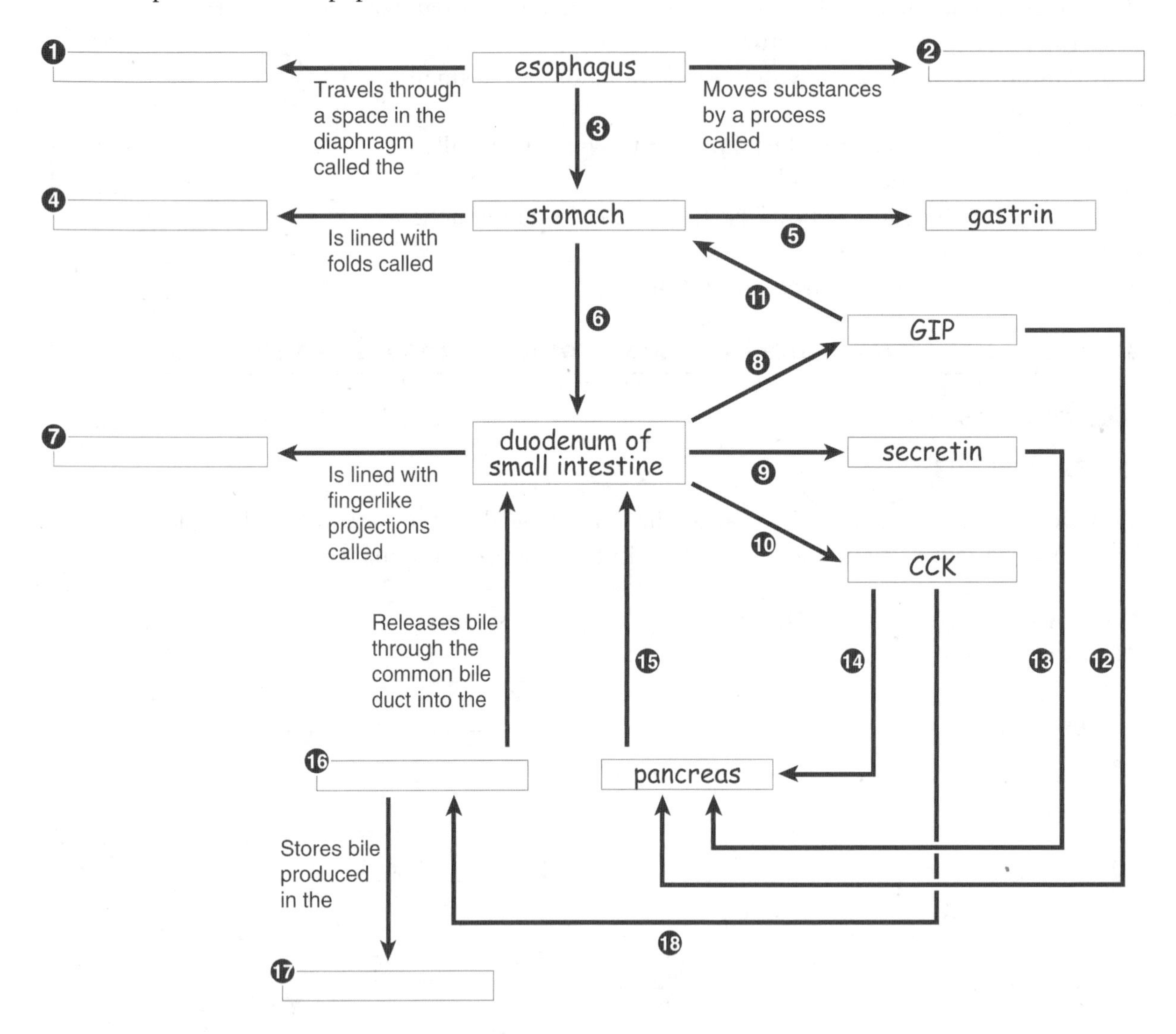

Optional Exercise: Make your own concept map, based on the three processes of digestion and ᵛ they apply to proteins, sugars, and fats. Choose your own terms to incorporate into your ʳ use the following list: digestion, absorption, elimination, stomach, small intestine, large ᶠats, carbohydrates, proteins, amylase, lipase, bile, hydrochloric acid, pepsin, trypsin, ¹ maltase.

Testing Your Knowledge

Building Understanding

I. Multiple Choice

Select the best answer and write the letter of your choice in the blank.

1. Which teeth would NOT be found in a 20-year-old man? 1. _______
   a. bicuspid
   b. cuspid
   c. deciduous
   d. incisor
2. Which of the following is the correct order of tissue from the innermost to the outermost layer in the wall of the digestive tract? 2. _______
   a. submucosa, serous membrane, smooth muscle, mucous membrane
   b. smooth muscle, serous membrane, mucous membrane, submucosa
   c. serous membrane, smooth muscle, submucosa, mucosa
   d. mucous membrane, submucosa, smooth muscle, serous membrane
3. The parotid salivary gland is located 3. _______
   a. inferior and anterior to the ear
   b. under the tongue
   c. in the cheek
   d. in the oropharynx
4. A highly contagious type of gingivitis caused by a spirochete or bacillus that is characterized by inflammation and ulceration is called 4. _______
   a. peritonitis
   b. enteritis
   c. leukoplakia
   d. Vincent disease
5. Which of the following is NOT a portion of the peritoneum? 5. _______
   a. mesocolon
   b. mesentery
   c. hiatus
   d. greater omentum
6. The active ingredients in gastric juice are 6. _______
   a. amylase and pepsin
   b. pepsin and hydrochloric acid
   c. maltase and secretin
   d. bile and trypsin
7. Gastric-inhibitory peptide stimulates the release of 7. _______
   a. gastric juice
   b. insulin
   c. bicarbonate
   d. none of the above
8. An instrument used to examine the lower portion of the colon is a(n) 8. _______
   a. sigmoidoscope
   b. bronchoscope
   c. catheter
   d. electrocardiogram

9. In the adult, fats are digested in the — 9. _______
   a. mouth
   b. stomach
   c. small intestine
   d. all of the above
10. Which of the following does NOT occur in the mouth? — 10. _______
   a. mastication
   b. digestion of starch
   c. absorption of nutrients
   d. ingestion
11. Which of the following is an enzyme? — 11. _______
   a. bile
   b. gastrin
   c. trypsin
   d. secretin
12. Which of the following is associated with the intestine? — 12. _______
   a. rugae
   b. lacteals
   c. LES
   d. greater curvature

II. Completion Exercise

Write the word or phrase that correctly completes each sentence.

1. The process by which ingested nutrients are broken down into smaller components is called ____________________.
2. Any inflammation of the liver is called ____________________.
3. The lower part of the colon bends into an S shape, so this part is called the ____________________.
4. A temporary storage section for indigestible and unabsorbable waste products of digestion is the ____________________.
5. The wavelike movement created by alternating muscle contractions is called ____________________.
6. A common cause of tooth loss is infection of the gums and bone around the teeth, a condition called ____________________.
7. The ascending and descending colon can be examined using a type of endoscope called a(n) ____________________.
8. Most of the digestive juices contain substances that cause the chemical breakdown of foods without entering into the reaction themselves. These catalytic agents are ____________________.

The portion of the peritoneum extending between the stomach and liver is called the ____________________.

cess of swallowing is called ____________________.

rgely composed of a calcified substance called ____________________.

passes through the diaphragm at a point called the ____________________.

hat stimulates gallbladder contraction is called ____________________.

ing folds in the plasma membrane of intestinal epithelial cells are called ____________.

15. The stomach enzyme involved in protein digestion is called ____________________.

Understanding Concepts

I. True/False

For each question, write *T* for true or *F* for false in the blank to the left of each number. If a statement is false, correct it by replacing the underlined term and write the correct statement in the blank below the question.

_______ 1. Ms. J is suffering from inflammation of the serous membrane lining the abdomen as a complication of appendicitis. This inflammation is called periodontitis.

____

_______ 2. There are 32 deciduous teeth.

____

_______ 3. The layer of the peritoneum attached to the liver is the visceral peritoneum.

____

_______ 4. Mr. D suffers from an ulcer because of a weakness in his LES. This ulcer is most likely in the duodenum.

____

_______ 5. Ms. Q is deficient in lactase. She will be unable to digest some carbohydrates.

____

_______ 6. Trypsin is secreted by the gastric glands.

____

_______ 7. Increased acidity in the chyme could be neutralized by the actions of the hormone gastrin.

____

_______ 8. Fats are absorbed into capillaries in the intestinal villi.

____

_______ 9. The middle section of the small intestine is called the jejunum.

____

_______ 10. Folds in the stomach wall are called villi.

____

_______ 11. The common bile duct delivers bile from the liver and gallbladder into the duodenum.

____

_______ 12. Amylase is involved in the digestion of carbohydrates.

____

_______ 13. The pancreas is responsible for the synthesis of urea.

____

II. Practical Applications

Study each discussion. Then write the appropriate word or phrase in the space provided.

1. Mr. P, age 42, came to the clinic complaining of pain in the "pit of the stomach." He was a tense man who divided up his long working hours with coffee and cigarette breaks. When asked about his alcohol consumption, Mr. C mentioned that he drank three or four beers and a glass of wine each night, and significantly more on the weekends. Endoscopy showed inflammation of the innermost layer of the stomach. A general term indicating inflammation of the stomach lining is _______________.
2. The name of the layer that is inflamed in Mr. P's stomach is the _______________.
3. A damaged area was also found in the most proximal part of the small intestine. This section of the small intestine is called the _______________.
4. Mr. P was diagnosed with an ulcer and given a prescription for antibiotics. The antibiotics will treat one of the causative factors in ulcer formation, a bacterium called _______________.
5. Mr. P also mentioned that his dentist had noticed a number of white patches in his mouth. These white patches characterize a condition that is common in smokers and may lead to cancer. This precancerous condition is called _______________.
6. The physician had almost completed her physical exam of Mr. P when she felt the edge of his liver about 5 cm (2 in) below the ribs. She also noticed that the whites of his eyes were tinged with yellow. This coloration of the sclera (and/or the skin) is termed _______________.
7. The yellowish coloration results from the release of a hepatic secretion into the bloodstream. This secretion, which is needed in digestion, is named _______________.
8. Based on her clinical findings and Mr. P's self-reported alcohol abuse, the physician suspected that Mr. P was suffering from a chronic liver disease in which active liver cells are replaced [illegible] connective tissue. This disease is called _______________.

III. Short Essays

1. Describe some features of the small intestine that increase the surface area for absorption of nutrients.

____

____

____

2. List four differences between the digestion and/or absorption of fats and proteins.

____

____

____

Conceptual Thinking

1. Ms. M is taking a drug that blocks the action of cholecystokinin. Discuss the impact of this drug on digestion.

____

____

____

2. Is bile an enzyme? Explain your answer.

____

____

____

Expanding Your Horizons

A growing number of young women (and even some young men) are "Dying to Be Thin." Due to eating disorders, they starve themselves and/or binge-and-purge, sometimes to death. You can learn more about eating disorders by viewing the PBS Nova program "Dying to Be Thin," which is available through the PBS website (http://shop.wgbh.org). The two journal articles listed below discuss eating disorders from a nurse's perspective.

Resources

1. Clark-Stone S, Joyce H. Understanding eating disorders. *Nurs Times* 2003; 99:20–23.
2. Murphy B, Manning Y. An introduction to anorexia nervosa and bulimia nervosa. *Nurs Stand* 2003; 18:45–52.

CHAPTER 20

Metabolism, Nutrition, and Body Temperature

Overview

The nutrients that reach the cells following digestion and absorption are used to maintain life. All the physical and chemical reactions that occur within the cells make up **metabolism**, which has two phases: a breakdown phase, or **catabolism**, and a building phase, or **anabolism.** Nutrients are oxidized to yield energy for the cells in the form of ATP using catabolic reactions. This process, termed **cellular respiration**, occurs in two steps: the first is **anaerobic** (does not require oxygen) and produces a small amount of energy; the second is **aerobic** (requires oxygen). This second step occurs within the cells' mitochondria. It yields a large amount of the energy contained in the nutrient plus carbon dioxide and water.

By the various metabolic pathways, the breakdown products of food can be built into substances needed by the body. The **essential** amino acids and fatty acids cannot be manufactured internally and must be ingested in food. **Minerals** and **vitamins** are also needed in the diet for health. A balanced diet includes carbohydrates, proteins, and fats consumed in amounts relative to individual activity levels. Excessive nutrient intake can result in obesity, whereas insufficient or inappropriate nutrient intake results in malnutrition.

The rate at which energy is released from nutrients is termed the **metabolic rate.** It is affected by many factors including age, size, sex, activity, and hormones. Some of the energy in nutrients is released as heat. Heat production is greatly increased during periods of increased muscular or glandular activity. Most heat is lost through he skin, but heat is also dissipated through exhaled air and eliminated waste prod- (urine and feces). The **hypothalamus** maintains body temperature at approxi- 7° C (98.6° F) by altering blood flow through the surface blood vessels and f sweat glands and muscles. Abnormalities of body temperature are a tic tool. The presence of **fever**—an abnormally high body temper- ates infection, but may also indicate a toxic reaction, a brain in- ders. The opposite of fever is **hypothermia**—an exceedingly ure—which most often occurs when the body is exposed to very perature. Hypothermia can cause serious damage to body tissues.

Addressing the Learning Outcomes

1. Differentiate between catabolism and anabolism.

EXERCISE 20-1.

INSTRUCTIONS

Which of the following statements apply to catabolism (C) and which apply to anabolism (A)?

1. The metabolic breakdown of complex compounds ________________
2. The metabolic building of simple compounds into substances needed by cells ________________
3. This process usually releases energy ________________

2. Differentiate between the anaerobic and aerobic phases of cellular respiration and give the end products and the relative amount of energy released by each.

EXERCISE 20-2.

INSTRUCTIONS

Which of the following statements apply to the anaerobic phase (AN) and which apply to the aerobic phase (AE) of cellular respiration?

1. This process generates 2 ATP per glucose molecule ____
2. This process generates about 30 ATP per glucose molecule ____
3. The end products are carbon dioxide and water ____
4. The end product is pyruvic acid ____
5. This process can occur in the absence of oxygen ____
6. This process requires oxygen ____
7. This process occurs first ____

3. Define metabolic rate and name several factors that affect the metabolic rate.

EXERCISE 20-3.

INSTRUCTIONS

Write a definition for each term in the blanks provided.

1. metabolism ____

____

2. metabolic rate ____

____

3. basal metabolism ____

____

4. Explain the roles of glucose and glycogen in metabolism (see Exercise 20-4).

5. Compare the energy contents of fats, proteins, and carbohydrates.

EXERCISE 20-4.

INSTRUCTIONS

Write the appropriate term in each blank.

glycolysis	lactic acid	pyruvic acid	glycerol
glycogen	deamination	fat	protein

1. The storage form of glucose ______________
2. A modification of amino acids that occurs before they can be oxidized for energy ______________
3. An organic product of glucose catabolism that can be completely oxidized within the mitochondria ______________
4. An organic substance produced when a muscle is generating energy in the absence of oxygen ______________
5. The nutrient type that generates the most energy per gram ______________
6. A product of fat digestion that can be used for energy ______________
7. The nutrient type that is used to generate energy only in emergencies ______________

6. Define *essential amino acid* (see Exercise 20-5).

7. Explain the roles of minerals and vitamins in nutrition and give examples of each.

EXERCISE 20-5.

INSTRUCTIONS

Write the appropriate term in each blank below.

essential amino acids	nonessential amino acids	essential fatty acids	antioxidants
trace elements	vitamins	minerals	

1. A class of substances that stabilizes free radicals ______________
2. Minerals required in extremely small amounts ______________
3. Complex organic molecules that are essential for metabolism ______________
4. Protein components that must be taken in as part of the diet ______________
5. Inorganic elements needed for body structure and many body functions ______________
6. Protein building blocks that can be manufactured by the body ______________
7. Linoleic acid and linolenic acid are examples of ______________

EXERCISE 20-6.

INSTRUCTIONS

Write the appropriate term in each blank below.

zinc	iodine	iron	potassium	calcium
folate	calciferol	riboflavin	vitamin A	vitamin K

1. The vitamin that prevents dry, scaly skin and night blindness ____________
2. The vitamin needed to prevent anemia, digestive disorders, and neural tube defects in the embryo ____________
3. Another name for vitamin D, the vitamin required for normal bone formation ____________
4. The mineral component of thyroid hormones ____________
5. A mineral important in blood clotting and muscle contraction ____________
6. The characteristic element in hemoglobin, the oxygen-carrying compound in the blood ____________
7. A mineral that promotes carbon dioxide transport and energy metabolism ____________
8. A vitamin involved in the synthesis of blood clotting factors that can be synthesized by colonic bacteria ____________

8. List the recommended percentages of carbohydrate, fat, and protein in the diet.

EXERCISE 20-7.

INSTRUCTIONS

Match each percentage to the corresponding nutrient type by writing the appropriate letter in each blank.

1. protein ___	a. 55% to 60%
2. fat ____	b. < 30%
3. carbohydrate ____	c. 15% to 20%

9. Distinguish between simple and complex carbohydrates, giving examples of each (see Exercise 20-8).

10. Compare saturated and unsaturated fats.

EXERCISE 20-8.

INSTRUCTIONS

Write the appropriate term in each blank.

trans-fatty acids	unsaturated fats	monosaccharides
saturated fats	polysaccharides	disaccharides

1. Fats that are usually of animal origin and are solid at room temperature ____________
2. Fats that are artificially saturated to prevent rancidity ____________
3. Carbohydrates with a low glycemic effect ____________
4. Glucose and fructose are examples of this type of nutrient ____________
5. Plant-derived fats that are usually liquid at room temperature ____________

11. List some adverse effects of alcohol consumption.

EXERCISE 20-9.

INSTRUCTIONS

Write three adverse effects of alcohol consumption in the spaces below.

1. ______________________________
2. ______________________________
3. ______________________________

12. Describe some nutritional disorders.

EXERCISE 20-10.

INSTRUCTIONS

Write the appropriate term in each blank.

underweight   overweight   obese   marasmus   kwashiorkor

1. A person with a body mass index greater than 30 ____________
2. A person with a body mass index less than 18.5 ____________
3. Severe malnutrition in infancy ____________
4. Protein deficiency, commonly observed in older children ____________

13. Explain how heat is produced and lost in the body.

EXERCISE 20-11.

INSTRUCTIONS

Write the appropriate term in each blank.

convection   evaporation   conduction   radiation

1. The direct transfer of heat to the surrounding air ____________
2. Heat loss resulting from the conversion of a liquid, such as perspiration, to a vapor ____________
3. Heat loss resulting from moving air ____________
4. Heat that travels from its source as heat waves ____________

14. Describe the role of the hypothalamus in regulating body temperature.

EXERCISE 20-12.

INSTRUCTIONS

Which of the following body changes would the hypothalamus induce when the body is cold (C), and which would it induce when the body is excessively hot (H)?

1. Constriction of the skin's blood vessels ______
2. Increased sweat gland activity ____
3. Dilation of the skin's blood vessels _____
4. Increased skeletal muscle contraction (shivering) ____

15. Explain the role of fever in disease.

EXERCISE 20-13.

INSTRUCTIONS

Write the appropriate term in each blank.

febrile crisis lysis antipyretic pyrogen

1. A substance that causes fever ______________
2. The term that describes a person who has a fever ______________
3. A sudden drop in temperature at the end of a period of fever ______________
4. A gradual fall in temperature at the end of a period of fever ______________
5. A class of drugs that treats fever ______________

16. Describe some adverse effects of excessive heat and cold.

EXERCISE 20-14.

INSTRUCTIONS

Write the appropriate term in each blank.

heat cramps heat exhaustion heat stroke hypothermia
frostbite constriction dilation

1. A response in the superficial blood vessels that increases heat loss ______________
2. An abnormally low body temperature, as may be caused by prolonged exposure to cold ______________
3. The final stages of excessive exposure to heat, characterized by central nervous system symptoms ______________
4. The change that occurs in the skin's blood vessels if too much heat is being lost from the body ______________
5. A condition that may follow heat cramps if adequate treatment is not given ______________
6. A term that describes regional tissue freezing ______________

17. Show how word parts are used to build words related to metabolism, nutrition, and body temperature.

EXERCISE 20-15.

INSTRUCTIONS

Complete the following table by writing the correct word part or meaning in the space provided. Write a word that contains each word part in the Example column.

Word Part	Meaning	Example
1. __________	heat	__________________
2. -lysis	_________	__________________
3. __________	sugar, sweet	__________________
4. pyr/o-	_________	__________________

Making the Connections

The following concept map deals with nutrition and metabolism. Each pair of terms is linked together by a connecting phrase into a sentence. The sentence should be read in the direction of the arrow. Complete the concept map by filling in the appropriate term or phrase. There is one right answer for each term (1 to 3, 8). However, there are many correct answers for the connecting phrases. Write the connecting phrases along the arrows if possible. If your phrases are too long, you may want to write them in the margins or on a separate sheet of paper.

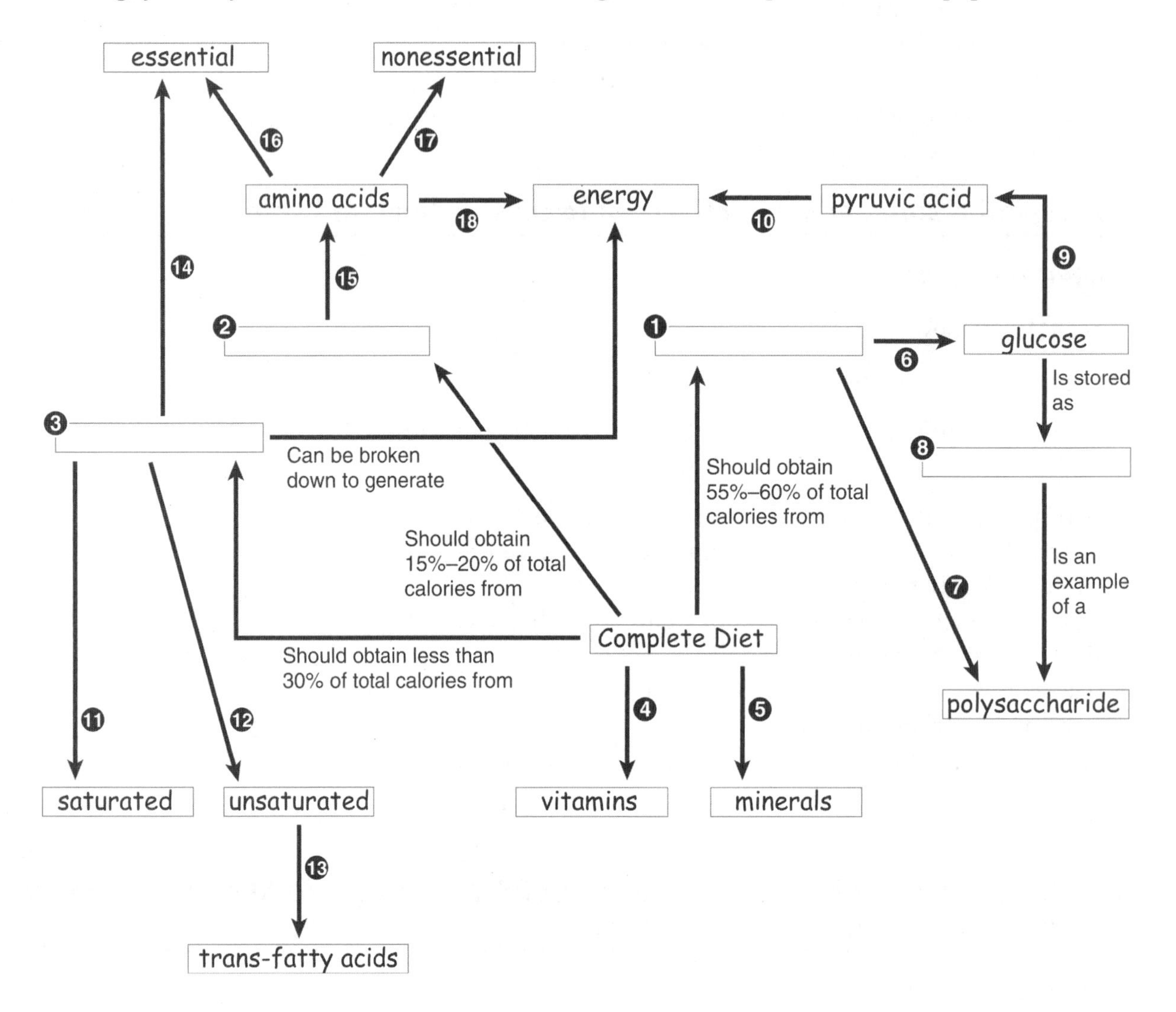

Optional Exercise: Make your own concept map, based on the regulation of body temperature. Choose your own terms to incorporate into your map, or use the following list: body temperature, fever, hypothermia, crisis, lysis, pyrogen, anti-pyretic, heat exhaustion, frostbite, hypothalamus, sweating, shivering, dilation, constriction, heat stroke, and heat cramps.

Testing Your Knowledge

Building Understanding

I. Multiple Choice

Select the best answer and write the letter of your choice in the blank.

1. The brain region involved in temperature regulation is the 1. _______
   a. hypothalamus
   b. cerebral cortex
   c. hippocampus
   d. thalamus
2. A complete protein contains 2. _______
   a. all the amino acids
   b. all the essential fatty acids
   c. a variety of minerals
   d. all the essential amino acids
3. Deamination is 3. _______
   a. an anabolic reaction
   b. the conversion of proteins into amino acids
   c. the conversion of glucose into glycogen
   d. the removal of a nitrogen group from an amino acid
4. If you have a fever, your temperature might realistically be 4. _______
   a. 96° F
   b. 39° F
   c. 39° C
   d. 96° C
5. The end product of the anaerobic phase of glucose metabolism is 5. _______
   a. glycogen
   b. pyruvic acid
   c. folic acid
   d. tocopherol
6. Trace elements in the diet are 6. _______
   a. sugars with a high glycemic effect
   b. vitamins needed in very small amounts
   c. minerals needed in large quantity
   d. minerals needed in extremely small amounts
7. Profuse sweating increases heat lost by the process of 7. _______
   a. radiation
   b. evaporation
   c. convection
   d. none of the above
8. A pyrogen is a 8. _______
   a. substance that induces fever
   b. cream used to treat frostbite
   c. drug that reduces body temperature
   d. substance that causes lysis

9. Unsaturated fats — 9. _______
   a. are generally healthier than saturated fats
   b. can be converted into trans fats
   c. contain double bonds between the carbon atoms
   d. all of the above
10. Which of the following is an example of an anabolic reaction? — 10. _______
   a. glycerol and fatty acids are used to form a fat
   b. starches and glycogen are converted into glucose
   c. a short peptide is converted into arginine and cysteine
   d. glucose is completely oxidized to carbon dioxide and water
11. A severe food shortage in a particular region will probably increase the prevalence of — 11. _______
   a. obesity
   b. type II diabetes
   c. kwashiorkor
   d. none of the above

II. Completion Exercise

Write the word or phrase that correctly completes each sentence.

1. Severe malnutrition beginning in infancy is called ____________________.
2. Tissue damage caused by exposure to cold is termed ____________________.
3. Shivering to generate additional body heat results from increased activity of the ____________________.
4. Organic substances needed in small amounts in the diet are the ____________________.
5. A gradual drop in temperature following a fever is known as ____________________.
6. The most important heat-regulating center is a section of the brain called the ____________________.
7. An individual subjected to excessive heat who is dizzy and no longer sweating is probably suffering from ____________________.
8. The series of catabolic reactions that results in the complete breakdown of nutrients is called ____________________.
9. Heat that is moved away from the skin by the wind is lost by the process of ____________________.
10. Glycolysis occurs in the part of the cell called the ____________________.
11. Fatty acids that must be consumed in the diet are called ____________________.

Understanding Concepts

I. True/False

For each question, write *T* for true or *F* for false in the blank to the left of each number. If a statement is false, correct it by replacing the underlined term and write the correct statement in the blank below the question.

_______ 1. A body temperature of 35° C would be considered <u>hypothermia</u>.

____

_______ 2. Grapeseed oil is liquid at room temperature. This oil is most likely a saturated fat.

____

_______ 3. The conversion of glycogen into glucose is an example of a catabolic reaction.

____

_______ 4. Most heat loss in the body occurs through the skin.

____

_______ 5. The element nitrogen is found in all sugars.

____

_______ 6. Mr. B is training for an upcoming road race by doing sprints. His leg muscles are working anaerobically. The pyruvic acid accumulating in his muscles may be responsible for the fatigue he is experiencing.

____

_______ 7. The rate at which energy is released from nutrients in Mr. B's cells during his sprints is called his basal metabolism.

____

_______ 8. Pernicious anemia is caused by a deficiency in vitamin B_{12}.

____

_______ 9. Mr. Q is suffering from a fever, but he has declined to take any medications. His blood contains large amounts of antipyretics that have induced the fever.

____

_______ 10. The most serious disorder resulting from excessive heat is called heat exhaustion.

____

_______ 11. A body mass index of 28 is indicative of obesity.

____

II. Practical Applications

Ms. S is researching penguin behavior at a remote location in Antarctica. She will be camping on the ice for two months. Study each discussion. Then write the appropriate word or phrase in the space provided.

1. Ms. S is spending her first night on the ice. She is careful to wear many layers of clothing in order to avoid a dangerous drop in body temperature, or ______________.
2. She is out for a moonlight walk to greet the penguins when she surprises an elephant seal stalking a penguin. Frightened, she sprints back to her tent. Her muscles are generating ATP by an oxygen-independent pathway. Each glucose molecule is generating a small number of ATP molecules, or to be exact ______________.
3. Ms. S realizes that she lost her face mask. She notices a whitish patch on her nose. Her nose is suffering from ______________.
4. The next morning, Ms. S is suffering from soreness in her leg muscles. She attributes the soreness to the accumulation of a byproduct of anaerobic metabolism called ______________.
5. This byproduct must be converted into another substance before it can be completely oxidized. This substance is called ______________.
6. After two weeks on the ice, Ms. S is out of fresh fruits and vegetables and the penguins have stolen her multivitamin supplements. She has been reading accounts of early explorers suffering from scurvy and fears she will experience the same fate. Scurvy is due to a deficiency of ______________.
7. Ms. S hikes to a distant penguin colony on her final day on the ice. She is dressed very warmly, and the sun is very bright. After several hours of hiking, Ms. S is sweating profusely. She is also experiencing a headache, tiredness, and nausea. She is probably suffering from ______________.
8. She removes some clothing to cool off. Some excess heat will be lost through the evaporation of sweat. Heat will also be lost directly from her skin to the surrounding air by the process called ______________.

III. Short Essays

1. Is alcohol a nutrient? Defend your answer.

   ____

   ____

   ____

2. A glucose molecule has been transported into a muscle cell. This cell has ample supplies of oxygen. Discuss the steps involved in using this glucose to produce energy. For each step, describe its location and oxygen requirements and name the substances produced.

   ____

   ____

   ____

Conceptual Thinking

1. Your friend wants to lose some weight. She is following a diet that contains 20% carbohydrates, 40% fat, and 40% protein. Why is this diet designed to limit fat deposition? You may have to review the actions of pancreatic hormones (see Chapter 12) to answer this question.

____

____

____

2. An invasion of influenza viruses has caused Mr. L to develop a fever. Using information from Chapters 17 and 20, describe the steps between the virus appearing in the bloodstream and Mr. L's increase in body temperature.

____

____

____

Expanding Your Horizons

How does your diet measure up? Go to http://www.mypyramid.gov to find out. By clicking on "MyPyramid Plan," you can get personalized dietary recommendations based on your age, gender, height, weight, and activity level. Some nongovernmental agencies, such as the Harvard School of Public Health, have criticized the government's efforts and proposed a different pyramid, the Healthy Eating Pyramid. You can read about their critique at http://www.hsph.harvard.edu/nutritionsource/pyramids.html/.

CHAPTER 21

Body Fluids

Overview

The majority (50% to 70%) of a person's body weight is **water.** This water is a solvent, a transport medium, and a participant in metabolic reactions. A variety of substances are dissolved in this water, including electrolytes, nutrients, gases, enzymes, hormones, and waste products. Body fluids are distributed in two main compartments: (a) The **intracellular fluid** compartment within the cells, and (b) The **extracellular fluid** compartment located outside the cells. The latter category includes blood plasma, interstitial fluid, lymph, and fluids in special compartments, such as the humors of the eye, cerebrospinal fluid, serous fluids, and synovial fluids.

Water balance is maintained by matching fluid intake with fluid output. Fluid intake is stimulated by the thirst center in the **hypothalamus,** but humans voluntarily control their fluid intake and may consume excess or insufficient fluids. Normally, the amount of fluid taken in with food and beverages equals the amount of fluid lost through the skin and the respiratory, digestive, and urinary tracts. When there is an imbalance between fluid intake and fluid output, serious disorders such as **edema, water intoxication,** and **dehydration** may develop.

The composition of intracellular and extracellular fluids is an important factor in homeostasis. These fluids must have the proper levels of electrolytes and must be kept at a constant pH. The kidneys are the main regulators of body fluids. Other factors that aid in regulation include hormones, buffers, and respiration. The normal pH of body fluids is a slightly alkaline 7.4. When regulating mechanisms fail to control shifts in pH, either **acidosis** or **alkalosis** results.

Fluid therapy is used to correct fluid and electrolyte imbalances and to give a patient nourishment.

Addressing the Learning Outcomes

1. Compare intracellular and extracellular fluids (see Exercises 21-1 and 21-3).

2. List four types of extracellular fluids.

EXERCISE 21-1: Main Fluid Compartments (Text Fig. 21-1)

INSTRUCTIONS

Label the different fluid compartments.

1. ______________________
2. ______________________
3. ______________________
4. ______________________

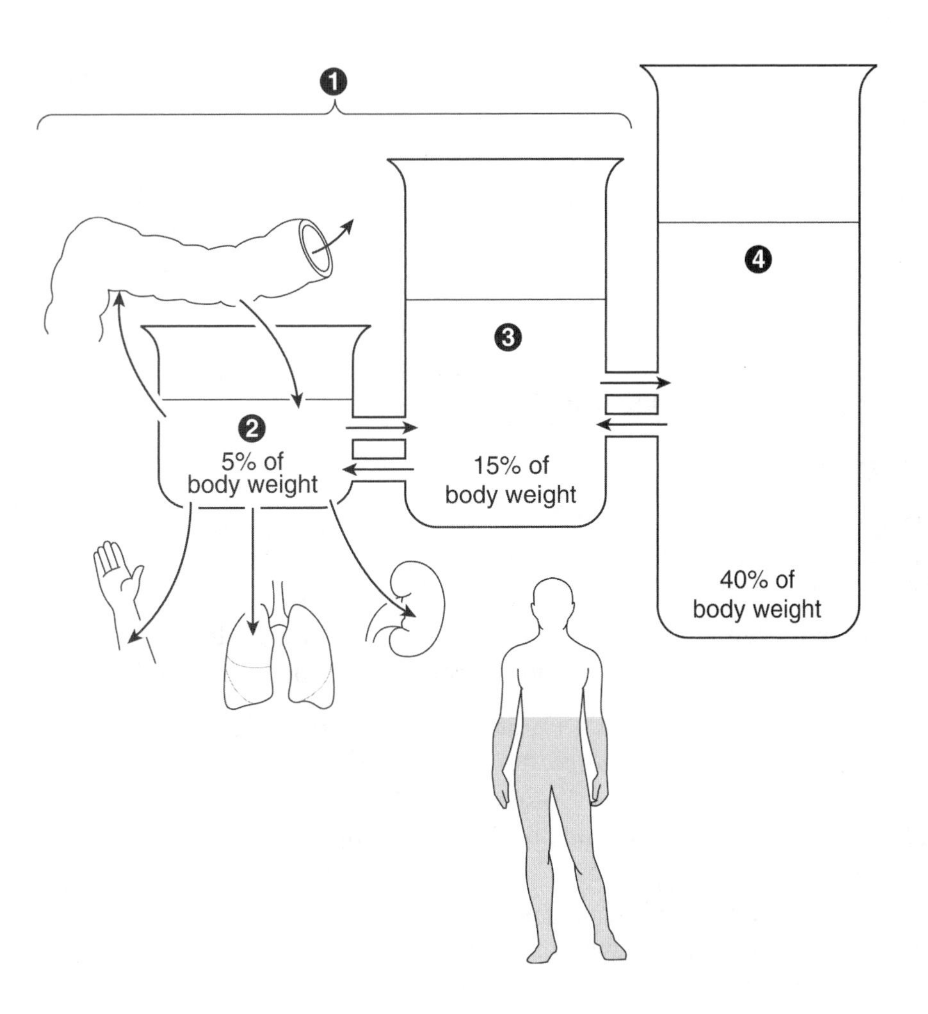

3. Name the systems that are involved in water balance.

EXERCISE 21-2: Daily Gain and Loss of Water (Text Fig. 21-2)

INSTRUCTIONS

Write the names of the different sources of water gain and loss on the appropriate numbered lines in different colors. Color the diagram with the appropriate colors.

1. ______________________
2. ______________________
3. ______________________
4. ______________________
5. ______________________
6. ______________________
7. ______________________

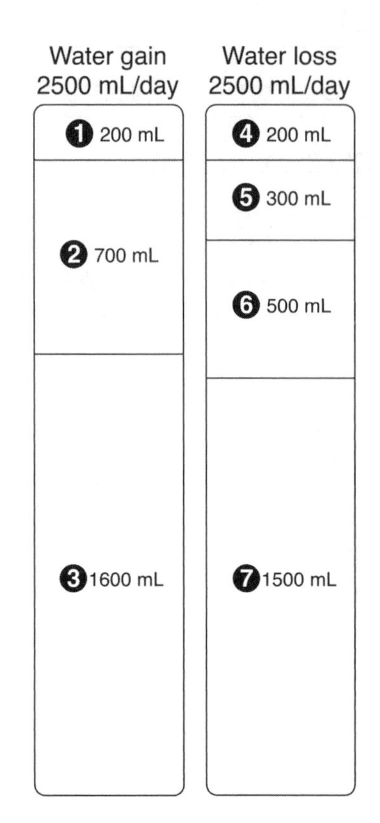

4. Explain how thirst is regulated.

EXERCISE 21-3.

INSTRUCTIONS

Write the appropriate term in each blank.

hypothalamus	brainstem	interstitial	intracellular	extracellular
water	blood plasma	body fluid concentration	body fluid volume	

1. The substance that makes up about 4% of a person's body weight ______________
2. The substance that makes up 50% to 70% of a person's body weight ______________
3. The part of the brain that controls the sense of thirst ______________
4. Term that specifically describes fluids in the microscopic spaces between cells ______________
5. The thirst center is stimulated when this is increased ______________
6. The thirst center is stimulated when this is decreased ______________
7. Term that describes the fluid within the body cells ______________

5. Define *electrolytes* and describe some of their functions.

EXERCISE 21-4.

INSTRUCTIONS

Write the appropriate term in each blank.

cation	anion	potassium	phosphate
electrolyte	sodium	calcium	chloride

1. A general term describing any positively charged ion ________
2. A general term describing any negatively charged ion ________
3. A component of stomach acid ________
4. The most abundant cation inside cells ________
5. A compound that forms ions in solution ________
6. A cation involved in bone formation ________
7. The most abundant cation in the fluid surrounding cells ________

6. Describe the role of hormones in electrolyte balance.

EXERCISE 21-5.

INSTRUCTIONS

Write the appropriate term in each blank.

parathyroid hormone	aldosterone	antidiuretic hormone
atrial natriuretic peptide	calcitonin	

1. A hormone secreted from the posterior pituitary that causes the kidney to reabsorb water ________
2. A hormone that is secreted when blood pressure rises too high ________
3. The adrenal hormone that promotes the reabsorption of sodium ________
4. A hormone that causes the kidney to reabsorb calcium ________
5. A hormone that promotes calcium deposition in bones ________

7. Describe three methods for regulating the pH of body fluids.

EXERCISE 21-6.

INSTRUCTIONS

Write the appropriate term in each blank.

buffer hydrogen ion kidney lungs carbon dioxide

1. Any substance that aids in maintaining a constant pH ________
2. Substance that directly determines the acidity or alkalinity of a fluid ________
3. The organ that regulates pH balance by altering urine acidity ________
4. The organ that regulates pH balance by altering carbon dioxide retention ________

8. Compare acidosis and alkalosis, including possible causes (see Exercise 21-7).

9. Describe three disorders involving body fluids.

EXERCISE 21-7.

INSTRUCTIONS

Write the appropriate term in each blank.

polydipsia	pulmonary edema	ascites	Addison disease	water intoxication
acidosis	alkalosis	respiratory	metabolic	effusion

1. A collection of fluid within the abdominal cavity __________
2. A condition that results from excess carbon dioxide in the blood __________
3. The accumulation of fluid in the lungs, such as may result from congestive heart failure __________
4. An abnormal increase in blood pH __________
5. Excessive thirst __________
6. A disorder resulting from aldosterone deficiency __________
7. The escape of fluid into a cavity or space __________
8. A condition in which body fluids are abnormally dilute __________

10. Specify some fluids used in therapy.

EXERCISE 21-8.

INSTRUCTIONS

Write the appropriate term in each blank.

dextrose	normal saline	Ringer lactate	25% serum albumin

1. A sugar that is often administered intravenously __________
2. An isotonic solution that is the first administered in emergencies __________
3. A hypertonic solution used to treat edema by drawing fluid out of the interstitial spaces __________
4. An isotonic solution that will help treat pH imbalances __________

11. Show how word parts are used to build words related to body fluids.

EXERCISE 21-9.

INSTRUCTIONS

Complete the following table by writing the correct word part or meaning in the space provided. Write a word that contains each word part in the Example column.

Word Part	Meaning	Example
1. ________	poison	________
2. intra-	________	________
3. ________	water	________
4. extra-	________	________
5. osmo-	________	________
6. ________	many	________
7. ________	condition, process	________
8. ________	partial, half	________

Making the Connections

The following concept map deals with the types of body fluids and the sources of fluid input and output. Each pair of terms is linked together by a connecting phrase into a sentence. The sentence should be read in the direction of the arrow. Complete the concept map by filling in the appropriate term or phrase. There is one right answer for each term. However, there are many correct answers for the connecting phrases (3, 5 to 7, 9, 11 to 13, 15 to 17).

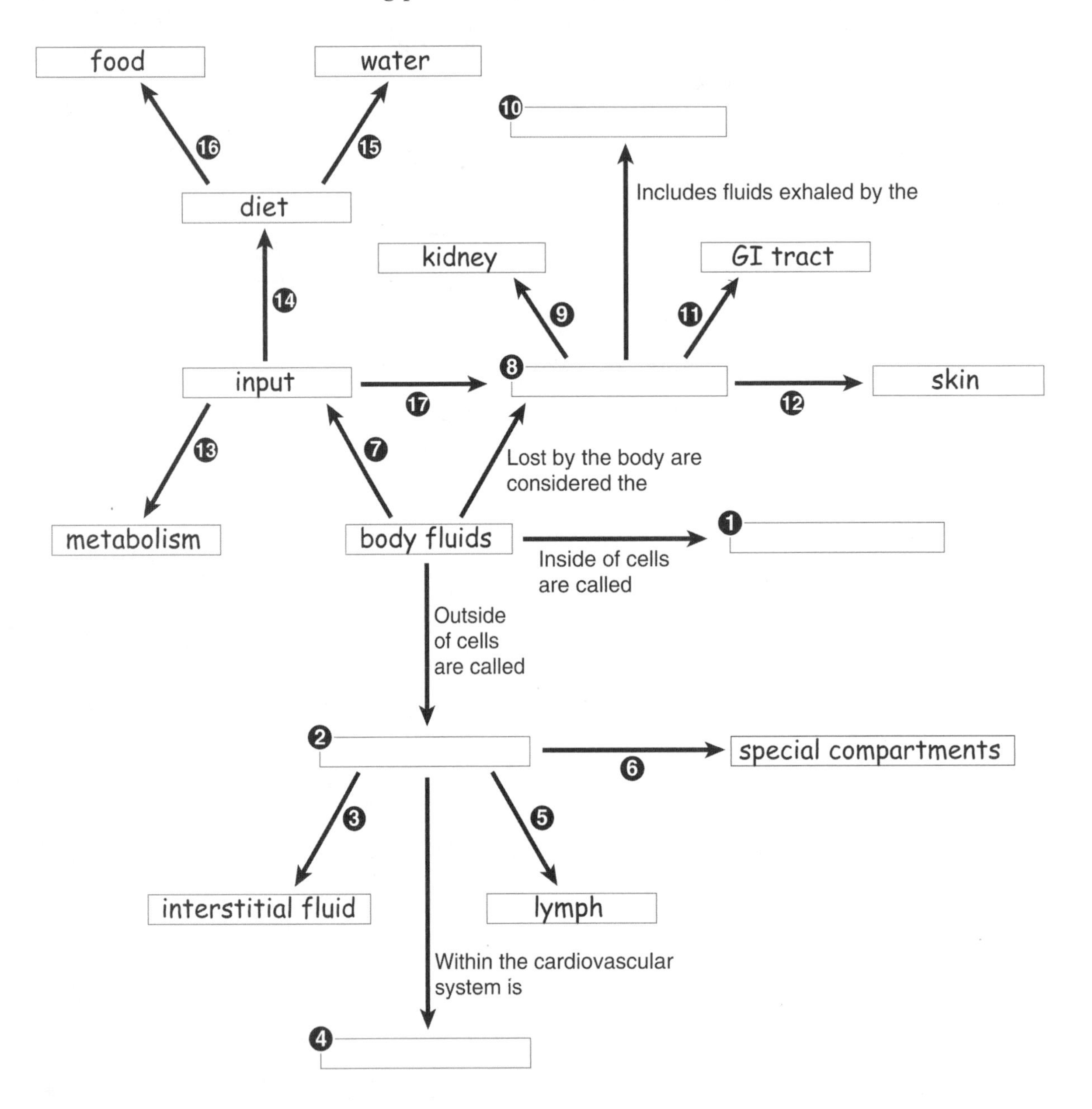

Optional Exercise: Make your own concept map, based on the regulation of acid–base balance. Choose your own terms to incorporate into your map, or use the following list: acidosis, alkalosis, metabolic, respiratory, carbon dioxide, buffer systems, and kidney.

Testing Your Knowledge

Building Understanding

I. Multiple Choice

Select the best answer and write the letter of your choice in the blank.

1. Aldosterone is produced by the 1. _______
   a. hypothalamus
   b. pituitary
   c. kidney
   d. adrenal cortex
2. Antidiuretic hormone 2. _______
   a. increases salt excretion by the kidney
   b. inhibits thirst
   c. decreases urine production
   d. increases calcium reabsorption by the kidney
3. Normal bone formation requires 3. _______
   a. calcium
   b. phosphate
   c. neither calcium nor phosphate
   d. both calcium and phosphate
4. Which of the following is NOT a buffer? 4. _______
   a. hemoglobin
   b. bicarbonate
   c. oxygen
   d. phosphate
5. Which of the following fluids is NOT in the extracellular compartment? 5. _______
   a. cytoplasm
   b. cerebrospinal fluid
   c. lymph
   d. interstitial fluid
6. Respiratory acidosis can result from 6. _______
   a. obstructive pulmonary disease
   b. kidney failure
   c. excessive exercise
   d. hyperventilation

III. Completion Exercise

Write the word or phrase that correctly completes each sentence.

1. A negatively charged electrolyte is called a(n) ___________________.
2. The receptors that detect an increase in body fluid concentration are called ___________________.
3. The electrolyte that plays the largest role in maintaining body fluid volume is ___________________.
4. Water balance is partly regulated by a thirst center located in a region of the brain called the ___________________.

5. Accumulation of excessive fluid in the intercellular spaces is ___________________.
6. A blood pH above 7.45 results in ___________________.
7. The escape of fluid into the pleural cavity is an example of a(n) ___________________.
8. The hormone that decreases blood volume is ___________________.

Understanding Concepts

I. True/False

For each question, write *T* for true or *F* for false in the blank to the left of each number. If a statement is false, correct it by replacing the underlined term and write the correct statement in the blank below the question.

_______ 1. Adherence to low-carbohydrate diets can result in metabolic acidosis.

____

_______ 2. The backup of fluid in the lungs that often accompanies congestive heart failure is called lymphedema.

____

_______ 3. The exhalation of carbon dioxide makes the blood more acidic.

____

_______ 4. Insufficient production of antidiuretic hormone characterizes Addison disease.

____

_______ 5. The fluids found in joint capsules and in the eyeball are examples of extracellular fluids.

____

_______ 6. The organ that excretes the largest amount of water per day is the skin.

____

II. Practical Applications

Study each discussion. Then write the appropriate word or phrase in the space provided.

1. Ms. S, age 26, was teaching English in rural India. She developed severe diarrhea, probably as a result of a questionable pakora from a road-side stand. On admission to a hospital, she was found to be suffering from a severe fluid deficit, a condition called _______________.
2. The physician's first concern was to increase Ms. S's plasma volume. Her fluid deficit was addressed by administering a 0.9% sodium chloride solution, commonly known as _______________.
3. This solution will not change the volume of Ms. S's cells, because it is _______________.
4. Next, the physician addressed Ms. S's acid–base balance. Prolonged diarrhea is often associated with a change in blood pH called _______________.

5. The pH of a blood sample was tested. The laboratory technician was looking for a pH value outside the normal range. The normal pH range for blood is ______________.
6. Ms. S's blood pH was found to be abnormal, and sodium bicarbonate was added to her IV. Bicarbonate is an example of a substance that helps maintain a constant pH. These substances are known as ______________.

III. Short Essays

1. Describe some circumstances under which fluid therapy might be given and cite some fluids that are used.

____

____

____

2. Compare respiratory acidosis and alkalosis, and cite some causes of each.

____

____

____

Conceptual Thinking

1. Ms. J is stranded on a desert island with limited water supplies. A. How will the volume and concentration of her body fluids change? B. How will her hypothalamus and pituitary gland respond to these changes? C. What will be the net effect of these responses?

____

____

____

Expanding Your Horizons

"Space. The final frontier." The classic statement from the show *Star Trek* remains valid in terms of human physiology. The effects of prolonged weightlessness and space travel on the human body remain an area of hot research. One effect of weightlessness is a dramatic redistribution of body fluids. You can read all about this and other effects of space travel in *Scientific American.*

Resources

1. White RJ. Weightlessness and the human body. *Sci Am* 1998; 279:58–63.

CHAPTER 22

The Urinary System

Overview

The urinary system comprises two **kidneys**, two **ureters**, one **urinary bladder**, and one **urethra**. This system is thought of as the body's main excretory mechanism; it is, in fact, often called the **excretory system**. The kidney, however, performs other essential functions; it aids in maintaining water and electrolyte balance and in regulating the acid–base balance (pH) of body fluids. The kidneys also secrete a hormone that stimulates red blood cell production and an enzyme that increases blood pressure.

The functional unit of the kidney is the **nephron**. It is the nephron that produces **urine** from substances filtered out of the blood through a cluster of capillaries, the **glomerulus**. The processes involved in urine formation in addition to filtration are tubular reabsorption, tubular secretion, and concentration of urine. Oxygenated blood is brought to the kidney by the **renal artery**. The arterial system branches through the kidney until the smallest subdivision, the **afferent arteriole**, carries blood into the glomerulus. Blood leaves the glomerulus by means of the **efferent arteriole** and eventually leaves the kidney by means of the **renal vein**. Before blood enters the venous network of the kidney, exchanges occur between the filtrate and the blood through the **peritubular capillaries** that surround each nephron.

Prolonged or serious diseases of the kidney have devastating effects on overall body function and health. Renal dialysis and kidney transplantation are effective methods for saving lives of people who otherwise would die of uremia and kidney failure.

Addressing the Learning Outcomes

1. List the systems that eliminate waste and name the substances eliminated by each.

EXERCISE 22-1.

INSTRUCTIONS

For each substance, determine the system or systems that eliminate it and write the appropriate letter(s) in the blank (U for urinary, D for digestive, R for respiratory, and I for integumentary). There may be more than one answer for each substance.

1. carbon dioxide ____
2. bile ____
3. water ____
4. salts ____
5. food residue ____
6. nitrogenous wastes ____

2. Describe the parts of the urinary system and give the functions of each.

EXERCISE 22-2: Male Urinary System (Text Fig. 22-1)

INSTRUCTIONS

1. Write the names of the arteries on the appropriate numbered lines in red, and color the arteries (except for the small box) on the diagram.
2. Write the names of the veins on the appropriate numbered lines in blue, and color the veins (except for the small box) on the diagram.
3. Write the names of the remaining structures on the appropriate numbered lines in different colors, and color the structures on the diagram.
4. Put arrows in the small boxes to indicate the direction of blood/urine movement.

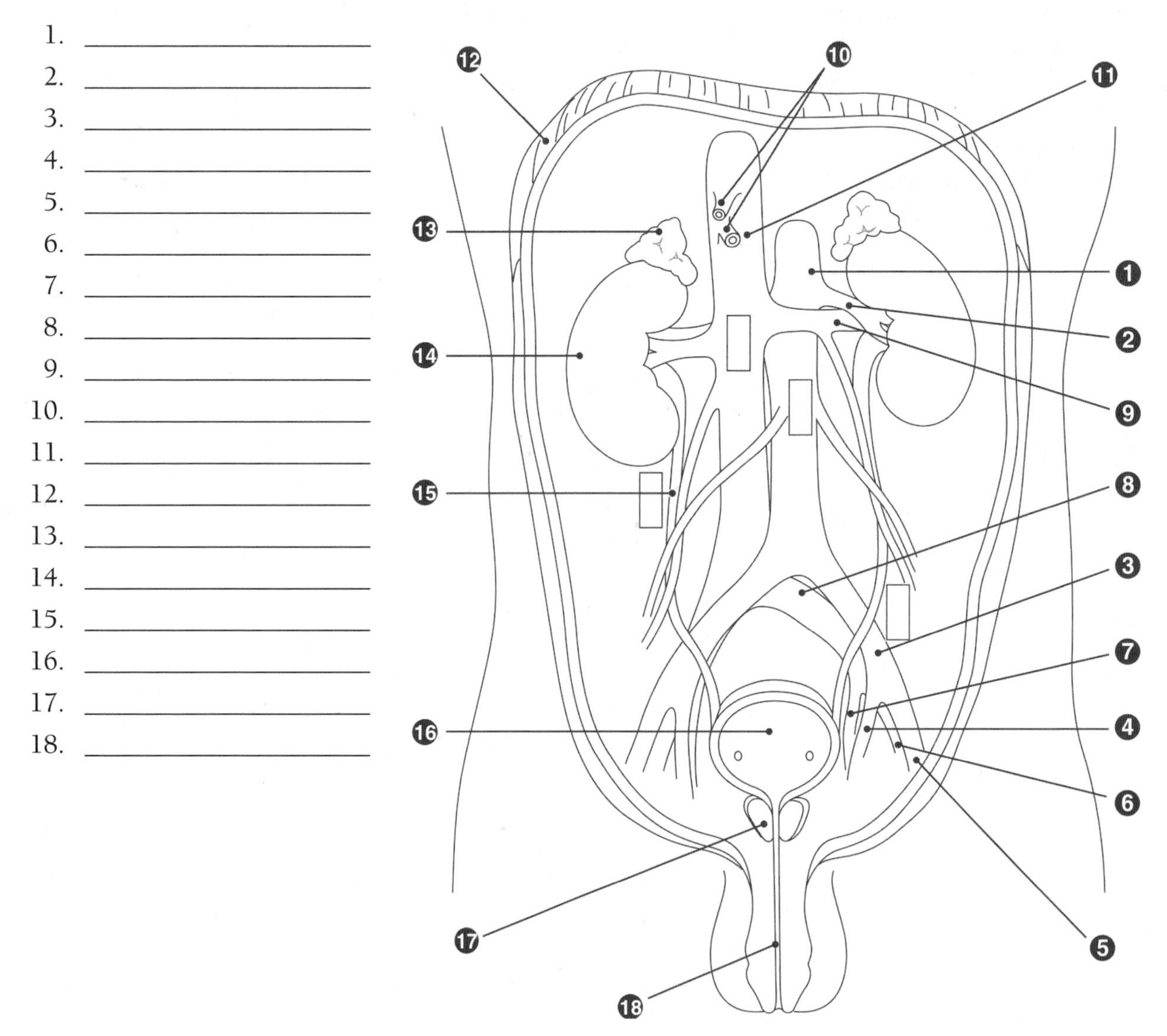

EXERCISE 22-3: Longitudinal Section through the Kidney (Text Figs. 22-2 and 22-3)

INSTRUCTIONS

1. Write the names of the blood vessels on lines 1 and 2 (vessel 1 drains from the abdominal aorta and vessel 2 drains into the inferior vena cava). Use red for the artery and blue for the vein.
2. Color the artery and arterioles on the diagram red and the veins and venules blue.
3. Write the names of the remaining structures on the appropriate numbered lines in different colors, and color the structures on the diagram. Use the same, dark color for structures 3 and 4. Use the same color for structures 6 and 7. Use yellow for structures 8 to 10.

1. ____________________
2. ____________________
3. ____________________
4. ____________________
5. ____________________
6. ____________________
7. ____________________
8. ____________________
9. ____________________
10. ____________________

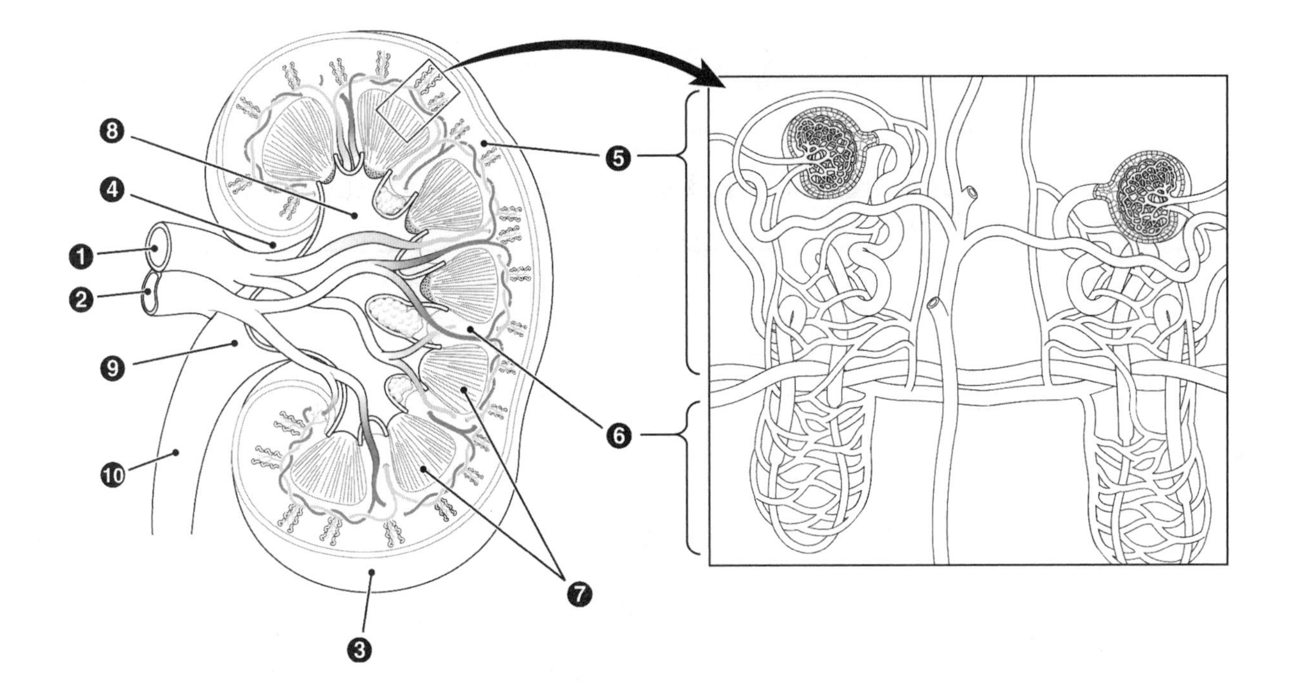

EXERCISE 22-4.

INSTRUCTIONS

Write the appropriate term in each blank.

ureter	urinary bladder	urethra	renal capsule	renal cortex
retroperitoneal space	nephron	renal pelvis	adipose capsule	renal medulla

1. A funnel-shaped basin that collects urine from collecting ducts ______________
2. The tube that carries urine from the bladder to the outside ______________
3. The area behind the peritoneum that contains the ureters and the kidneys ______________
4. A microscopic functional unit of the kidney ______________
5. A tube connecting a kidney with the bladder ______________
6. The crescent of fat that helps to support the kidney ______________
7. The inner region of the kidney ______________
8. The outer region of the kidney ______________

3. List the activities of the kidneys in maintaining homeostasis.

EXERCISE 22-5.

INSTRUCTIONS

In the spaces below, list five body parameters that the kidney helps to maintain within tight limits.

1. ____
2. ____
3. ____
4. ____
5. ____

4. Trace the path of a drop of blood as it flows through the kidney (see Exercise 22-6).

5. Describe a nephron.

EXERCISE 22-6: A Nephron and Its Blood Supply (Text Fig. 22-4)

INSTRUCTIONS

1. Follow the passage of blood by labeling structures 1 to 6. Use different shades of red for the arteries/arterioles, purple for the capillaries, and blue for the venules/veins. Lines indicate the boundaries between different vessels.
2. Follow the passage of filtrate through the nephron by labeling structures 7 to 13. Use different colors (perhaps shades of yellow and orange) for the different structures. Lines indicate the boundaries between the different nephron parts.
3. Write the names of the remaining structures on the appropriate numbered lines in different colors, and color the structures on the diagram.

1. ____________________
2. ____________________
3. ____________________
4. ____________________
5. ____________________
6. ____________________
7. ____________________
8. ____________________
9. ____________________
10. ____________________
11. ____________________
12. ____________________
13. ____________________

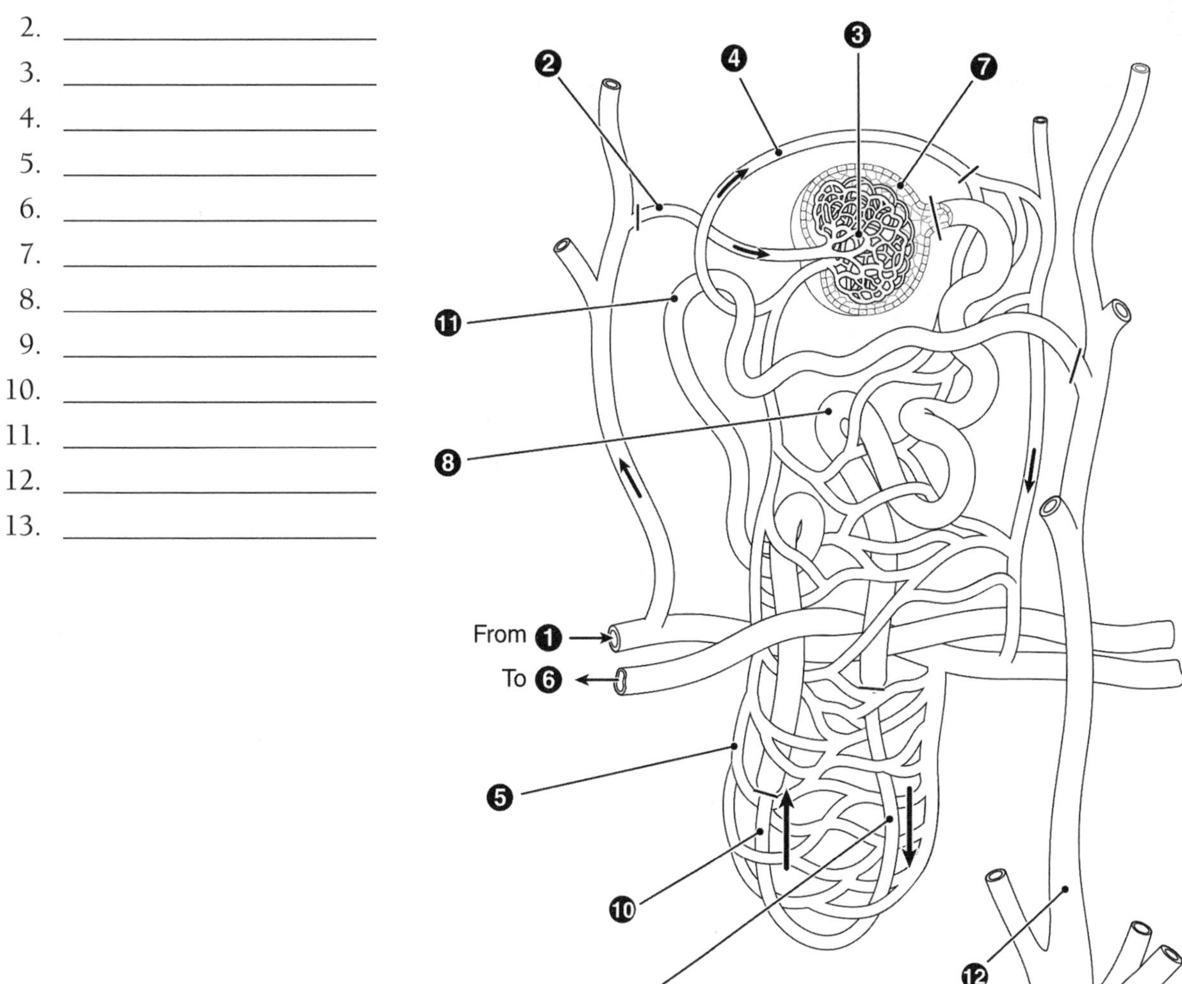

EXERCISE 22-7.

INSTRUCTIONS

Write the appropriate term in each blank.

afferent arteriole
proximal convoluted tubule
peritubular capillaries
nephron loop
glomerulus
glomerular capsule
efferent arteriole
renal artery
collecting duct
renal vein

1. A hollow bulb at the proximal end of the nephron ________________
2. The blood vessels connecting the afferent and efferent arterioles ________________
3. The portion of the nephron receiving filtrate from the glomerular capsule ________________
4. The vessel that branches to form the glomerulus ________________
5. The vessel that drains the kidney ________________
6. The blood vessels that exchange substances with the nephron ________________
7. A tube that receives urine from the distal convoluted tubule ________________
8. The portion of the nephron that dips into the medulla ________________

6. Name the four processes involved in urine formation and describe the action of each.

EXERCISE 22-8: Filtration Process (Text Fig. 22-6)

INSTRUCTIONS

1. Write the names of the different structures on the appropriate numbered lines in different colors, and color the structures on the diagram. Use different shades of red for structures 1 to 3 and yellow for part 7. Do not color over the symbols. Some structures have appeared on earlier diagrams. You may want to use the same color scheme.
2. Color the symbols beside "soluble molecules," "proteins," and "blood cells" in different, dark colors and color the corresponding symbols on the diagram.

1. ____________________
2. ____________________
3. ____________________
4. ____________________
5. ____________________
6. ____________________
7. ____________________

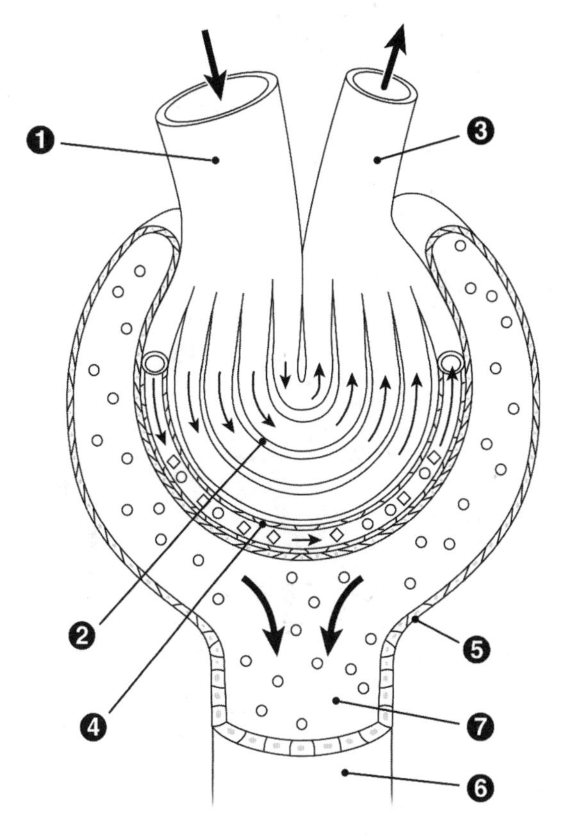

EXERCISE 22-9: Summary of Urine Formation (Text Fig. 22-8)

INSTRUCTIONS

1. Label the structures and fluids by writing the appropriate terms on lines 5 to 11.
2. Write the name of the hormone that controls water reabsorption in the distal tubule and collecting duct on line 12.
3. In boxes 1 to 4, summarize the process that is occurring.

5. ______________________
6. ______________________
7. ______________________
8. ______________________
9. ______________________
10. ______________________
11. ______________________
12. ______________________

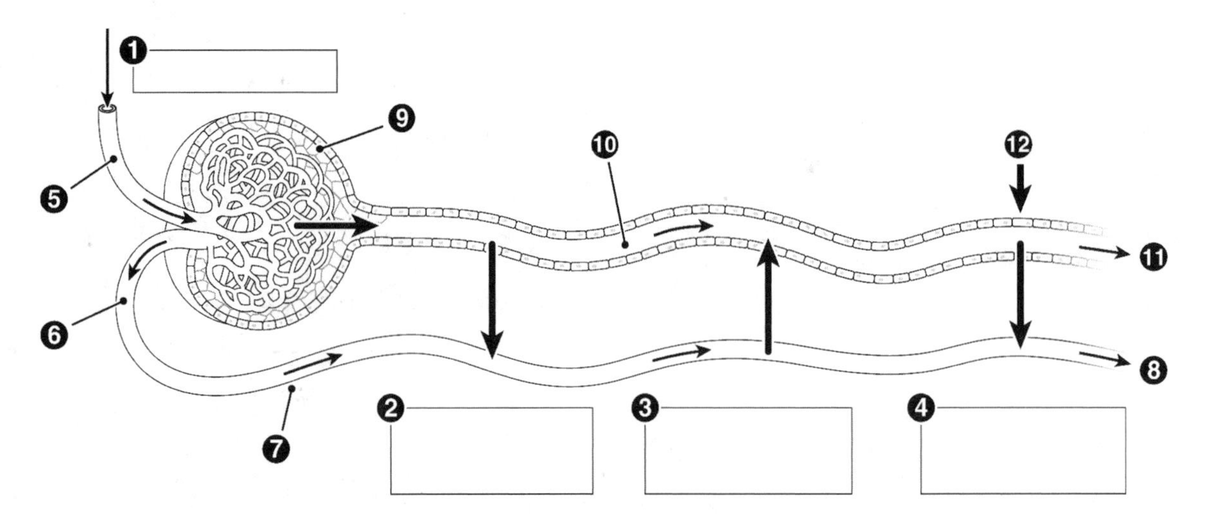

7. Identify the role of antidiuretic hormone (ADH) in urine formation.

EXERCISE 22-10.

INSTRUCTIONS

Which of the following processes will ONLY occur if ADH is present? Write the correct answer here. ____

A. filtration in the glomerulus
B. establishment of the interstitial fluid concentration gradient
C. water reabsorption from the collecting duct back into the blood
D. water reabsorption from the descending nephron loop back into the blood

8. Describe the components and functions of the juxtaglomerular (JG) apparatus.

EXERCISE 22-:11 Structure of the Juxtaglomerular (JG) Apparatus (Text Fig. 22-9)

INSTRUCTIONS

Write the names of the different structures on the appropriate numbered lines in different colors, and color the structures on the diagram. Use different shades of red for structures 1 to 3. Use a dark color to outline the box surrounding structure 8. Some structures have appeared on earlier diagrams. You may want to use the same color scheme.

1. ____________________
2. ____________________
3. ____________________
4. ____________________
5. ____________________
6. ____________________
7. ____________________
8. ____________________

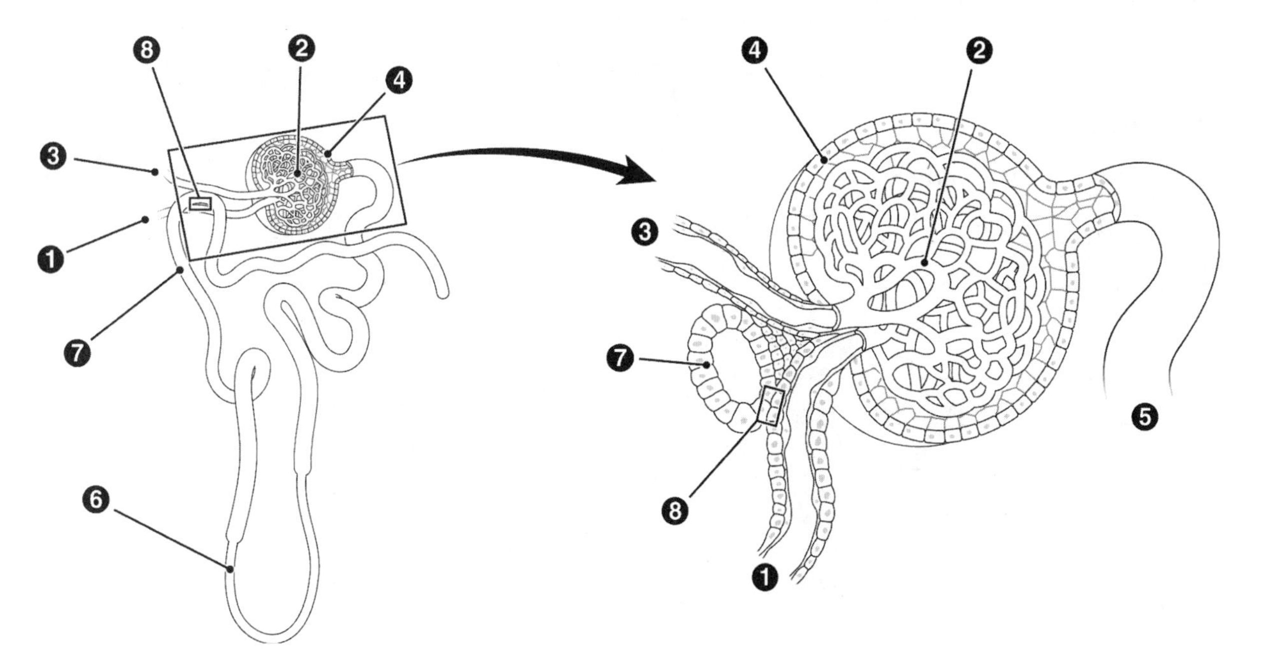

EXERCISE 22-12.

INSTRUCTIONS

Write the appropriate term in each blank.

renin	juxtaglomerular apparatus	urea	EPO	angiotensin
filtration	tubular reabsorption	tubular secretion	antidiuretic hormone	aldosterone

1. An enzyme produced by the kidney ________________
2. The hormone that increases sodium reabsorption in the DCT ________________
3. The process that returns useful substances in the filtrate to the bloodstream ________________
4. The process by which substances leave the glomerulus and enter the glomerular capsule ________________
5. The structure in the kidney that produces renin ________________
6. The hormone produced in the kidney that stimulates erythrocyte production by the bone marrow ________________
7. The process by which the renal tubule actively moves substances from the blood into the nephron to be excreted ________________
8. The hormone that increases the permeability of the DCT and collecting duct to water ________________

9. Describe the process of micturition.

EXERCISE 22-13: Interior of the Male Urinary Bladder (Text Fig. 22-10)

INSTRUCTIONS

Label the indicated parts.

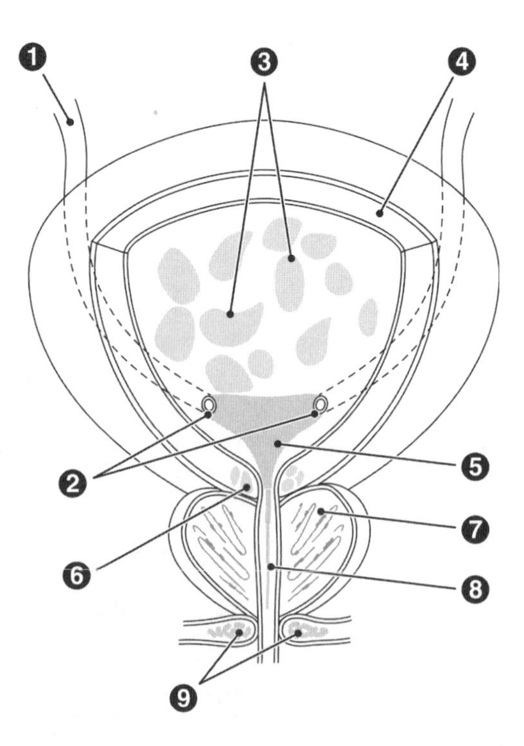

EXERCISE 22-14.

INSTRUCTIONS

In the blank following each statement, write T if the statement is true and F if it is false.

1. The internal urethral sphincter is formed of skeletal muscle. ____
2. The external urethral sphincter is formed by the pelvic floor muscles. ____
3. When the bladder fills with liquid, stretch receptors are activated. ____
4. When the bladder fills with liquid, muscles in the bladder wall contract. ____
5. Urination will occur when the external urethral sphincter is contracted. ____

10. Name three normal and six abnormal constituents of urine.

EXERCISE 22-15.

INSTRUCTIONS

State whether each of the following is a normal (N) or abnormal (A) constituent of urine.

1. glucose ____
2. blood ____
3. pigment ____
4. nitrogenous waste products ____
5. albumin ____
6. casts ____
7. white blood cells ____
8. electrolytes ____

11. List the common disorders of the urinary system.

EXERCISE 22-16.

INSTRUCTIONS

Write the appropriate term in each blank.

specific gravity	glycosuria	albuminuria	hematuria
pyelonephritis	acute glomerulonephritis	hydronephrosis	white blood cells

1. The presence of this material in the urine indicates pyuria ________________
2. An indication of the amount of dissolved substances in the urine ________________
3. Inflammation of the renal pelvis and the kidney tissue itself ________________
4. An inflammation of part of the nephron subsequent to a streptococcal infection ________________
5. The presence of an abundant blood protein in the urine ________________
6. The presence of blood in the urine ________________
7. Accumulation of fluid in the renal pelvis and calyces ________________

EXERCISE 22-17.

INSTRUCTIONS

Write the appropriate term in each blank.

nocturia	acute renal failure	chronic renal failure	edema	anemia
uremia	renal calculi	polyuria	cystitis	urethritis

1. A gradual loss of nephrons ________________
2. Elimination of very large amounts of urine ________________
3. Kidney stones; solids formed when uric acid or calcium salts precipitate out of the urine ________________
4. A sudden decrease in kidney function resulting from a medical or surgical emergency ________________
5. Elimination of urine during the night ________________
6. The general condition caused by accumulation of nitrogenous waste products in the blood ________________
7. Excessive accumulation of fluid in the body tissues, as may be caused by kidney failure ________________
8. Inflammation of the urinary bladder ________________

12. List six signs of chronic renal failure.

EXERCISE 22-18.

INSTRUCTIONS

List the six most important signs and symptoms of kidney failure in the spaces below.

1. ________________________________
2. ________________________________
3. ________________________________
4. ________________________________
5. ________________________________
6. ________________________________

13. Explain the principle and the purpose of kidney dialysis.

EXERCISE 22-19.

INSTRUCTIONS

In the blank following each statement, write T if the statement is true and F if it is false.

1. Dialysis separates dissolved molecules based on their ability to pass through a semipermeable membrane. ____
2. During dialysis, substances move from the region of low concentration to the region of high concentration. ____
3. In hemodialysis, dialysis fluid is pumped into the peritoneal space. ____
4. In dialysis, accumulated waste products move from the blood into the dialysis solution. ____

14. Show how word parts are used to build words related to the urinary system.

EXERCISE 22-20.

INSTRUCTIONS

Complete the following table by writing the correct word part or meaning in the space provided. Write a word that contains each word part in the Example column.

Word Part	Meaning	Example
1. ________	night	________________
2. dia-	________	________________
3. ________	renal pelvis	________________
4. trans-	________	________________
5. nephr/o-	________	________________
6. ________	sac, bladder	________________
7. ________	backward, behind	________________
8. ________	next to	________________
9. ren/o	________	________________
10. -cele	________	________________

Making the Connections

The following concept map deals with the organization of the kidney. Each pair of terms is linked together by a connecting phrase into a sentence. The sentence should be read in the direction of the arrow. Complete the concept map by filling in the appropriate term or phrase. There is one right answer for each term. However, there are many correct answers for the connecting phrases (3, 6, and 7). Three phrases (13 and 15 and 16) involve 3 terms. For each of these phrases, build a single sentence linking the three terms together. For instance, use terms 4 and 8 and filtration to build phrase 13; term 12, peritubular capillaries, and secretion to build phrase 15; and term 9, reabsorption, and peritubular capillaries to build phrase 16.

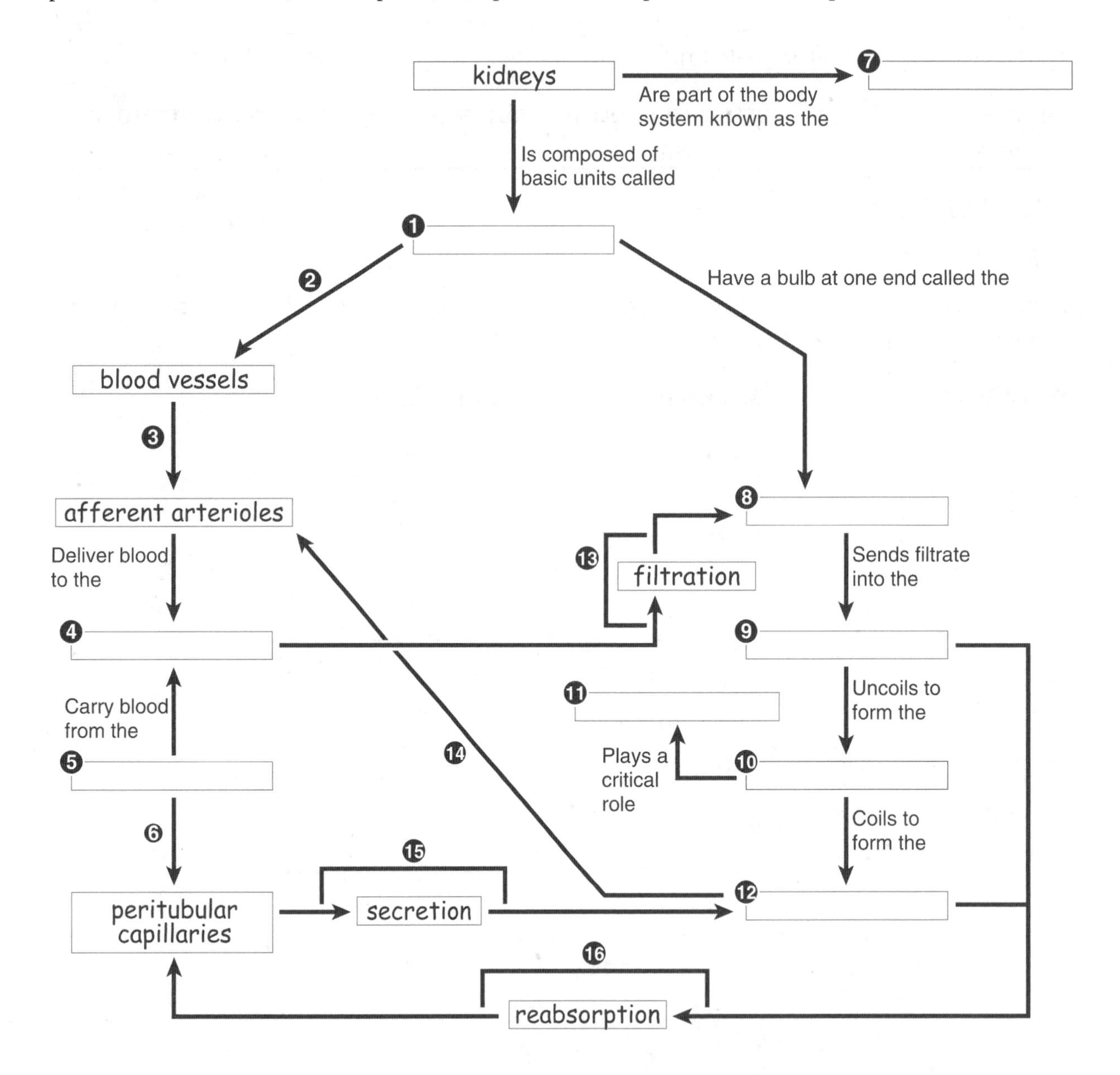

Choose your own terms to incorporate into your map, or use the following list: urinary system, kidneys, nephrons, ureters, urinary bladder, urethra, adipose capsule, renal capsule, renal pelvis, rugae, trigone, urinary meatus, penis, micturition, internal urethral sphincter, and external urethral sphincter.

Testing Your Knowledge

Building Understanding

I. Multiple Choice

Select the best answer and write the letter of your choice in the blank.

1. Select the correct order of urine flow from its source to the outside of the body. 1. _______
   a. urethra, bladder, kidney, ureter
   b. bladder, kidney, urethra, ureter
   c. kidney, ureter, bladder, urethra
   d. kidney, urethra, bladder, ureter
2. Select the correct order of filtrate flow through the nephron 2. _______
   a. nephron loop, distal convoluted tubule, proximal convoluted tubule, collecting duct
   b. glomerular capsule, proximal convoluted tubule, nephron loop, distal convoluted tubule
   c. proximal convoluted tubule, distal convoluted tubule, loop of Henle, glomerular capsule
   d. glomerular capsule, distal convoluted tubule, proximal convoluted tubule, collecting duct
3. The enzyme renin raises blood pressure by activating 3. _______
   a. urea
   b. glomerular filtration
   c. angiotensin
   d. erythropoietin
4. The juxtaglomerular apparatus consists of cells in the 4. _______
   a. proximal convoluted tubule and efferent arteriole
   b. renal artery and afferent arteriole
   c. collecting tubules and renal vein
   d. distal convoluted tubule and afferent arteriole
5. Which of the following is NOT a function of the kidneys? 5. _______
   a. red blood cell destruction
   b. blood pressure regulation
   c. elimination of nitrogenous wastes
   d. modification of body fluid composition
6. The movement of a substance from the distal convoluted tubule to the peritubular capillaries occurs by the process of 6. _______
   a. secretion
   b. filtration
   c. reabsorption
   d. excretion
7. Urine does NOT usually contain 7. _______
   a. sodium
   b. hydrogen
   c. casts
   d. urea

8. The process of expelling urine is called 8. _______
   a. reabsorption
   b. micturition
   c. dehydration
   d. defecation
9. A cystoscope is used to diagnose 9. _______
   a. interstitial cystitis
   b. polycystic kidney
   c. staghorn calculi
   d. renal ptosis
10. Substances leave the patient's blood during dialysis by the process of 10. _______
    a. active transport
    b. diffusion
    c. filtration
    d. all of the above

II. Completion Exercise

Write the word or phrase that correctly completes each sentence.

1. The congenital anomaly in which the urethra opens on the undersurface of the penis is known as ____________________.
2. One indication of nephritis is the presence in the urine of solid materials molded in the kidney tubules. They are called ____________________.
3. Inflammation of the bladder is called ____________________.
4. When the bladder is empty, its lining is thrown into the folds known as ____________________.
5. The vessel that brings blood to the glomerulus is the ____________________.
6. Large kidney stones that fill the renal pelvis and extend into the calyces are called ____________________.
7. The most distal portion of the nephron is the ____________________.
8. The device that employs shock waves to destroy kidney stones is called a(n) ____________________.
9. Fluid passes from capillaries into the glomerular capsule of the nephron by the process of ____________________.

Understanding Concepts

I. True/False

For each question, write *T* for true or *F* for false in the blank to the left of each number. If a statement is false, correct it by replacing the underlined term and write the correct statement in the blank below the question.

_______ 1. Polyuria could be caused by a deficiency of the hypothalamic hormone erythropoietin.

____

_______ 2. The movement of hydrogen ions from the peritubular capillaries into the distal convoluted tubule is an example of tubular secretion.

____

_______ 3. The type of dialysis which does NOT use dialysis tubing is called hemodialysis.

____

_______ 4. Urine with a specific gravity of 0.05 is more concentrated than urine with a specific gravity of 0.04.

____

_______ 5. Aldosterone tends to decrease the amount of urine produced.

____

_______ 6. Adults can control the internal urethral sphincter.

____

_______ 7. The triangular-shaped region in the floor of the bladder is called the renal pelvis.

____

_______ 8. Most water loss occurs through the actions of the digestive system.

____

_______ 9. The presence of glucose in the urine, known as pyuria, is indicative of diabetes mellitus.

____

II. Practical Applications

Study each discussion. Then write the appropriate word or phrase in the space provided.

➤ Group A

1. K, age 11, was brought to the emergency room by her mother, Ms. L, because her urine was red. She had not consumed beets recently, so the nurse suspected that the red coloration was due to the presence of blood in the urine, a condition called ______________.
2. The presence of blood was confirmed by a laboratory study of her urine. This test is called a(n) ______________.
3. Ms. L mentioned that K had recently recovered from "strep throat." Tests indicated that K was suffering from the most common disease of the kidneys, namely ______________.
4. Ms. L was quite concerned by this diagnosis, because she had had several similar episodes during her own childhood. She was assured that her daughter would recover with appropriate treatment, but was advised to have her own urine tested as a precaution. The tests demonstrated protein in her urine, consistent with damage to the kidney tubules. The presence of protein in the urine is called ______________.
5. Other substances were also found in the mother's urine, including ketones and glucose. Ms. L may be suffering from an endocrine disease called ______________.

➤ Group B

1. Young S, age 1, was brought to the hospital with an enlarged abdomen. His blood pressure was shown to be extremely high, a disorder called ______________.
2. High blood pressure can result from increased renin production. Renin is produced by a specialized region of the kidney called the ______________.
3. Tests showed that S was suffering from a slow, progressive, irreversible loss of nephrons. This disease is called ______________.
4. By the age of 5, S's kidneys were no longer functioning to any appreciable degree. S must visit a clinic three times weekly, where excess waste products are removed from his blood by diffusion. This process is called ______________.
5. Two years later, S received a kidney from his mother. The process by which an organ is transferred from one individual to another is called ______________.

III. Short Essays

1. Name five signs of chronic renal disease. Explain why each sign indicates that the kidney is no longer functioning properly.

____

____

____

____

____

2. Mr. B has uncontrolled diabetes mellitus. Name two abnormal constituents you would expect to find in his urine, and explain why each would be present.

____

____

____

Conceptual Thinking

1. Ms. W has just eaten a large bag of salty popcorn. Her blood is now too salty and must be diluted. Is it possible for the kidney to increase water reabsorption without increasing salt absorption? Explain.

____

____

____

2. Mr. R is taking penicillin to cure a throat infection. He must take the drug frequently, because the kidney clears penicillin very efficiently. That is, all of the penicillin that enters the renal artery leaves the kidney in the urine. However, only some penicillin molecules will be filtered into the glomerular capsule. How can the kidney excrete all of the penicillin it receives?

____

____

____

Expanding Your Horizons

Renal failure is a very common disorder, because the kidneys are sensitive to toxins and physical trauma. Patients with chronic renal failure face a life of illness and frequent dialysis, unless they can receive a kidney transplant. Kidney transplants are quite common (over 15,000 were performed in 2001 in the United States alone). In some cases, the patient receives a kidney from a live, related donor or an acceptable tissue match. You can learn more about live related kidney donations on the Web site of the University of Southern California Kidney Transplant Program (http://www.kidneytransplant.org/) or by watching the Discovery Channel documentary "Kidney Transplant" (1995).

The ideal treatment for chronic renal failure would be an artificial kidney, which would abolish the problems of kidney availability and donor incompatibility. Artificial nephrons have been synthesized that are capable of urine production, but a fully functioning artificial kidney is likely more than a decade away. You can read about some of the trials and tribulations of this line of research in the following article.

Resources

1. Soares C. Body building. *Sci Am* 2004; 290:20, 22.

UNIT VII Perpetuation of Life

CHAPTER 23

The Male and Female Reproductive Systems

Overview

Reproduction is the process by which life continues. Human reproduction is **sexual**, that is, it requires the union of two different **germ cells** or **gametes.** (Some simple forms of life can reproduce without a partner in the process of **asexual** reproduction.) These germ cells, the **spermatozoon** in males and the **ovum** in females, are formed by **meiosis**, a type of cell division in which the chromosome number is reduced to one half. When fertilization occurs and the gametes combine, the original chromosome number is restored.

The reproductive glands or **gonads** manufacture the gametes and also produce hormones. These activities are continuous in the male but cyclic in the female. The male gonad is the **testis.** The remainder of the male reproductive tract consists of passageways for storage and transport of spermatozoa; the male organ of copulation, the **penis**; and several glands that contribute to the production of **semen.**

The female gonad is the **ovary.** The ovum released each month at the time of **ovulation** travels through the **oviducts** to the **uterus**, where the egg, if fertilized, develops. If no fertilization occurs, the ovum, along with the built-up lining of the uterus, is eliminated through the **vagina** as the **menstrual flow.**

Reproduction is under the control of hormones from the **anterior pituitary**, which in turn is controlled by the **hypothalamus** of the brain. These organs respond to **feedback** mechanisms, which maintain proper hormone levels.

Aging causes changes in both the male and female reproductive systems. A gradual decrease in male hormone production begins as early as age 20 and continues throughout life. In the female, a more sudden decrease in activity occurs between ages 45 and 55 and ends in **menopause**, the cessation of menstruation and the childbearing years.

The reproductive tract is subject to infection and inflammation, including **sexually transmitted infections (STIs)** and inflammation secondary to other infections such as mumps. Infections, congenital malformations, and hormonal imbalances can cause **infertility** or **sterility.**

Addressing the Learning Outcomes

Note that the learning outcomes address male and female reproductive physiology together, while the textbook covers them in separate sections. Use these exercises to compare the anatomy and physiology of the male and the female.

1. Name the male and female gonads and describe the function of each. (Note that the female gonad is shown in Exercise 23-6.)

EXERCISE 23-1: Structure of the Testis (Text Fig. 23-2)

INSTRUCTIONS

Label the indicated parts. (Hint: The vein is more branched than the artery.)

1. ____________________
2. ____________________
3. ____________________
4. ____________________
5. ____________________
6. ____________________
7. ____________________
8. ____________________
9. ____________________
10. ____________________
11. ____________________
12. ____________________

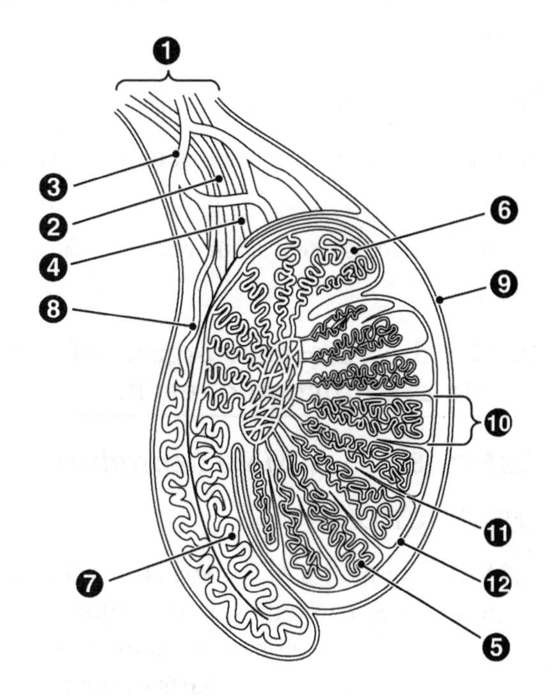

EXERCISE 23-2.

INSTRUCTIONS

Write the appropriate term in each blank.

interstitial cell	sustentacular cell	spermatozoon
seminiferous tubule	ovarian follicle	ovum

1. A cell that nourishes and protects developing spermatozoa ________________
2. A cell that secretes testosterone ________________
3. The germ cell of the male ________________
4. The germ cell of the female ________________
5. The cluster of cells that surrounds the female germ cell ________________

2. State the purpose of meiosis.

EXERCISE 23-3.

INSTRUCTIONS

In the blank following each statement, write T if the statement is true and F if it is false.

1. Meiosis only occurs in germ cells. _____
2. After meiosis occurs, the cells will have the same number of chromosomes as the parent cell. _____

3. List the accessory organs of the male and female reproductive tracts and cite the function of each.

EXERCISE 23-4: Male Reproductive System (Text Fig. 23-1)

INSTRUCTIONS

1. Write the names of the structures that are not part of the reproductive system on the appropriate lines 1 to 6 in different colors. Use the same color for structures 1 and 2 and for structures 4 and 5. Color the structures on the diagram.
2. Write the names of the parts of the male reproductive system on the appropriate lines 7 to 20 in different colors. Use the same color for structures 10 and 11. Color the structures on the diagram.

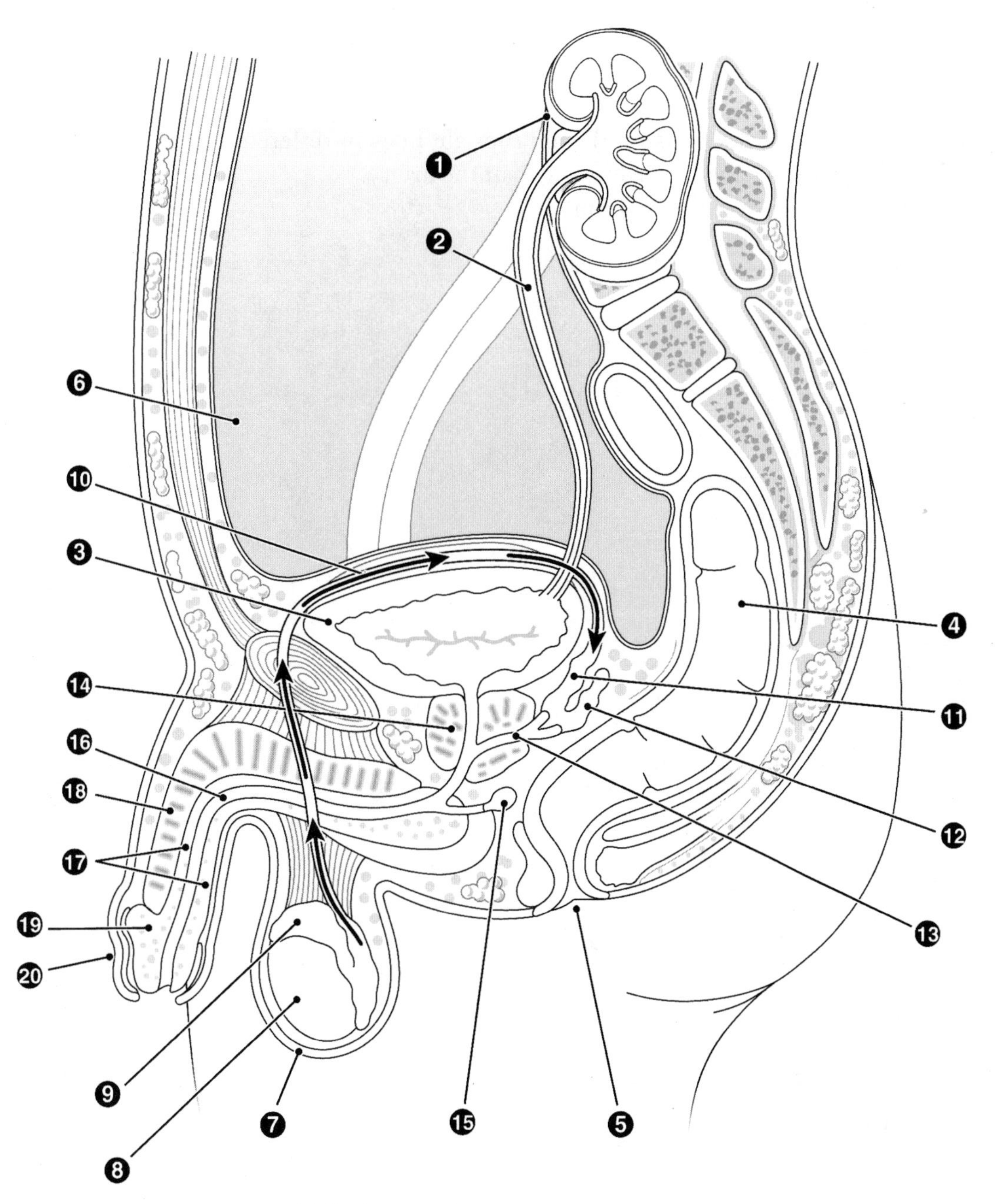

1. ______________________
2. ______________________
3. ______________________
4. ______________________
5. ______________________
6. ______________________
7. ______________________
8. ______________________
9. ______________________
10. ______________________
11. ______________________
12. ______________________
13. ______________________
14. ______________________
15. ______________________
16. ______________________
17. ______________________
18. ______________________
19. ______________________
20. ______________________

EXERCISE 23-5: Cross-section of the Penis (Text Fig. 23-5)

INSTRUCTIONS

1. Write the names of the parts on the appropriate lines in different colors. (Hint: the nerve is solid and the veins have larger lumens than the artery.)
2. Color the structures on the diagram.

1. ______________________
2. ______________________
3. ______________________
4. ______________________
5. ______________________
6. ______________________
7. ______________________
8. ______________________
9. ______________________
10. ______________________

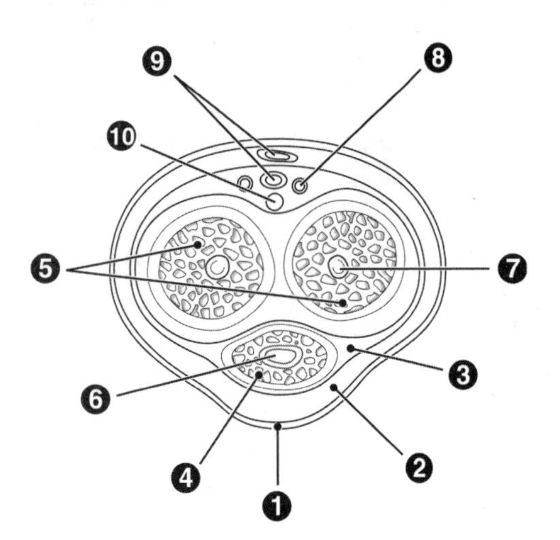

EXERCISE 23-6: Female Reproductive System (Text Fig. 23-8)

INSTRUCTIONS

1. Write the names of the parts on the appropriate lines in different colors. Use the same color for parts 8 and 9, for parts 10 to 13, and for parts 14 and 15.
2. Color the structures on the diagram.

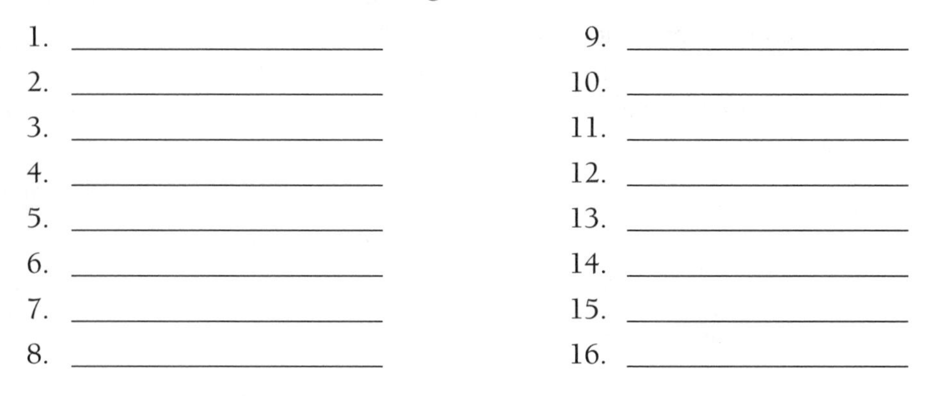

1. ____________________ 9. ____________________
2. ____________________ 10. ____________________
3. ____________________ 11. ____________________
4. ____________________ 12. ____________________
5. ____________________ 13. ____________________
6. ____________________ 14. ____________________
7. ____________________ 15. ____________________
8. ____________________ 16. ____________________

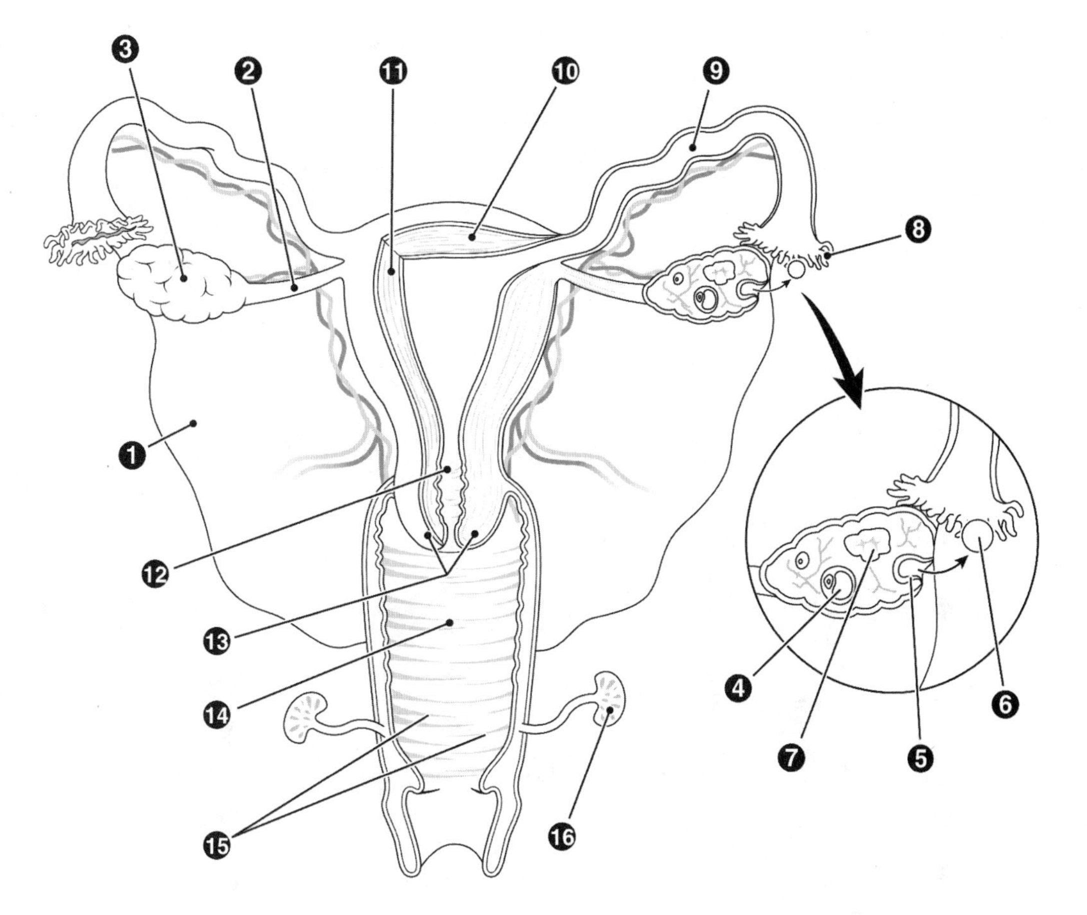

EXERCISE 23-7: Female Reproductive System (Sagittal Section) (Text Fig. 23-11)

INSTRUCTIONS

1. Write the names of the structures that are not part of the reproductive system on the appropriate lines 1 to 8 in different colors. Use the same color for structures 1 and 2, for structures 3 and 4, and for structures 5 and 6. Color the structures on the diagram.
2. Write the names of the parts of the female reproductive system and supporting ligaments on the appropriate lines 9 to 19 in different colors. Use the same color for structures 15 and 16. Color the structures on the diagram.

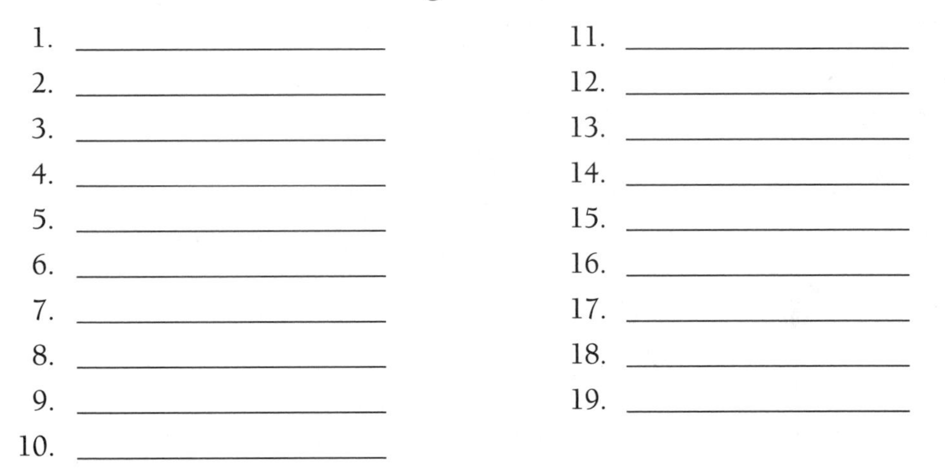

1. ____________________
2. ____________________
3. ____________________
4. ____________________
5. ____________________
6. ____________________
7. ____________________
8. ____________________
9. ____________________
10. ____________________
11. ____________________
12. ____________________
13. ____________________
14. ____________________
15. ____________________
16. ____________________
17. ____________________
18. ____________________
19. ____________________

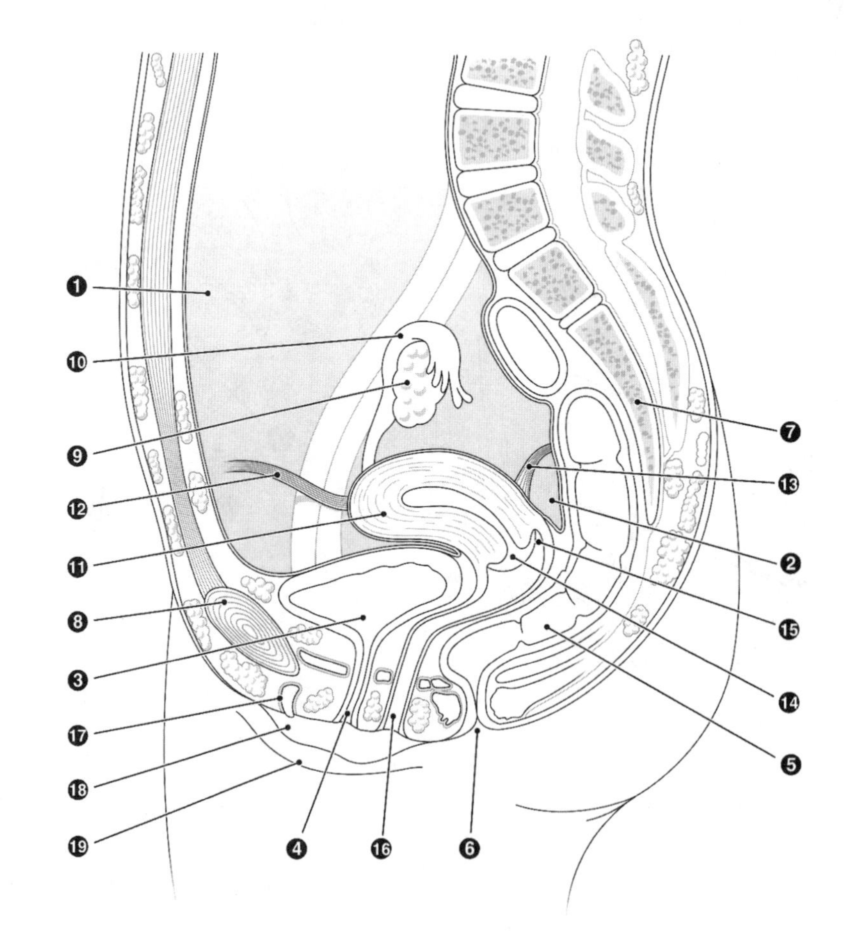

EXERCISE 23-8: External Parts of the Female Reproductive System (Text Fig. 23-12)

INSTRUCTIONS

Label the indicated parts.

1. ______________________
2. ______________________
3. ______________________
4. ______________________
5. ______________________
6. ______________________
7. ______________________
8. ______________________
9. ______________________

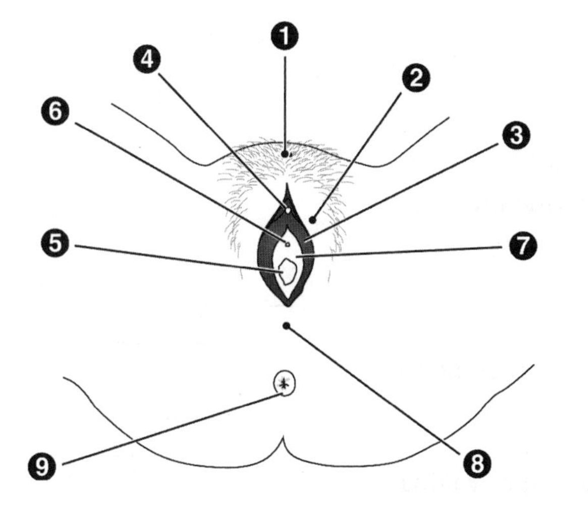

EXERCISE 23-9.

INSTRUCTIONS

Write the appropriate term in each blank.

epididymis	vas deferens	seminal vesicle	prostate gland	ejaculatory duct
fimbriae	oviduct	bulbourethral gland	greater vestibular gland	

1. One of the two glands that secrete a thick, yellow, alkaline secretion ________________
2. The gland that secretes a thin, alkaline secretion and that can contract to aid in ejaculation ________________
3. The coiled tube in which spermatozoa are stored as they mature and become motile ________________
4. The tube that transports the female germ cells ________________
5. Fringelike extensions that sweep the ovum into the uterine tube ________________
6. A gland in the female reproductive tract that secretes into the vagina ________________
7. A duct in the male that empties into the urethra ________________
8. A gland located inferior to the prostate that secretes mucus during sexual stimulation ________________

EXERCISE 23-10.

INSTRUCTIONS

Write the appropriate term in each blank.

cervix	myometrium	fundus	vestibule
endometrium	fornix	hymen	perineum

1. The Bartholin glands secrete into this area ________________
2. The vaginal region around the cervix ________________
3. The specialized tissue that lines the uterus ________________
4. The small, rounded part of the uterus located above the openings of the oviducts ________________
5. A fold of membrane found at the opening of the vagina ________________
6. The necklike part of the uterus that dips into the upper vagina ________________
7. The muscular wall of the uterus ________________

4. Describe the composition and function of semen.

EXERCISE 23-11.

INSTRUCTIONS

In the spaces below, list 5 functions of semen.

1. ____
2. ____
3. ____
4. ____
5. ____

5. Draw and label a spermatozoon.

EXERCISE 23-12: Diagram of a Human Spermatozoon (Text Fig. 23-4)

INSTRUCTIONS

1. Write the names of the parts on the appropriate lines in different colors. Use black for structures 1 to 3, because they will not be colored.
2. Color the structures on the diagram.

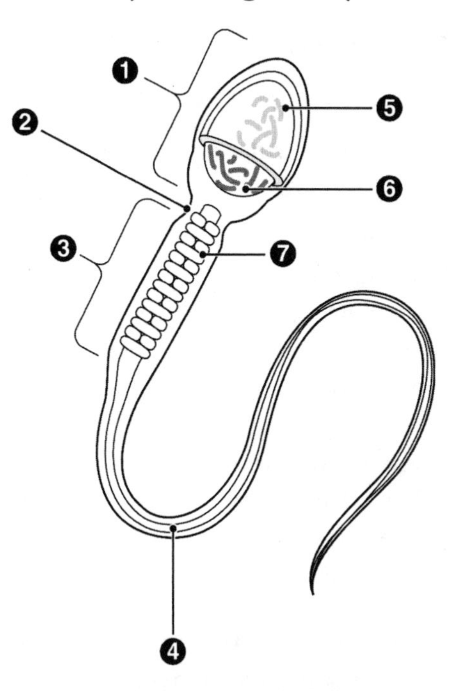

1. ____________________
2. ____________________
3. ____________________
4. ____________________
5. ____________________
6. ____________________
7. ____________________

6. List in the correct order the hormones produced during the menstrual cycle and cite the source of each.

EXERCISE 23-13: The Menstrual Cycle (Text Fig. 23-13)

INSTRUCTIONS

1. Identify the hormone responsible for each of the numbered lines in graph A, and write the hormone names on the appropriate lines in different colors. Trace over the lines in the graph with the appropriate color.
2. Write the names of the phases of the ovarian cycle (part B) and the uterine cycle (part C) on the appropriate lines.

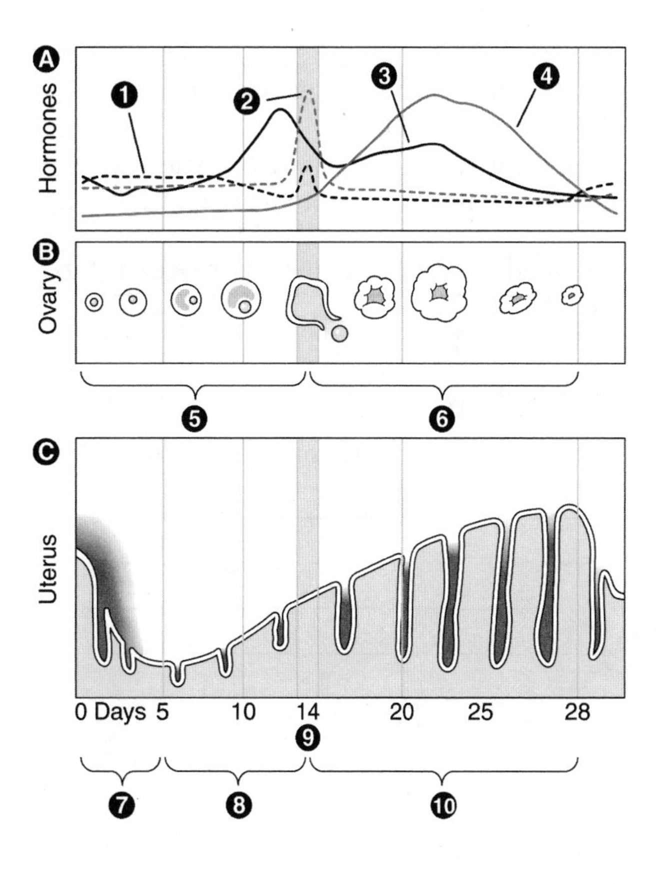

1. ____________________
2. ____________________
3. ____________________
4. ____________________
5. ____________________
6. ____________________
7. ____________________
8. ____________________
9. ____________________
10. ____________________

7. Describe the functions of the main male and female sex hormones.

EXERCISE 23-14.

INSTRUCTIONS

Write the appropriate term in each blank.

luteinizing hormone	follicle stimulating hormone	testosterone	estrogen
corpus luteum	ovulation	menstruation	progesterone

1. A hormone that is produced only during the luteal phase of the menstrual cycle ________
2. An ovarian hormone that is produced during the follicular and luteal phases of the menstrual cycle ________
3. A hormone produced by the testicular interstitial cells ________
4. Discharge of an ovum from the surface of the ovary ________
5. The hormone that stimulates Sertoli cells ________
6. The structure formed by the ruptured follicle after ovulation ________
7. The phase of the menstrual cycle when the endometrium is degenerating ________

8. Explain how negative feedback regulates reproductive function in both males and females.

EXERCISE 23-15.

INSTRUCTIONS

For each statement, choose the correct term from the two in brackets.

1. Testosterone ________ [stimulates, inhibits] the production of FSH and LH.
2. Just before ovulation, high estrogen levels ________ [stimulate, inhibit] the production of LH.
3. During the rest of the menstrual cycle, estrogen and/or progesterone ________ [stimulate, inhibit] the production of FSH and LH.

9. Describe the changes that occur during and after menopause.

EXERCISE 23-16.

INSTRUCTIONS

Write 5 different changes that may be associated with menopause in the spaces below.

1. ________
2. ________
3. ________
4. ________
5. ________

10. Cite the main methods of birth control in use.

EXERCISE 23-17.

INSTRUCTIONS

Write the appropriate term in each blank.

IUD	birth control patch	birth control ring	tubal ligation
male condom	diaphragm	vasectomy	

1. A method used to administer estrogen and progesterone through the skin ________________
2. A device implanted into the uterus that prevents fertilization and implantation ________________
3. A rubber cap fitted over the cervix ________________
4. The birth control method that is also highly effective against sexually transmitted infections ________________
5. A method used to administer birth control hormones internally ________________
6. A surgical method of birth control in females ________________

11. Briefly describe the major disorders of the male and female reproductive tracts.

EXERCISE 23-18.

INSTRUCTIONS

Write the appropriate term in each blank.

cryptorchidism	phimosis	prostatitis	orchitis	oligospermia
syphilis	epididymitis	genital herpes	salpingitis	

1. A sexually transmitted infection caused by a virus ________________
2. The condition that results when the testes fail to descend into the scrotal sac during fetal life ________________
3. A deficiency in the number of sperm cells in the semen ________________
4. Inflammation of the testis ________________
5. A sexually transmitted infection caused by a spirochete ________________
6. Tightness of the foreskin, which prevents it from being drawn back ________________
7. Infection of the prostate gland ________________
8. Inflammatory infection of the oviducts ________________

EXERCISE 23-19.

INSTRUCTIONS

Write the appropriate term in each blank.

amenorrhea	mastectomy	dysmenorrhea	fibroid	hysterectomy
premenstrual syndrome	gonorrhea	mammogram	endometriosis	

1. Depression, nervousness, and irritability preceding menses ______________
2. Surgical removal of the breast ______________
3. Absence of the menstrual flow for any reason ______________
4. An alternate name for myoma ______________
5. Surgical removal of the uterus ______________
6. Painful or difficult menstruation ______________
7. A condition in which the tissue normally lining the uterus grows in other sites, such as the external surface of the ovary ______________

12. Show how word parts are used to build words related to the reproductive systems.

EXERCISE 23-20.

INSTRUCTIONS

Complete the following table by writing the correct word part or meaning in the space provided. Write a word that contains each word part in the Example column.

Word Part	Meaning	Example
1. __________	extreme end	__________________
2. hyster/o-	_________	__________________
3. __________	egg	__________________
4. metr/o	_________	__________________
5. mast/o	_________	__________________
6. test/o	_________	__________________
7. __________	few, deficiency	__________________
8. orchid/o, orchi/o	_________	__________________
9. mamm/o	_________	__________________
10. salping/o	_________	__________________

Making the Connections

The following concept map deals with the organization of the reproductive system. Each pair of terms is linked together by a connecting phrase into a sentence. The sentence should be read in the direction of the arrow. Complete the concept map by filling in the appropriate term or phrase. There is one right answer for each term (1, 2, 4, 14). However, there are many correct answers for the connecting phrases.

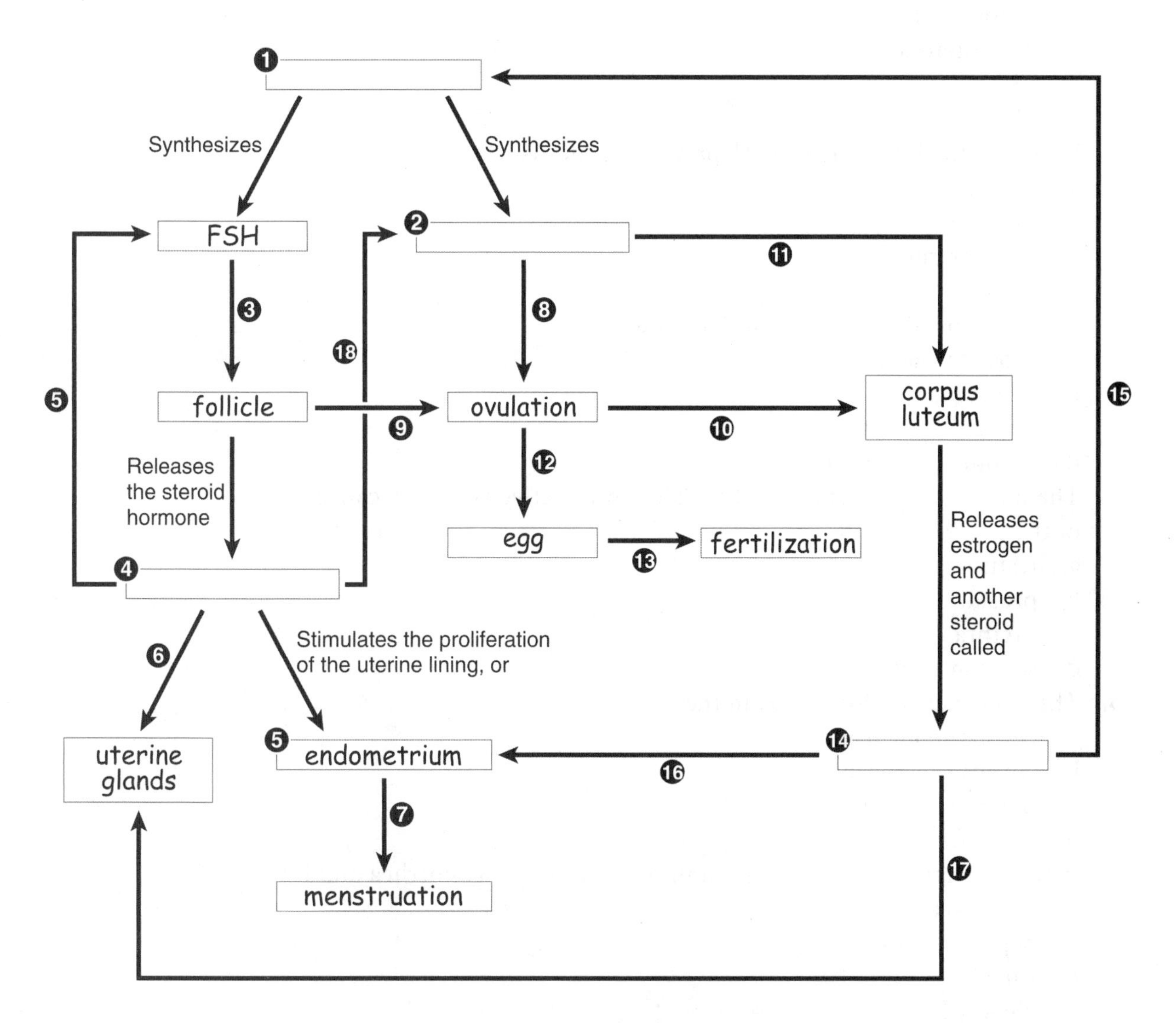

Optional Exercise: Make your own concept map, based on male reproductive anatomy and physiology. Choose your own terms to incorporate into your map, or use the following list: spermatozoon, testosterone, testis, FSH, LH, interstitial cell, epididymis, vas deferens, seminiferous tubule, Sertoli cell, prostate gland, bulbourethral gland, and penis.

Testing Your Knowledge

Building Understanding

I. Multiple Choice

Select the best answer and write the letter of your choice in the blank.

1. The corpus luteum secretes 1. _______
   a. testosterone
   b. progesterone
   c. FSH
   d. LH
2. Which of the following is NOT part of the uterus? 2. _______
   a. cervix
   b. fimbriae
   c. endometrium
   d. corpus
3. Spermatozoa develop in the walls of the 3. _______
   a. prostate gland
   b. penis
   c. seminal vesicles
   d. seminiferous tubules
4. The most common cancer of the male reproductive system is cancer of the 4. _______
   a. testis
   b. prostate
   c. urethra
   d. seminal vesicle
5. The glans penis is formed from the 5. _______
   a. corpus spongiosum
   b. corpus cavernosum
   c. pubic symphysis
   d. vas deferens
6. Pelvic inflammatory disease is commonly caused by gonorrhea and by 6. _______
   a. rickettsia
   b. papilloma virus
   c. rubella
   d. chlamydia
7. The ovum develops within the 7. _______
   a. corpus luteum
   b. ovarian follicle
   c. interstitial cells
   d. Sertoli cells
8. The Pap smear is a test for 8. _______
   a. pregnancy
   b. breast cancer
   c. cervical cancer
   d. infertility

9. Which of the following glands is found in the female? 9. _______
   a. greater vestibular glands
   b. bulbourethral glands
   c. Cowper glands
   d. prostate gland
10. Semen does NOT contain 10. _______
    a. seminal fluid
    b. a high concentration of hydrogen ions
    c. sugar
    d. sperm

II. Completion Exercise

Write the word or phrase that correctly completes each sentence.

1. The enzyme-containing cap on the head of a spermatozoon is called the _______________.
2. The hormone that promotes development of spermatozoa in the male and development of ova in the female is _______________.
3. The process of cell division that reduces the chromosome number by half is _______________.
4. The testes are contained in an external sac called the _______________.
5. The main male sex hormone is _______________.
6. The pelvic floor in both males and females is called the _______________.
7. The hormone that stimulates ovulation is _______________.
8. In the male, the tube that carries urine away from the bladder also carries sperm cells. This tube is the _______________.
9. An x-ray study of the breasts for the detection of cancer is a(n) _______________.
10. The surgical removal of the foreskin of the penis is called _______________.

Understanding Concepts

I. True/False

For each question, write *T* for true or *F* for false in the blank to the left of each number. If a statement is false, correct it by replacing the underlined term and write the correct statement in the blank below the question.

_______ 1. A low ability to reproduce is called sterility.

____

_______ 2. The urethra is contained in the corpus spongiosum of the penis.

____

_______ 3. When estrogen and progesterone levels are high, luteinizing hormone secretion is inhibited.

____

_______ 4. Gametes are produced by the process of mitosis.

____

_______ 5. Absence of menstrual flow is amenorrhea.

____

_______ 6. Progesterone levels are highest during the follicular phase of the menstrual cycle.

____

_______ 7. Progesterone is the first ovarian hormone produced in the menstrual cycle.

____

_______ 8. A deficiency in luteinizing hormone would result in testosterone deficiency.

____

II. Practical Applications

Study each discussion. Then write the appropriate word or phrase in the space provided.

➤ Group A

1. Ms. J, age 22, complained of painful menstrual cramps and excessive menstrual flow. The medical term for painful menstruation is ____________________.
2. Menstruation is the shedding of the uterine layer called the ____________________.
3. The physician suggested anti-inflammatory medication, but Ms. J had previously tried medication to no avail. An alternate approach was tried, in which the passageway between the vagina and uterus was artificially dilated. The part of the uterus containing this passageway is the ____________________.
4. The excessive bleeding was also a disturbing symptom. Ultrasound revealed the presence of numerous small, benign uterine tumors. These tumors are called myomas or ______________.

➤ Group B

1. Mr. and Ms. S, both aged 35, had been trying to conceive for two years with no success. Analysis of Mr. S's semen revealed a low sperm count. This condition is called ____________.
2. The fertility specialist asked Mr. S about any illnesses or infections that might be responsible for his low sperm count. Mr. S mentioned that his testes were very swollen when he had a mumps infection at the age of 17. Inflammation of the testis is called ____________________.
3. Since Mr. S's testes were of normal size, the mumps infection was probably not the cause of his infertility. Blood tests were performed, revealing a deficiency in the pituitary hormone that acts on sustentacular cells. This hormone is called ____________________.
4. Conversely, testosterone production was normal. The cells that produce testosterone are called ____________________.
5. Mr. S's hormone deficiency was successfully treated, and Mr. and Ms. S became the proud parents of triplets. Mr. S went to the clinic shortly after the birth, and his ductus deferens were cut. This method of birth control is called a(n) ____________________.

III. Short Essays

1. Discuss changes occurring in the ovary and in the uterus during the follicular phase of the menstrual cycle.

   ____

   ____

   ____

2. Name the microorganisms that commonly cause infections of the reproductive tract in males and females.

   ____

   ____

   ____

Conceptual Thinking

1. Mr. S is taking synthetic testosterone in order to improve his wrestling performance. To his alarm, he notices that his testicles are shrinking. Explain why this is happening.

   ____

   ____

   ____

2. Compare and contrast the role of FSH in the regulation of the male and female gonad.

   ____

   ____

   ____

Expanding Your Horizons

Have you heard of kamikaze sperm? Based on studies in animals, some animal behavior researchers have hypothesized that semen contains a population of spermatozoa that is capable of destroying spermatozoa from a competing male. A "sperm team" contains blockers, which are misshapen spermatozoa with multiple tails and/or heads. Other sperm, the kamikaze spermatozoa, seek out and destroy spermatozoa from other males. A very small population of egg-getting spermatozoa attempts to fertilize the ovum. Although some aspects of this hypothesis have been challenged, the presence of sperm subpopulations is now well established. You can read more about the kamikaze sperm hypothesis in the following scientific articles.

Resources

1. Baker RR, Bellis MA. Elaboration of the Kamikaze sperm hypothesis: a reply to Harcourt. *Anim Behav* 1989; 37:865–867.
2. Moore HD, Martin M, Birkhead TR. No evidence for killer sperm or other selective interactions between human spermatozoa in ejaculates of different males in vitro. *Proc R Soc Lond B Biol Sci* 1999; 266:2343–2350.

CHAPTER 24

Development and Birth

Overview

Pregnancy begins with fertilization of an ovum by a spermatozoon to form a **zygote.** Over the next 38 weeks of **gestation**, the offspring develops first as an **embryo** and then as a **fetus.** During this period, it is nourished and maintained by the **placenta**, formed from tissues of both the mother and the embryo. The placenta secretes a number of hormones, including progesterone, estrogen, human chorionic gonadotropin, human placental lactogen, and relaxin. These hormones induce changes in the uterus and breasts to support the pregnancy and prepare for childbirth and milk production.

Childbirth or **parturition** occurs in four stages, beginning with contractions of the uterus and dilation of the cervix. Subsequent stages include delivery of the infant, delivery of the afterbirth, and control of bleeding. Milk production, or **lactation**, is stimulated by the hormones prolactin and oxytocin. Milk removal from the breasts stimulates further production.

Although pregnancy and birth are much safer than in previous centuries, they are not without risk. Infectious diseases, endocrine abnormalities, tumors, and structural problems can harm the fetus and/or the mother.

Addressing the Learning Outcomes

1. Describe fertilization and the early development of the fertilized egg.

EXERCISE 24-1.

INSTRUCTIONS

Write the appropriate term in each blank.

gestation zygote embryo implantation fetus ovum

1. The fertilized egg ____________
2. The developing offspring from the third month until birth ____________
3. The entire period of development in the uterus ____________
4. Attachment of the fertilized egg to the lining of the uterus ____________
5. The developing offspring from implantation until the third month ____________

2. Describe the structure and function of the placenta (also see Exercise 24-3).

EXERCISE 24-2.

INSTRUCTIONS

Write the appropriate term in each blank.

placenta venous sinuses placental villi umbilical cord
human placental lactogen human chorionic gonadotropin relaxin

1. The hormone that loosens the pubic symphysis and softens the cervix for parturition ____________
2. A hormone that prepares the breasts for lactation and alters maternal metabolism ____________
3. A placental hormone that stimulates progesterone synthesis and is detected by pregnancy tests ____________
4. The structure that serves as the organ for nutrition, respiration, and excretion for the fetus ____________
5. The portion of the placenta containing fetal capillaries ____________
6. Channels in the placenta containing maternal blood ____________

3. Describe how fetal circulation differs from adult circulation.

EXERCISE 24-3: Fetal Circulation (Text Fig. 24-2)

INSTRUCTIONS

1. On the diagram on the next page, label the parts of the fetal circulation and placenta. (Hint: you can differentiate between the uterine arterioles and uterine venules by following the path of the vessels into the umbilical arteries and vein.)
2. Color the boxes in the legend using the following color scheme: color the oxygen-rich blood box red, the oxygen-poor blood box blue, and the mixed blood box purple.
3. As much as possible, color the blood vessels on the diagram with the appropriate color, based on the oxygen content of the blood.

1. ______________________
2. ______________________
3. ______________________
4. ______________________
5. ______________________
6. ______________________
7. ______________________
8. ______________________
9. ______________________
10. ______________________

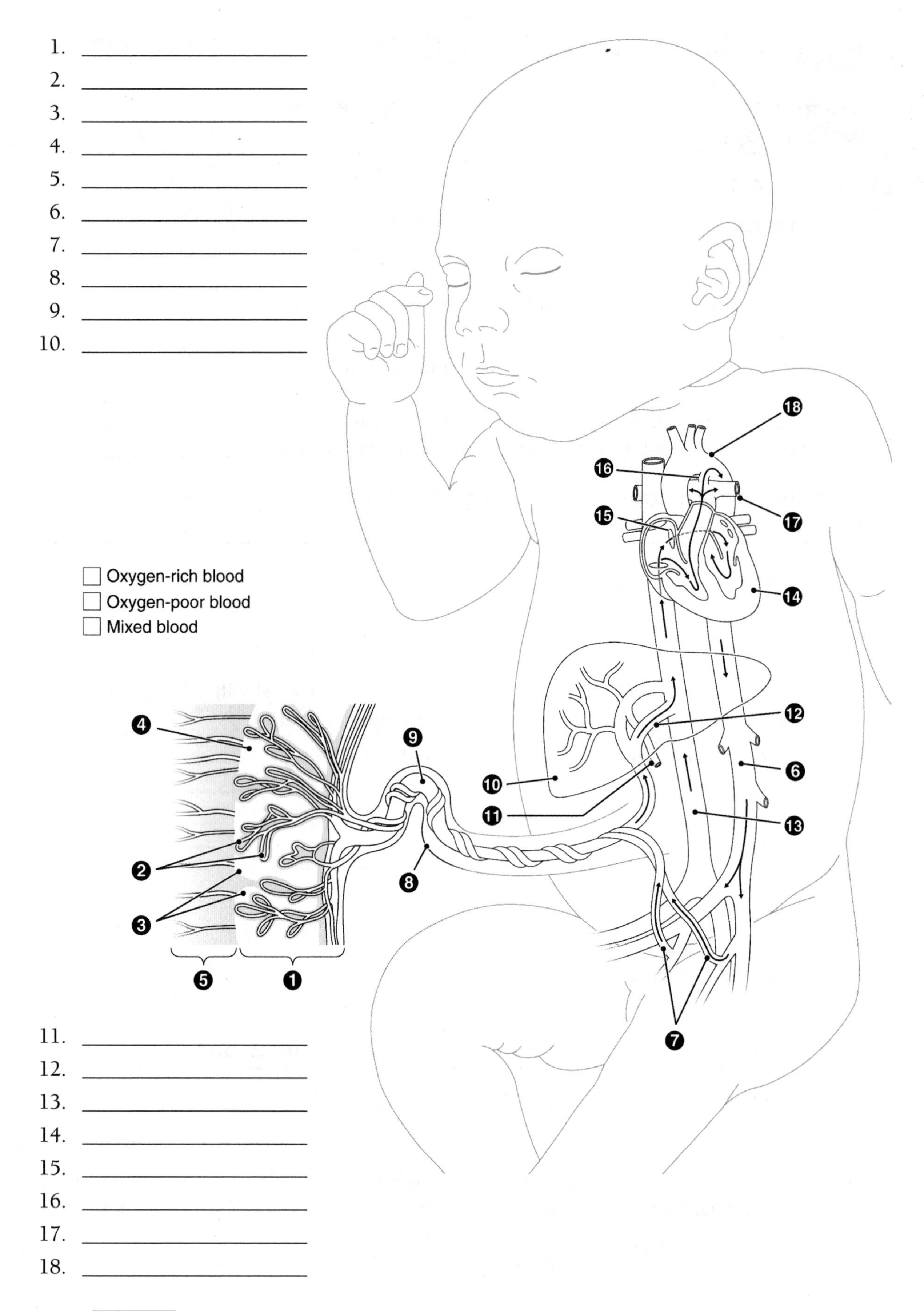

11. ______________________
12. ______________________
13. ______________________
14. ______________________
15. ______________________
16. ______________________
17. ______________________
18. ______________________

4. Briefly describe changes that occur in the fetus and the mother during pregnancy.

EXERCISE 24-4: Midsagittal Section of a Pregnant Uterus (Text Fig. 24-5)

INSTRUCTIONS

Write the name of each labeled part on the numbered lines.

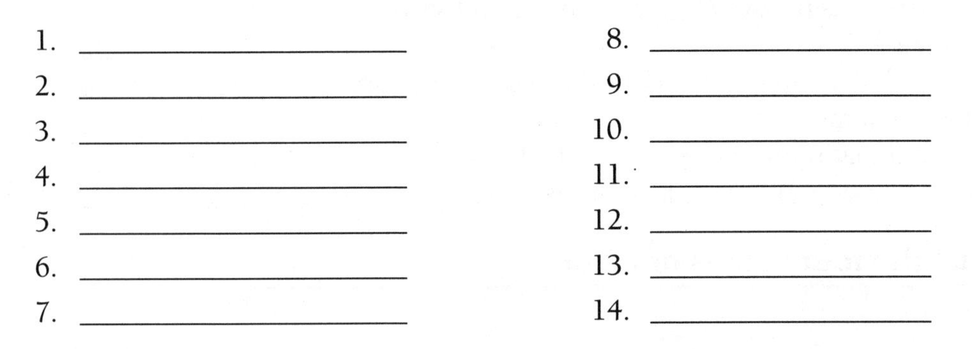

1. ____________________
2. ____________________
3. ____________________
4. ____________________
5. ____________________
6. ____________________
7. ____________________
8. ____________________
9. ____________________
10. ____________________
11. ____________________
12. ____________________
13. ____________________
14. ____________________

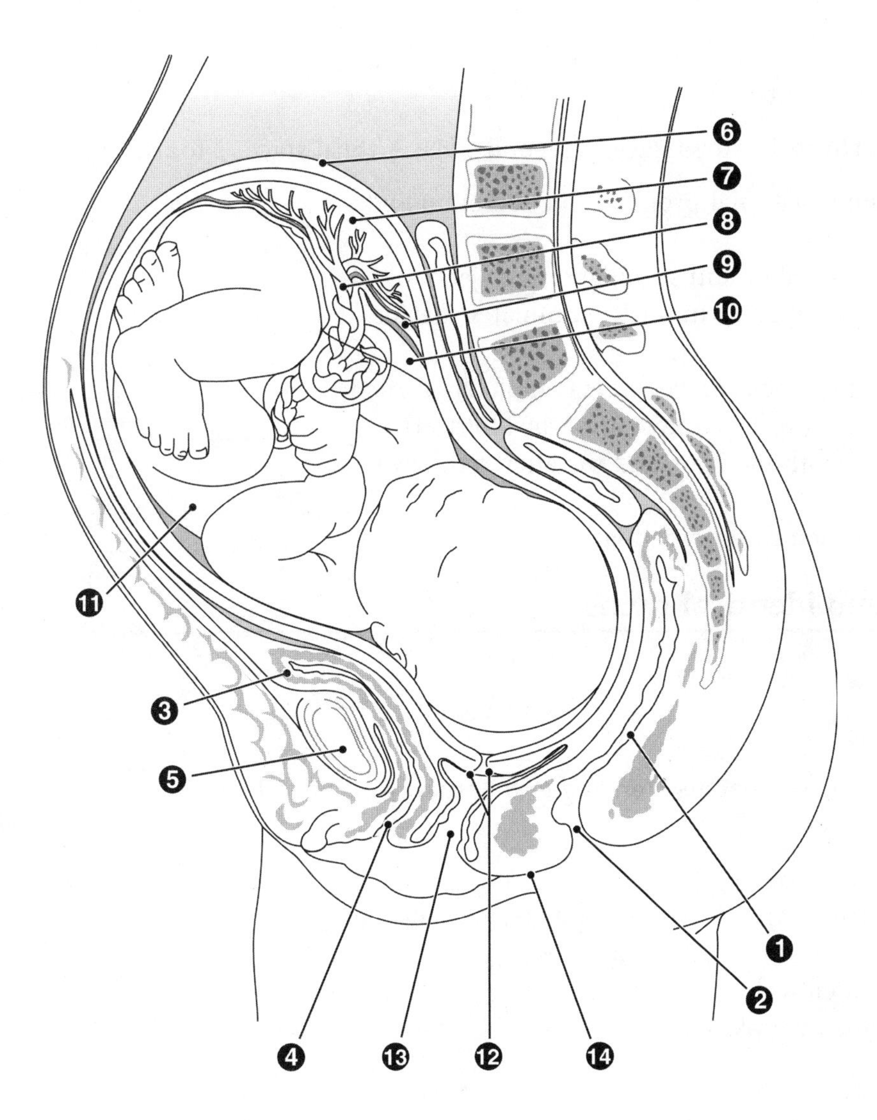

EXERCISE 24-5.

INSTRUCTIONS

Write the appropriate term in each blank.

vernix caseosa	ultrasonography	amniotic sac
ductus arteriosus	foramen ovale	ductus venosus

1. The structure that surrounds the developing offspring and serves as a protective cushion ____________
2. A small vessel joining the pulmonary artery to the descending aorta ____________
3. A small hole in the fetal atrial septum ____________
4. The cheeselike material that protects the skin of the fetus ____________
5. A technique commonly used to monitor fetal development ____________

5. Briefly describe the four stages of labor.

EXERCISE 24-6.

INSTRUCTIONS

Write the appropriate term in each blank.

oxytocin prostaglandin cortisol first stage second stage third stage fourth stage

1. A fetal hormone that inhibits maternal progesterone secretion and stimulates uterine contractions ____________
2. The hormone that can initiate labor and stimulate milk ejection ____________
3. A substance produced by the myometrium that stimulates uterine contractions ____________
4. The stage of labor that begins when the cervix is completely dilated ____________
5. The stage of labor that ends with the expulsion of the afterbirth ____________
6. The stage of labor in which both the baby and the afterbirth have been expelled ____________
7. The stage of labor during which the cervix dilates ____________

6. Compare fraternal and identical twins.

EXERCISE 24-7.

INSTRUCTIONS

Write "F" in the blank following statements applying to fraternal twins and "I" in the blank following statements applying to identical twins.

1. Twins formed from a single zygote ____
2. Twins formed from the fertilization of two ova by two spermatozoa ____
3. Twins that may be of the same sex or different sexes ____
4. Twins that are genetically identical ____
5. Twins that always have their own placenta ____

7. Cite the advantages of breastfeeding.

EXERCISE 24-8: Section of the Breast (Text Fig. 24-7)

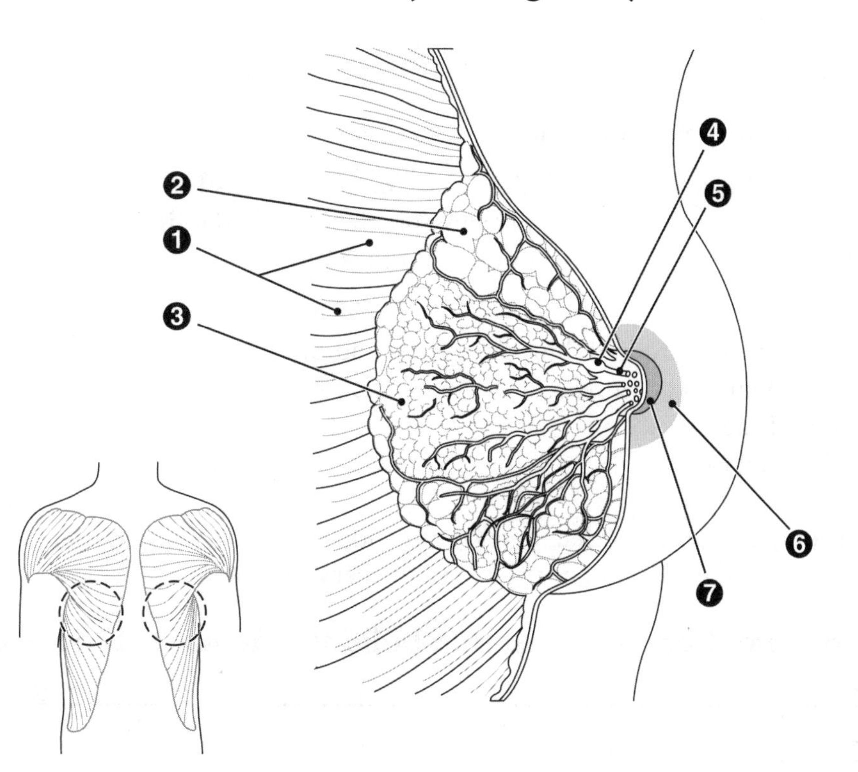

INSTRUCTIONS

Write the name of each labeled part on the numbered lines.

1. ____________________
2. ____________________
3. ____________________
4. ____________________
5. ____________________
6. ____________________
7. ____________________

EXERCISE 24-9.

INSTRUCTIONS

Briefly summarize four advantages of breastfeeding in the spaces below.

1. ____
2. ____
3. ____
4. ____

8. Describe several disorders associated with pregnancy, childbirth, and lactation.

EXERCISE 24-10.

INSTRUCTIONS

Write the appropriate term in each blank.

pregnancy-induced hypertension	choriocarcinoma	mastitis
episiotomy	placenta previa	hydatidiform mole
ectopic	puerperal infection	
abruptio placentae		

1. A disorder associated with high blood pressure and proteinuria ____________
2. Term describing an infection related to childbirth ____________
3. Inflammation of the breast ____________
4. A benign placental tumor ____________
5. A malignant placental tumor ____________
6. Term describing a pregnancy that develops outside the uterine cavity ____________
7. The attachment of the placenta near or over the cervical opening ____________

9. Show how word parts are used to build words related to development and birth.

EXERCISE 24-11.

INSTRUCTIONS

Complete the following table by writing the correct word part or meaning in the space provided. Write a word that contains each word part in the Example column.

Word Part	Meaning	Example
1. ________	outside, external	________
2. zyg/o	________	________
3. ________	labor	________
4. ox/y	________	________
5. chori/o	________	________
6. ________	body	________

Making the Connections

The following concept map deals with different aspects of pregnancy, parturition, and lactation. Each pair of terms is linked together by a connecting phrase into a sentence. The sentence should be read in the direction of the arrow. Complete the concept map by filling in the appropriate term or phrase. There is one right answer for each term. However, there are many correct answers for the connecting phrases.

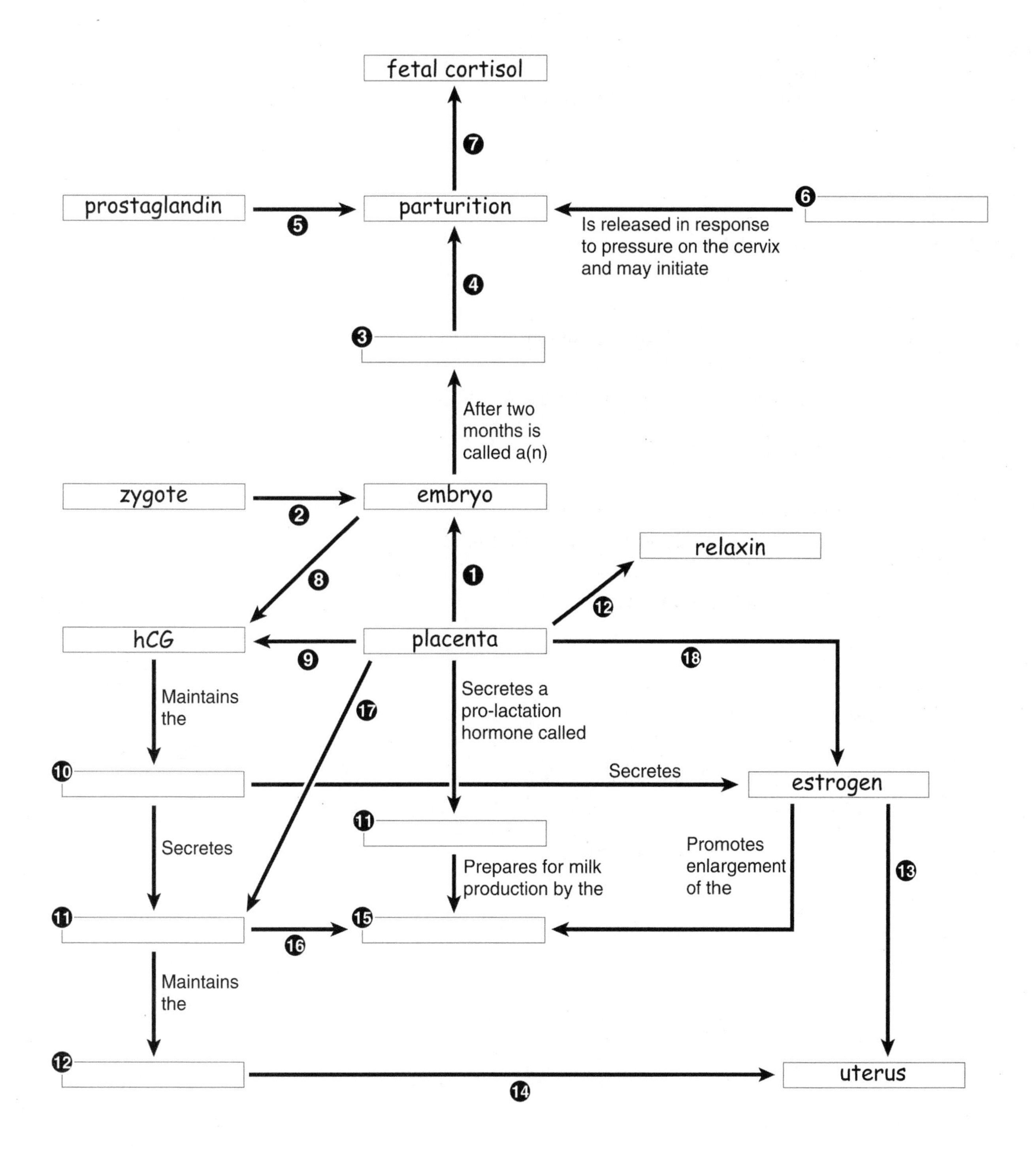

Testing Your Knowledge

Building Understanding

I. Multiple Choice

Select the best answer and write the letter of your choice in the blank.

1. Contractions of the myometrium are stimulated by 1. _______
   a. progesterone
   b. estrogen
   c. prolactin
   d. prostaglandin
2. The fetal circulation is designed to bypass the 2. _______
   a. lungs
   b. placenta
   c. spleen
   d. brain
3. The second stage of labor includes 3. _______
   a. the onset of contractions
   b. expulsion of the afterbirth
   c. passage of the fetus through the vagina
   d. expulsion of the placenta
4. Premature separation of the placenta from the uterine wall is 4. _______
   a. ectopic pregnancy
   b. placenta previa
   c. eclampsia
   d. abruptio placentae
5. The term *viable* is used to describe a fetus that is 5. _______
   a. spontaneously aborted
   b. developing outside the uterus
   c. capable of living outside the uterus
   d. stillborn
6. Maternal blood is found in the 6. _______
   a. umbilical cord
   b. placental villi
   c. venous sinuses of the placenta
   d. all of the above

II. Completion Exercise

Write the word or phrase that correctly completes each sentence.

1. A surgical cut and repair of the perineum to prevent tearing is called a(n) ___________________.
2. By the end of the first month of embryonic life, the beginnings of the extremities may be seen. These are four small swellings called ___________________.
3. The mammary glands of the female provide nourishment for the newborn through the secretion of milk; this is a process called ___________________.
4. The stage of labor during which the afterbirth is expelled from the uterus is the ___________________.
5. Twins that develop from the same fertilized egg are called ___________________.
6. The normal site of fertilization is the ___________________.
7. The clear liquid that flows from the uterus when the mother's "water breaks" is technically called ___________________.
8. Untreated PIH may progress to a more severe form of the disorder, called ___________________.

Understanding Concepts

I. True/False

For each question, write *T* for true or *F* for false in the blank to the left of each number. If a statement is false, correct it by replacing the underlined term and write the correct statement in the blank below the question.

_______ 1. Ms. J is 7 weeks pregnant. Her uterus contains an embryo.

____

_______ 2. The baby is delivered during the fourth stage of labor.

____

_______ 3. Fraternal twins are genetically distinct.

____

_______ 4. Blood is carried from the placenta to the fetus in the umbilical vein.

____

_______ 5. Oxytocin secreted from the fetal adrenal gland may help induce labor.

____

_______ 6. A hydatidiform mole is considered to be a malignant tumor.

____

II. Practical Applications

Study each discussion. Then write the appropriate word or phrase in the space provided.

1. Mr. and Ms. L had been trying to conceive for 2 years. Finally, Ms. L realized her period was late and purchased a pregnancy detection kit. These kits test for the presence of a hormone produced exclusively by embryonic tissues that helps maintain the corpus luteum. This hormone is called ______________.
2. The pregnancy test was positive. Based on the date of her last menstrual period, Ms. L was determined to be 6-weeks pregnant. The gestational age of Ms. L's new offspring at this time would be ______________.
3. During a prenatal appointment 14 weeks later, Ms. L was able to see her future offspring using a technique for visualizing soft tissues without the use of x-rays. This technique is called ______________.
4. The radiologist noted that the placenta was located near the cervix. Ms. L was told that her placenta would probably migrate upward as her pregnancy progressed. However, if the placenta remained at its current position she was at risk for a syndrome called ______________.
5. Happily, later exams showed that the placenta had migrated upward toward the fundus of the uterus. However, urinalysis revealed the presence of protein in Ms. L's urine, and her blood pressure was higher than normal. Ms. L may be suffering from a serious disorder called ______________.
6. Ms. L's condition did not change, and at 39 weeks the decision was made to deliver the baby through an incision made in the abdominal wall and the wall of the uterus. This operation is called a(n) ______________.
7. The operation went smoothly, and baby L was born. Ms. L immediately began to breastfeed her infant. The first secretion from her breasts was not milk, but rather ______________.

III. Short Essays

1. Compare and contrast human placental lactogen (hPL) and human chorionic gonadotropin (hCG). Discuss the synthesis and action of each hormone.

   ____

   ____

   ____

2. Describe the site of synthesis and actions of the hormones involved in lactation and preparing the breasts for lactation.

   ____

   ____

   ____

Conceptual Thinking

1. Trace the path of a blood cell from the placenta to the fetal heart and back to the placenta. Assume that this blood cell passes through the foramen ovale.

   ____

   ____

   ____

2. Ms. J has suffered from repeated miscarriages. Blood tests reveal a deficiency in progesterone. Could this deficiency be implicated in Ms. J's miscarriages?

   ____

   ____

   ____

Expanding Your Horizons

Have you ever wondered why animals can give birth alone, but humans require all sorts of technological equipment? Women rarely give birth alone, even in societies with few technological advances. The difficulties of human birth reflect some of our evolutionary adaptations. For instance, walking upright requires the pelvis to be backward, and the large human brain size results in large fetal skulls. You can read more about "The Evolution of Birth" in *Scientific American* article.

Resources

1. Rosenberg KR, Trevathan WR. *Sci Am* 2001; 285:72–77.

CHAPTER 25

Heredity and Hereditary Diseases

Overview

The scientific study of heredity has advanced with amazing speed in the past 50 years. Nevertheless, many mysteries remain. Gregor Mendel was the first person known to have carried out formal experiments in genetics. He identified independent units of heredity, which he called factors and which we now call **genes.**

The chromosomes in the nucleus of each cell are composed of a complex molecule, **DNA.** This material makes up the many thousands of genes that determine a person's traits and are passed on to offspring at the time of fertilization. Genes direct the formation of **enzymes**, which in turn make possible all the chemical reactions of metabolism. Defective genes, produced by **mutation**, may disrupt normal enzyme activity and result in hereditary disorders such as sickle cell anemia, albinism, and phenylketonuria. Some human traits are determined by a single pair of genes (one gene from each parent), but most are controlled by multiple pairs of genes acting together.

Genes may be classified as **dominant** or **recessive.** If one parent contributes a dominant gene, then any offspring who receives that gene will show the trait (e.g., Huntington disease). Traits carried by recessive genes may remain hidden for generations and be revealed only if they are contributed by both parents (e.g., albinism, cystic fibrosis, PKU, and sickle cell anemia). In some cases, treatment begun early may help prevent problems associated with a genetic disorder. Genetic counseling should be sought by all potential parents whose relatives are known to have an inheritable disorder.

Addressing the Learning Outcomes

1. Briefly describe the mechanism of gene function.

EXERCISE 25-1.

INSTRUCTIONS

In the blank following each statement, write "T" if the statement is true and "F" if it is false.

1. Genes are segments of proteins. ____
2. Genes contain the blueprints to make proteins. ____
3. Genes are composed of DNA. ____
4. Genes code for specific traits. ____
5. Humans have 46 autosomes. ____
6. Humans have 46 chromosomes. ____

2. Explain the difference between dominant and recessive genes.

EXERCISE 25-2.

INSTRUCTIONS

Write the appropriate term in each blank.

heterozygous	homozygous	autosome	sex chromosome
recessive	dominant	allele	

1. One member of a gene pair that controls a specific trait ________________
2. Term describing a gene that expresses its effect only if homozygous ________________
3. A gene pair consisting of two dominant or two recessive alleles ________________
4. Any chromosome except the X and Y chromosomes ________________
5. Term describing a gene pair composed of two different alleles ________________
6. Term describing a gene that expresses its effect if homozygous or heterozygous ________________

3. Compare phenotype and genotype and give examples of each.

EXERCISE 25-3.

INSTRUCTIONS

In the blank following each statement, write "P" if the statement applies to *phenotype* and "G" if it applies to *genotype*.

1. The genetic makeup of an individual ____
2. The characteristics that can be observed and/or measured ____
3. Eye color ____
4. Homozygous dominant ____
5. Blood type ____
6. Heterozygous ____

4. Describe what is meant by a carrier of a genetic trait.

EXERCISE 25-4.

INSTRUCTIONS

Circle the correct answer.

A carrier is:

1. an individual heterozygous for a recessive trait.
2. an individual heterozygous for a dominant trait.
3. an individual that shows the symptoms of a disease.
4. an individual homozygous for either a dominant or a recessive trait.

5. Define *meiosis* and explain its function in reproduction.

EXERCISE 25-5.

INSTRUCTIONS

In the blank following each statement, write "T" if the statement is true and "F" if it is false.

1. Following meiosis, each reproductive cell contains 46 chromosomes. ____
2. Following meiosis, each reproductive cell contains 23 chromosomes. ____
3. Some reproductive cells contain all of the maternal chromosomes, while other reproductive cells contain all of the paternal chromosomes. ____
4. Each reproductive cell contains a mix of maternal and paternal chromosomes. ____

6. Explain how sex is determined in humans.

EXERCISE 25-6.

INSTRUCTIONS

For each of the following examples, state if the genotype would result in a female phenotype (F) or a male phenotype (M), assuming that development proceeds normally.

1. Union between an X sperm and a X ovum ____
2. Union between an Y sperm and an X ovum ____

7. Describe what is meant by the term sex-linked and list several sex-linked traits.

EXERCISE 25-7.

INSTRUCTIONS

In the blank following each statement, write "S" if the statement refers to sex-linked traits or "A" if it applies to autosomal (non–sex-linked) traits.

1. A trait carried on the Y chromosome ____
2. A trait carried on the X chromosome ____
3. A trait carried on chromosomes other than the X or Y chromosomes ____
4. A trait for which males or females can be carriers ____
5. A trait for which only females can be carriers ____

8. List several factors that may influence the expression of a gene.

EXERCISE 25-8.

INSTRUCTIONS

In the spaces below, list three factors that influence gene expression.

1. ______________________________
2. ______________________________
3. ______________________________

9. Define *mutation*.

EXERCISE 25-9.

INSTRUCTIONS

In the spaces below, define *mutation* and *mutagen*. In your definition of mutation, describe some different types of mutations.

1. mutation: ____

____

2. mutagen: ____

____

10. Differentiate among congenital, genetic, and hereditary disorders and give several examples of each.

EXERCISE 25-10.

INSTRUCTIONS

Write the appropriate term in each blank.

multifactorial	XX	sex-linked	mutation	mutagen
hereditary	XY	congenital	genetic	

1. The genotype of a female ______________
2. Term for a trait carried on the X chromosome, such as hemophilia ______________
3. A specific term describing a trait determined by multiple gene pairs acting together ______________
4. Term describing any disease caused by a change in the DNA sequence, whether or not it is inherited from a parent ______________
5. Term describing any trait present at birth ______________
6. Term that indicates specifically that a trait is inherited from a parent ______________
7. A substance that induces a change in DNA ______________
8. The genotype of a male ______________

EXERCISE 25-11.

INSTRUCTIONS

Write the appropriate term in each blank

Marfan syndrome	Huntington disease	PKU	trisomy 21
spina bifida	cystic fibrosis	Tay-Sachs disease	osteogenesis imperfecta

1. A neurodegenerative disease that is genetic and hereditary, but is not evident until midlife ________
2. A genetic disorder that is genetic but not always hereditary ________
3. An inherited disorder that prevents the metabolism of phenylalanine ________
4. A congenital disorder that is not genetic ________
5. A genetic, hereditary disease affecting all connective tissue, resulting in problems in many body systems ________
6. A genetic disease in which fat is abnormally deposited in CNS neurons ________
7. A genetic disease that results in blockages of the bronchi, intestine, and pancreatic ducts ________

11. List several factors that may cause genetic disorders.

EXERCISE 25-12.

INSTRUCTIONS

List three types of mutagens that may produce genetic disorders.

1. ________
2. ________
3. ________

12. Define *karyotype* and explain how karyotypes are used in genetic counseling (see Exercise 25-13).

13. Briefly describe several methods used to treat genetic disorders.

EXERCISE 25-13.

INSTRUCTIONS

Write the appropriate term in each blank.

pedigree amniocentesis Klinefelter syndrome phenylketonuria Marfan syndrome maple syrup urine disease chorionic villus sampling karyotype

1. A study of the chromosomes, as done on fetal cells, used to determine fetal sex and screen for significant chromosomal abnormalities ____________
2. Removal of fluid from the sac surrounding the fetus ____________
3. Removal of projections from a placental membrane ____________
4. A complete, detailed family tree that can detect carriers ____________
5. A genetic disorder with the genotype XXY that is treated with hormones and psychotherapy ____________
6. A genetic disease treated with large doses of thiamin ____________
7. An inherited disease treated by dietary management that is detected by a blood test ____________

14. Show how word parts are used to build words related to heredity.

EXERCISE 25-14.

INSTRUCTIONS

Complete the following table by writing the correct word part or meaning in the space provided. Write a word that contains each word part in the Example column.

Word Part	Meaning	Example
1. ________	nucleus	____________
2. con-	________	____________
3. ________	color	____________
4. cele-	________	____________
5. phen/o	________	____________
6. ________	self	____________
7. ________	tapping, perforation	____________
8. ________	other, different	____________
9. homo-	________	____________
10. dactyl/o	________	____________

Making the Connections

The following concept map deals with some aspects of heredity. Each pair of terms is linked together by a connecting phrase into a sentence. The sentence should be read in the direction of the arrow. Complete the concept map by filling in the appropriate term or phrase. There is one right answer for each term (4 to 6, 11, 13). However, there are many correct answers for the connecting phrases.

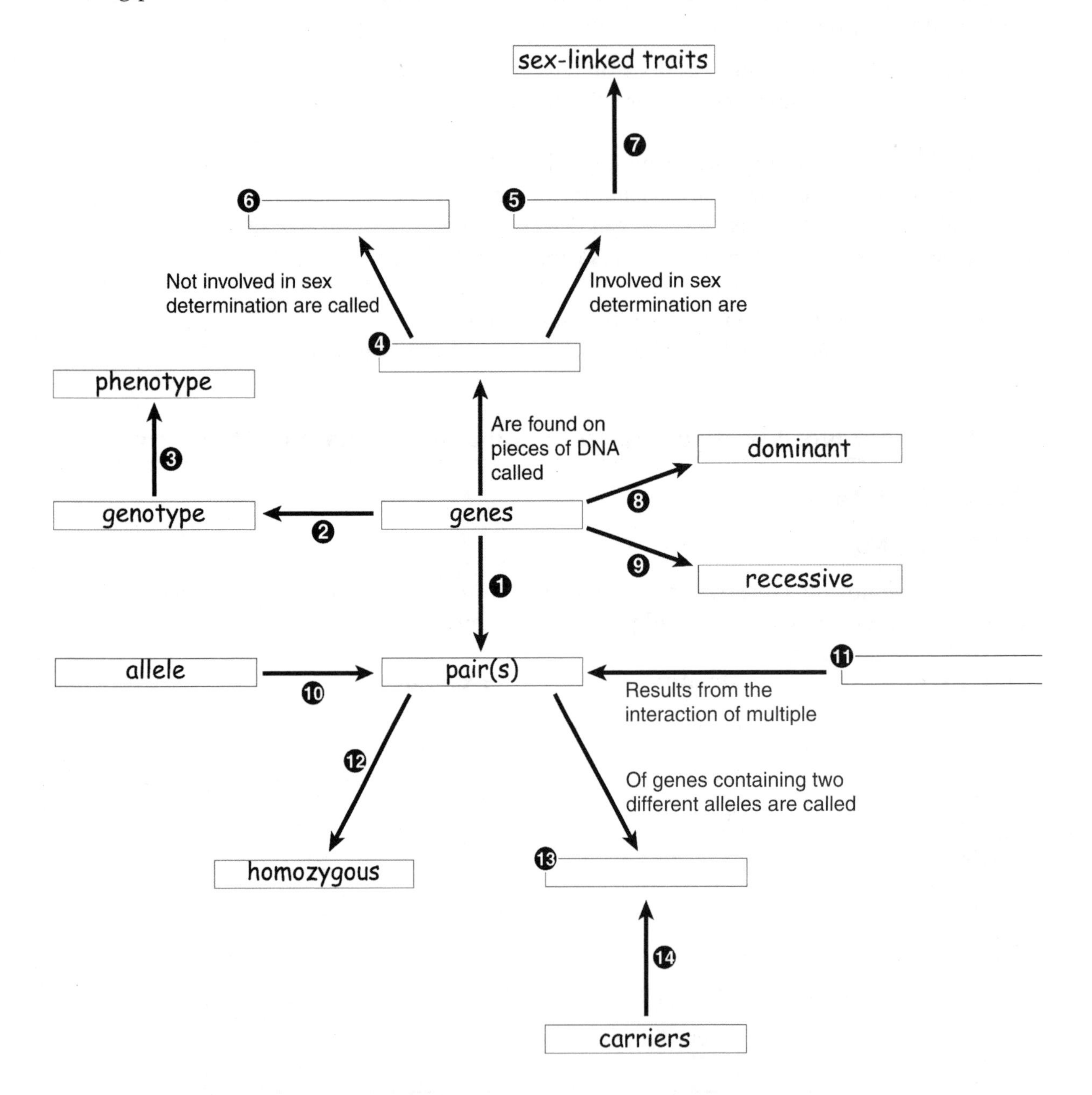

Optional Exercise: Make your own concept map, based on aspects of genetic diseases. Choose your own terms to incorporate into your map, or use the following list: karyotype, amniocentesis, chromosomes, chorionic villus sampling, hereditary, congenital, genetic, genetic engineering, pedigree, genetic screening, Huntington disease, Down syndrome, cystic fibrosis, and phenylketonuria.

Testing Your Knowledge

Building Understanding

I. Multiple Choice

Select the best answer and write the letter of your choice in the blank.

1. Two parents with normal pigmentation can give birth to a child with albinism. Albinism is NOT a 1. _______
   a. recessive disease
   b. dominant trait
   c. congenital disorder
   d. genetic disorder
2. Klinefelter syndrome is a 2. _______
   a. disorder in which masses grow along the nerves
   b. disorder that involves the sex chromosomes
   c. condition of having extra fingers
   d. form of dwarfism
3. Gender in humans is determined by 3. _______
   a. the number of X chromosomes
   b. the sex chromosome carried by the ovum
   c. the number of autosomes
   d. the sex chromosome carried by the spermatozoon
4. The name *karyotype* for the genetic test refers to the fact that the chromosomes of the cell are located in the 4. _______
   a. nucleus
   b. cytoplasm
   c. lysosome
   d. endoplasmic reticulum
5. Traits that are determined by more than one more gene pair are termed 5. _______
   a. sex-linked
   b. recessive
   c. multifactorial
   d. dominant

II. Completion Exercise

Write the word or phrase that correctly completes each sentence.

1. The larger sex chromosome is called the ____________________.
2. A change in a gene or chromosome is called a(n) ____________________.
3. The number of autosomes in the human genome is ____________________.
4. The process of cell division that halves the chromosome number is called ________________.
5. An agent that induces a change in chromosome structure is called a(n) ________________.
6. Incomplete closure of the spine results in a disorder called ____________________.

Understanding Concepts

I. True/False

For each question, write *T* for true or *F* for false in the blank to the left of each number. If a statement is false, correct it by replacing the underlined term and write the correct statement in the blank below the question.

_______ 1. An individual with Down syndrome received 23 chromosomes from one parent and 23 chromosomes from the other parent.

____

_______ 2. Hereditary diseases are always evident at birth.

____

_______ 3. Sex-linked traits appear almost exclusively in males.

____

_______ 4. The sex of the offspring is determined by the sex chromosome carried in the spermatozoon.

____

_______ 5. Carriers for a particular gene are always homozygous for the gene.

____

_______ 6. A recessive trait is expressed in individuals heterozygous for the recessive gene.

____

II. Practical Applications

Study each discussion. Then write the appropriate word or phrase in the space provided.

➤ Group A

1. Mr. and Ms. J have consulted a genetic counselor. Ms. J is 8-weeks pregnant, and cystic fibrosis runs in both of their families. Cystic fibrosis is a disease in which an individual may carry the disease gene but not have cystic fibrosis. A term to describe this type of trait is _______________________.
2. First, the genetic counselor prepared a diagram summarizing the incidence of cystic fibrosis in the two families and the relationships between family members. This type of diagram is called a(n) _______________________.
3. Mr. and Ms. J were then screened for the presence of the cystic fibrosis gene. It was determined that both Mr. and Ms. J have the gene, even though they do not have cystic fibrosis. For the cystic fibrosis trait, they are both considered to be _______________________.

4. Ms. J wanted to know if her baby would have cystic fibrosis. Cells were taken from the hair-like projections of the membrane that surrounds the embryo. This technique is called ________________.
5. The fetus was shown to carry one normal allele and one cystic fibrosis allele. The fact that the alleles are different means that they can be described as ________________.
6. As part of the prenatal screening, the chromosomes were obtained from cells in metaphase and photographed for analysis. This form of evaluation is called a(n) ________________.
7. The analysis revealed the presence of two XX chromosomes. The gender of the baby is therefore ________________.

➤ Group B

1. Mr. and Ms. B have a child with three copies of chromosome 21. The child has a chromosomal abnormality known as ________________.
2. This abnormality, like any change in DNA, is known as a(n) ________________.
3. Chromosome 21 is one of 22 pairs of chromosomes known as ________________.
4. The remainder of the child's chromosomes are normal. The total number of chromosomes in the child's cells is ________________.
5. The chromosome abnormality arose during the production of gametes. The form of cell division that occurs exclusively in germ cell precursors is called ________________.

III. *Short Essays*

1. Name three methods used to treat genetic disorders. If possible, provide specific examples of diseases treated by each method.

____

____

____

2. Some traits in a population show a range instead of two clearly alternate forms. List some of these traits, and explain what causes this variety.

____

____

____

Conceptual Thinking

1. Are identical twins identical individuals? Defend your answer, using the terms "genotype" and "phenotype."

____

____

____

2. Young S, female, is afflicted with Edwards syndrome, or trisomy 18. She was born with a small jaw, misshapen ears, malfunctioning kidneys, and a heart defect. None of her relatives suffer from this disorder, which resulted from the presence of an extra chromosome 18 in the spermatozoon. Is trisomy 18 (a) congenital, (b) hereditary, or (c) genetic? Defend your answers.

____

____

____

Expanding Your Horizons

The movie *Gattaca* (1997) describes a futuristic world in which genetic screening and engineering are the norm. Everyone has a genetic description (including lifespan) available to potential mates, and virtually everyone (except the hero) is genetically engineered. Genetic screening determines one's future on Gattaca, and genetic engineering has essentially abolished independent thought and creativity. Science fiction aside, genetic screening and engineering may soon become a technological reality in our society. What are the implications of genetic screening and engineering? Will a poor genetic outlook affect health insurance coverage and job prospects? Will society genetically engineer away creativity? The articles listed below can provide you with more information about the use and abuse of genetic testing.

Resources

1. Hodge JG, Jr. Ethical issues concerning genetic testing and screening in public health. *Am J Med Genet* 2004; 125C:66–70.
2. Sermon K, Van Steirteghem A, Liebaers I. Preimplantation genetic diagnosis. *Lancet* 2004; 363:1633–1641.